Intelligent Environmental Data Monitoring for Pollution Management

Intelligent Environmental Data Monitoring for Pollution Management

Edited by

Siddhartha Bhattacharyya
CHRIST (Deemed to be University), Bangalore, India

Naba Kumar Mondal
The University of Burdwan, Burdwan, India

Jan Platos
VŠB - Technical University of Ostrava, Czech Republic

Václav Snášel
VŠB - Technical University of Ostrava, Czech Republic

Pavel Krömer
VŠB - Technical University of Ostrava, Czech Republic

Series Editor

Fatos Xhafa
Universitat Politècnica de Catalunya, Spain

Academic Press is an imprint of Elsevier
125 London Wall, London EC2Y 5AS, United Kingdom
525 B Street, Suite 1650, San Diego, CA 92101, United States
50 Hampshire Street, 5th Floor, Cambridge, MA 02139, United States
The Boulevard, Langford Lane, Kidlington, Oxford OX5 1GB, United Kingdom

Notices
Knowledge and best practice in this field are constantly changing. As new research and experience broaden our understanding, changes in research methods, professional practices, or medical treatment may become necessary.

Practitioners and researchers must always rely on their own experience and knowledge in evaluating and using any information, methods, compounds, or experiments described herein. In using such information or methods they should be mindful of their own safety and the safety of others, including parties for whom they have a professional responsibility.

To the fullest extent of the law, neither the Publisher nor the authors, contributors, or editors, assume any liability for any injury and/or damage to persons or property as a matter of products liability, negligence or otherwise, or from any use or operation of any methods, products, instructions, or ideas contained in the material herein.

Library of Congress Cataloging-in-Publication Data
A catalog record for this book is available from the Library of Congress

British Library Cataloguing-in-Publication Data
A catalogue record for this book is available from the British Library

ISBN 978-0-12-819671-7

For information on all Academic Press publications
visit our website at https://www.elsevier.com/books-and-journals

Publisher: Candice Janco
Acquisitions Editor: Marisa LaFleur
Editorial Project Manager: Lena Sparks
Production Project Manager: Bharatwaj Varatharajan
Cover Designer: Alan Studholme

Typeset by SPi Global, India

Dedication

Siddhartha Bhattacharyya would like to dedicate this book to Dr Fr Benny Thomas, Professor, Department of Computer Science and Engineering & Information Technology and Dr Fr Joseph Varghese, Personnel Officer and Professor, Department of Mathematics, CHRIST (Deemed to be University), India.

Naba Kumar Mondal would like to dedicate this book to his sweet and loving father and mother, whose affection, love, encouragement, and prayer of day and night made him achieve such success and honor.

Jan Platos would like to dedicate this book to his wife Daniela and his daughters Emma and Margaret.

Contents

Contributors

Jinat Aktar Centre for the Environment, Indian Institute of Technology Guwahati, Guwahati, Assam, India

Anindya Banerjee Department of Electronics and Communication Engineering, Kalyani Government Engineering College, Kalyani, West Bengal, India

Jatindra N. Bhakta Department of Ecological Studies & International Center for Ecological Engineering, University of Kalyani, Kalyani, West Bengal, India

Mariem Brahem DAVID Lab, University of Versailles Saint-Quentin-en-Yvelines, Paris-Saclay University, Versailles, France

Mohamed Chachoua LASTIG, University Gustave Eiffel, EIVP, Paris, France

Adriane Chapman University of Southampton, Southampton, United Kingdom

Soumya Chattoraj University Institute of Technology, General Science and Humanities, The University of Burdwan, Bardhaman, West Bengal, India

Priyanka Debnath Environmental Chemistry Laboratory, Department of Environmental Science, The University of Burdwan, Bardhaman, West Bengal, India

T.T. Dhivyaprabha Department of Computer Science, Avinashilingam Institute for Home Science and Higher Education for Women, Coimbatore, India

Apurba Ratan Ghosh Department of Environmental Science, The University of Burdwan, Burdwan, West Bengal, India

Manash Gope Department of Chemistry, National Institute of Technology Durgapur (NITD), Durgapur, West Bengal, India

Metehan Guzel Department of Computer Engineering, Gazi University, Ankara, Turkey

Hafsa E.L. Hafyani DAVID Lab, University of Versailles Saint-Quentin-en-Yvelines, Paris-Saclay University, Versailles, France

G. Jayashree Northeastern University, Boston, MA, United States

Zoubida Kedad DAVID Lab, University of Versailles Saint-Quentin-en-Yvelines, Paris-Saclay University, Versailles, France

Ibrahim Kok Department of Computer Science, Gazi University, Ankara, Turkey

M. Krishnaveni Department of Computer Science, Avinashilingam Institute for Home Science and Higher Education for Women, Coimbatore, India

Ahmad Ktaish DAVID Lab, University of Versailles Saint-Quentin-en-Yvelines, Paris-Saclay University, Versailles, France

Bradley McLaughlin University of Southampton, Southampton, United Kingdom

Souheir Mehanna DAVID Lab, University of Versailles Saint-Quentin-en-Yvelines, Paris-Saclay University, Versailles, France

Arghadip Mondal Environmental Chemistry Laboratory, Department of Environmental Science, The University of Burdwan, Bardhaman, West Bengal, India

Naba Kumar Mondal Environmental Chemistry Laboratory, Department of Environmental Science, The University of Burdwan, Bardhaman, West Bengal, India

Suat Ozdemir Department of Computer Engineering, Hacettepe University, Ankara, Turkey

Anirudha Paul Department of Ecological Studies & International Center for Ecological Engineering, University of Kalyani, Kalyani, West Bengal, India

Avedananda Ray Department of Agricultural and Environmental Sciences, Tennessee State University, Nashville, TN, United States

Cyril Ray The Naval School, IRENAV, Brest, France

Mouni Roy Department of Chemistry, National Institute of Technology Durgapur (NITD), Durgapur, West Bengal; Department of Chemistry, Banasthali University, Banasthali, Rajasthan, India

Ritabrata Roy Hydro-Informatics Engineering Department, National Institute of Technology Agartala, Agartala, Tripura, India

Rajnarayan Saha Department of Chemistry, National Institute of Technology Durgapur (NITD), Durgapur, West Bengal, India

Palas Samanta Department of Environmental Science, Sukanta Mahavidyalaya, Dhupguri, West Bengal, India

N. Santhi Department of Bioinformatics, Avinashilingam Institute for Home Science and Higher Education for Women, Coimbatore, India

Tarun Sasmal Department of Geography, Panskura Banamali College, Purba Medinipur, West Bengal, India

Kamalesh Sen Environmental Chemistry Laboratory, Department of Environmental Science, The University of Burdwan, Bardhaman, West Bengal, India

Tarakeshwar Senapati Department of Environmental Science, Directorate of Distance Education, Vidyasagar University, Midnapore, West Bengal, India

Ramesh Sivanpillai Department of Botany, University of Wyoming, Laramie, WY, United States

Stephen Snow University of Queensland, Brisbane, QLD, Australia

P. Subashini Department of Computer Science, Avinashilingam Institute for Home Science and Higher Education for Women, Coimbatore, India

Yehia Taher DAVID Lab, University of Versailles Saint-Quentin-en-Yvelines, Paris-Saclay University, Versailles, France

Laurent Yeh DAVID Lab, University of Versailles Saint-Quentin-en-Yvelines, Paris-Saclay University, Versailles, France

Karine Zeitouni DAVID Lab, University of Versailles Saint-Quentin-en-Yvelines, Paris-Saclay University, Versailles, France

Preface

Efficient monitoring of environmental data for the purpose of pollution management has been at the forefront of affairs given the vast amount of pollutants injected into the environment due to industrial affluent. The state-of-the-art methods rely on time-intensive and costly batch procedures. As a result, environmental scientists and researchers are engaged in devising newer, cost-effective methods. The latest additions in this direction are inspired by computational intelligence that relies on intelligent tools like neural networks and evolutionary intelligence.

This book focuses on evolving novel intelligent algorithms and their applications in the area of environmental data-centric systems guided by batch process-oriented data. It may be noted that there are many potential applications and prototypes in this field, yet very few real-life applications are known so far. This is mainly due to the fact that most pollution management procedures still rely on the time-consuming batch processes and field study. Thus, this book will usher in a new era as far as environmental pollution management is concerned. This book reviews the fundamental concepts of gathering, processing, and analyzing data from batch processes, supplemented by a review of intelligent tools and techniques that can be used in this direction. This book also covers novel intelligent algorithms for the purpose of effective environmental pollution data management at par with the standards laid down by the World Health Organization.

The book is organized into 13 well-versed chapters on different aspects of environmental data monitoring, both traditional and intelligent.

The batch adsorption process remains one of the most practical approaches used to adsorb pollutants from a liquid solution for the purification of water. Chapter 1 presents a literature survey of different adsorption types with their characteristics, including an overview of different factors influencing the batch process. This chapter also helps to understand the removal mechanism of the batch process through bulk solution transport, film diffusion transport, pore transport, and adsorption on available adsorption sites.

The term "heavy metal" is commonly envisaged to be those whose density surpasses $5 \, g/cm^3$. An enormous number of elements belong to this group, but Cd, Cr, Cu, Ni, As, Pb, Hg, and Zn are those of relevance in the environmental context. Heavy metals cause serious health issues such as organ injury, nervous system impairment, reduced growth and development, cancer, and in extreme cases, death. Therefore, it is essential to treat metal-contaminated wastewater prior to its discharge into the environment. Among the various types of treatment processes in vogue, adsorption is acknowledged a financially cheap practice for heavy metal expulsion from wastewater as it is economical, simple to deal with, and exceedingly productive. Chapter 2 explains the adsorption technique by using biochar, which is a carbon-rich item obtained when biomass, for example, wood, fertilizer, or leaves, is heated in a shut vessel with next to zero accessible oxygen. In a more methodical term, biochar is obtained by thermal disintegration of organic material under partial or no source of oxygen (O_2) and at comparatively low temperatures ($<750°C$). The use of biochar in industrial wastewater can potentially diminish the heavy metal's mobility due to the porous assemblage, large surface area, and high adsorption capacity of biochar.

Nanoscience is an interdisciplinary subject that has been one of the most dynamic disciplines in material science. Key features of nanoparticles are clusters of atoms in the size range of 1–100 nm. Metal nanoparticles can be synthesized by physical, chemical, and biological routes. But the green or biological route for nanoparticle synthesis from biological origin is of more interest than the other two ways because of its environmental friendliness, economically cheap, feasibility, and applications in various fields. Several analytical tools are used for structure determination and characterization of synthesized nanoparticles. Chapter 3 focuses on the power of biomolecules for the synthesis of silver, gold, zinc, and copper nanoparticles and their applications toward the effect on mosquito larvicidal mortality.

The generation of heavy metal(loids) and their derivative compounds (such as oxides, carbides, sulfides, etc.) in increasing rates by various anthropogenic activities is of major global concern. To control these problems, the removal of metal(loids) from effluents is essential. Although several physicochemical methods (adsorption, electrodialysis, floatation, ion exchange) are available, the effectiveness of these methods are poor and of high cost. Therefore, eco-friendly and green technology is required in this regard. Among various green synthesis techniques, biosorption drew great attention for scientific research as a novel method for industrial wastewater treatment to reduce or neutralize metal content. Chapter 4 reviews the state-of-art works carried out in this regard by different scientists. Different parameters of biosorbent materials along with optimum treatment conditions have also been considered in this chapter. This can be helpful for future studies in exploring novel biosorbents to treat industrial wastewater using biosorption technology.

Glyphosate [N-(phosphonomethyl) glycine] pollution, mainly due to industrial drainage and unnecessarily used for weed control of agricultural and residential purposes, creates ecosystem and environmental toxins. There are very important research fields needed for decontamination in a sustainable way. In Chapter 5, a review on the adsorption process of glyphosate from aqueous solution is reported with influencing mechanisms like kinetics, isotherms, and thermodynamics. Reported results are depicted by consequence factors, namely pH, contact time, initial taken concentration, doses, etc.

Today, dyes, highly colored organic compounds, find widespread applications in industries like textile, food technology, leather tanning, paper, etc., for coloring products, light-harvesting solar-cells, photoelectrochemical cells, etc. However, the large usage of these dyes leads to their release as effluents, which eventually contaminate drinking water sources, hampering the ecological system. The contaminated wastewater effluent exhibits detrimental effects related to people and environmental concerns. Thus, a requirement for a suitable, cost-effective treatment method for contaminated wastewater seems to be indispensable. Chapter 6 aims to compose various information about dyes, dye effluents, and existing treatment technologies for wastewater.

Water is an essential natural resource for the survival of living beings in the world. Freshwater bodies significantly support ecological and economic activities in each state or country, especially India, which has a large number of rivers, lakes, ponds, and wetlands. Notably, lake water is critically polluted due to industrial effluents, encroachment, eutrophication, garbage dumping, poor drainage, and climate changes that have a severe effect on water quality. Hence, monitoring and assessment of water quality over time is important for minimizing the negative effects to the ecosystem, and it greatly helps to sustain valuable aquatic species. In Chapter 7, an intelligent estimation model is developed using the Kalman filter integrated with a Synergistic Fibroblast Optimization (SFO) algorithm for forecasting water quality parameters. The developed predictive model is evaluated with real-time water samples collected from Ukkadam Periyakulam Lake, Coimbatore, and the promising results provide qualitative information to enhance water quality in this lake.

Many countries are suffering from air pollution due to imbalanced urbanization, unregulated increase in transportation, and inorganic industrialization. Air pollution directly affects the ecosystem and climate thereby compromising the well-being of citizens and cities. For this reason, predicting air pollution in advance has great importance in supporting proactive plans and environmental management actions for decision-makers. In the literature, many research efforts focus on predicting air pollutant concentrations. In Chapter 8, the authors present recent artificial intelligence-based air pollution prediction approaches. In this context, the authors present a comprehensive review of the literature and have categorized the proposed prediction methods based on feature selection methods, air pollutant types, and learning models.

Chapter 9 examines the equilibrium adsorption of carbaryl insecticide by several easily available natural adsorbents like Alluvial soil, Pistia Stratiotes biodust, Lemna major biodust, Neem Bark dust (NBD), and Eggshell powder to remove carbaryl from an aqueous environment. The batch experiments are archived out as different operating parameters. For overall validity of the process, response surface (RSM) and artificial neural network (ANN) models are used. The maximum amount of carbaryl adsorbed onto different adsorbents is presented in this study. Adsorption isotherms, kinetics, and thermodynamics of the process have also been provided. Finally, results show that NBD has the maximum adsorption capacity of carbaryl. As NBD is easily prepared, it can be a used as an effective adsorbent for removal of the insecticide from contaminated water.

Biochemical oxygen demand (BOD), representing the biodegradable organic load in water, is a prime parameter to assess water quality. Estimation of BOD requires prolonged incubation of water samples. Thus, it is a time-consuming as well as energy-consuming process. Therefore, it is not possible to respond rapidly for mitigation if the BOD level goes beyond the permissible limit. In Chapter 10, a study is presented on River Damodar; the BOD value is predicted from electrical conductivity (EC), turbidity, and chloride, using an artificial neural network (ANN)-based empirical model. Because predictor parameters of these models are rapidly measurable, the BOD value can also be quickly predicted. As all the predictor parameters are highly correlated (correlation coefficient, 0.9 or more) with BOD, the models are valid for prediction of BOD. Additionally, for further validation of the models, a portion of field data (20%) was used for model testing. As the model predictions are close to the actual values ($r = 0.98$, MAE = 0.43, RMSE = 0.57), the model, developed in this study, can be considered as successful in BOD prediction.

The effective monitoring and management of indoor-sourced pollutants is vital, given that poor indoor air quality (IAQ) reduces the academic performance of school students and contributes to short- and long-term health effects. Despite this, 66% of classrooms are found to exceed UK government IAQ guidelines, and there is not yet a requirement for real-time IAQ monitoring in UK classrooms. Chapter 11 describes the design and deployment of a visual iPad display of classroom IAQ called Airlert. Findings inform a discussion of which visual aspects of IAQ feedback devices hold the best potential for empowering teachers to improve understanding of IAQ in their classroom, to employ healthier ventilation practices. Implications for the design of IAQ feedback devices and necessary data management considerations are discussed, along with suggestions for future research.

The advent of the new generation of low-cost, lightweight, and connected sensors makes a paradigm shift in environmental studies. In particular, nomadic sensors allow for a very precise personalized measurement, by continuously quantifying the individual exposure to air pollution components. Moreover, a broad dissemination among volunteers of these devices, or their deployment on vehicle fleets, is becoming a credible solution. Another major interest of such sensor deployment is to densify the air quality monitoring network, indoor, as in the outdoor, which is today restricted to sparse nodes. However, this high spatiotemporal resolution raises several issues related to their analysis. After an overview of the projects relying on this technology, Chapter 12 points out the remaining challenges to be addressed. Part of these challenges constitute the research program of the ongoing project Polluscope in France.

In Chapter 13, the authors aim to tabulate the statistical repertoire used to predict adsorption of contaminant from an aqueous solution. The effects of various parameters like pH, temperature, adsorption time, initial concentration, and adsorbent dosage have been analyzed, and their optimization by an ANN, RSM, and Path analysis model has been presented. A comparison of predictions of the experimental data with the adsorption efficiency shows the sustainability of the different models of the adsorption process under different conditions.

This book is envisaged to offer benefits to researchers and several categories of students for some part of their curriculum. The editors would feel good about their attempt if the academic and scientific community finds benefit from this novel venture.

Siddhartha Bhattacharyya
CHRIST (Deemed to be University), Bangalore, India

Naba Kumar Mondal
The University of Burdwan, Burdwan, India

Jan Platos
VŠB - Technical University of Ostrava, Czech Republic

Václav Snášel
VŠB - Technical University of Ostrava, Czech Republic

Pavel Krömer
VŠB - Technical University of Ostrava, Czech Republic

1

Batch adsorption process in water treatment

Jinat Aktar

CENTRE FOR THE ENVIRONMENT, INDIAN INSTITUTE OF TECHNOLOGY GUWAHATI, GUWAHATI, ASSAM, INDIA

1 Introduction

Wastewater gets released from different types of industries and is a significant source of different heavy metals, dyes, detergents, and other contaminants that can consume dissolved oxygen of accommodating water stream, can alter chemical and biological characteristics of the water, and poses environmental hazards by endangering ecosystems and human health [1]. Hence, wastewater treatment is prerequisite before its discharge into the ecosystem. Although there are multiple water treatment options are being used frequently, adsorption process is still unrivaled because of its effectiveness, low energy consumption, and easiness to perform; therefore, the versatility of adsorption makes it extensively used while removing contaminants from wastewater [2]. Mainly, the adsorption process is used for the removal of a highly toxic or low concentrated compound, which is not readily treated by biological processes [3]. The batch process is promptly used for adsorption treatment of wastewater. The batch process occurs in a closed system containing the optimum amount of adsorbent in contact with the predetermined volume of adsorbate solution. In a closed vessel, agitation is provided by rotating stirrers for the full mixing of adsorbent materials with the contaminated solution. Well-designed batch processes are highly efficient, which results in a high-quality recyclable effluent after treatment (Fig. 1). Moreover, batch adsorption process can be cost effective if low-cost adsorbents are used or regeneration is feasible [4]. Furthermore, the batch adsorption process can also be used for abatement of pollutants at source and quality improvement of industrial and other wastewater [5].

Literature in Table 1 shows that different heavy metals like cadmium, copper, lead, chromium present in wastewater are treated by the batch process (Table 1). The batch process can be used for the treatment of textile discharge containing different dyes, detergents, and other contaminants [5,34] and also used in the pharmaceutical industrial water treatment like acetaminophen and ibuprofen removal [2].

The purpose of this study is to develop the detailed view of the batch process and its affecting factors, mechanism of pollutant removal in the batch process, the significance of various mathematical models to predict the performance of the batch process, and

Intelligent Environmental Data Monitoring for Pollution Management. https://doi.org/10.1016/B978-0-12-819671-7.00001-4

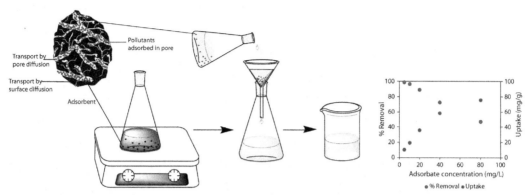

FIG. 1 Scheme of adsorption of pollutants by the batch process.

analyze the surface nature and pore size of the adsorbent. This information could help to optimize the factors for the efficient removal of pollutants in the batch process.

This book chapter is organized in multiple sections: beginning with introduction, which discussed the problem caused by wastewater, role of adsorptive removal while treating wastewater, various adsorbents used for removal of particular contaminants from wastewater. Further, Section 2 comprised the concepts of batch experiments along with its working principle(s), which were discussed in detail. Moreover, types of adsorption were also covered in brief in this section. Section 3 focused mainly on factors influencing adsorption process. The detailed information discussed here may help in optimizing the factors to achieve maximum removal of pollutants from wastewater. One of the most important aspects of adsorption process is mechanism followed during the adsorption, which is discussed in Section 4. Here, bulk diffusion transport, pore transport, film diffusion transport, and surface diffusion were summarized. The investigation of theoretical uptake capacity of the material is obligatory; therefore, various kinds of adsorption isotherms, including Langmuir, Freundlich, Sips, Brunauer Emmett and Teller (BET), and Extended Langmuir models are discussed in Section 5. The mechanism of adsorption could be well understood by adsorption kinetics; therefore to study the contaminant removal rate, the pseudo-first-order, pseudo-second-order, and Elovich model, intraparticle diffusion model and kinetics of finite bath and infinite bath experiments are illustrated in Section 6. In Section 7, significance of thermodynamics in adsorption reactions is mentioned. Lastly, chapter concludes with providing briefs regarding regeneration capacity of adsorbent and its recyclability after adsorption in Section 8. Various desorption studies are also presented in tabular format.

2 Batch experiments

In the batch tank, adsorbent is mixed with wastewater for a predetermined time and pH; after that, the final concentration of adsorbate is measured. After the batch process,

Table 1   Use of batch process for different pollutant removal with different types of adsorbent.

Adsorbent	Adsorbate	Initial concentration (mg/L)	Removal percentage (%)	Reference
Modified Fe_3O_4 NPs	Fluoride	20	85	[6]
Fe_3O_4/Al_2O_3 NPs	Fluoride	4	90	[7]
Al_2O_3-Fe_3O_4-expanded graphite nanosandwich adsorbent	Fluoride	5	98	[8]
Graphene oxide	Mercury (II)	100	94	[9]
Alumina NPs	Mercury (II)	90	100	[10]
Sorghum biomass	Total Arsenic	1	85	[11]
Graphene oxide composite	Arsenic	0.001	99	[12]
Orange waste	Cadmium	100	98	[13]
Bamboo charcoal	Cadmium	40	80	[14]
Activated alumina	Cadmium	50	95	[15]
Banana peel	Cadmium	150	98	[16]
Activated alumina	Lead	10	95	[15]
Hydroxyapatite/magnetite	Lead	10	99	[17]
Alumina modified with 2,4-dinitrophenylhydrazine	Lead	50	95	[18]
Chitosan-based mesoporous Ti-Al binary metal oxide	Fluoride	5	80	[19]
Nettle and Thyme leaf Iron nano	Cephalexin antibiotic	25	85	[20]
Iron nanoparticles	Chromium (VI)	5	100	[21]
aniline-formaldehyde condensate polymer	Chromium (VI)	10	95	[22]
Magnetic activated carbon	Chromium (VI)	2	75	[23]
Urtica dioica-leaf iron nano	Chromium (VI)	150	92	[24]
Amine-based polymer	Chromium (III)	50	80	[25]
Activated carbon from coconut coil	Methylene Blue dye	80	82	[26]
Red mud	Congo red	10	75	[27]
Rice husk carbon	Malachite green	20	85	[28]
Dry *Annona squmosa* seed	Malachite green	50	78	[29]
Zn-Al-Cl-layered hydroxide	Nitrate	50	80	[30]
Magnolia champaca leaf nanoparticle	Phosphate	92	98	[31]
Sawdust	Copper (II)	170	60	[32]
Aniline formaldehyde condensate polymer	Copper (II)	200	85	[33]

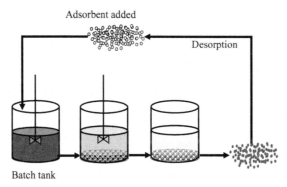

FIG. 2 The framework of pollutant removal by the batch process.

separation of the solution is done by either centrifugation, sedimentation, or filtration process. In the Industrial area, multiple batches or crossflow systems are established, and large quantities of adsorbent are required. In industry, the batch process is easily scaled up-scale or scale-down, depending on requirements. At the end of the steps, a finite amount of the final product is produced, and after desorption, the adsorbent can be used for another batch process sequence. In the batch tank, adsorbent comes with contact with pollutants present in solution and adsorption occurs. Two types of adsorption may happen: physical adsorption and Chemisorption. In physical adsorption, Van der Waals forces work as a force of attraction, and the resulting adsorption is reversible. In Chemisorption, strong chemical bonding acts as a force of attraction, and this adsorption is irreversible. In Fig. 2, the framework of batch process is described properly. In a properly designed batch tank, the adsorbent can be reused after the desorption process.

3 Factors affecting adsorption process

pH: pH is one of the most significant parameters that can directly affect the removal efficiency of pollutants or adsorbate by adsorbents [35]. pH can affect the ionization as well as the speciation of adsorbate in solution and also the surface nature of adsorbent material. In solution, hydrogen ions (H^+) and hydroxide ions (OH^-) interact with activated site of adsorbents. Hence, the adsorption process gets affected by the pH of the solution.

Fig. 3 shows variation in chemical speciation of chromium with respect to pH, which also affects the batch process. Redox potential (Eh) plays an essential role in the conversion of different valence states of chromium [37]. In lower pH, Cr(VI) has a high positive redox potential value, which acts as a strong oxidizing agent. $Cr(OH)^{2+}$ is the principal species at pH 5, although $Cr(OH)_3$ present at pH 8. The chromate ion is present at pH > 7, though CrO_4^{-2} is dominating species at pH < 6 [38]. In case of Cu(II) removal, the effect pH was measured in 2 to 6 pH range, because above pH 6 precipitation of $Cu(OH)_2$ occurs, which causes sharp drop of Cu(II) removal [33,39].

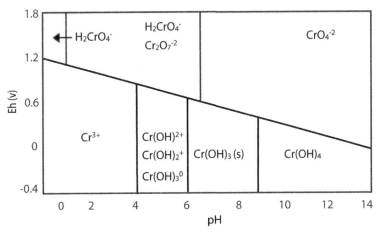

FIG. 3 Speciation of chromium in water environment [36,37].

pHzpc: Surface charge of adsorbent plays an important role in the batch processes and also helps to understand the sorption mechanism. The pH at which the surface charge of the adsorbent is zero, considered as the point of zero charges (pH_{pzc}). This allows hypothesizing that below the point of zero charges, the material surface is positive, and it can adsorb negatively charged pollutants [40].

Among several techniques of the determination of pH_{pzc} of adsorbent, immersion technique and mass technique have been used widely [22]. The plot shown in Fig. 4 for initial pH versus delta pH and the point zero charge value by the immersion process was found to be at pH 4.7. The pH_{zpc} value of the mass titration technique was 4.6. The values of both techniques are quite close; therefore, the value of pH_{pzc} was relevant.

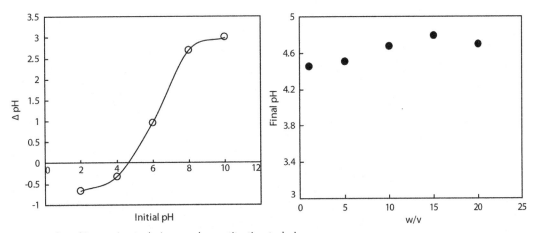

FIG. 4 A plot of immersion technique and mass titration technique.

The electrostatic attraction between oppositely charged adsorbent species increases the rate of adsorption.

Adsorbent dose: The active adsorption sites increase with increase in the adsorption dose, which helps in more removal of contaminant(s). However, with the increase of the adsorption dose, the total uptake of pollutants (q_e) per unit mass of an adsorbent reduces due to the unsaturated site that exists in adsorption process. Literature showed increase of removal percentage of copper (II) from 47% to 87% with the rising dose from 0.5 to 8 g/L; however, the uptake capacity decreased from 9.6 to 1.1 mg/g [39]. Baral et al. examined the effect of sawdust dose on chromium (VI) removal and found that the percentage removal increased from 20% to 100% with dose of 0.2 to 1.6 g/L, though the Chromium adsorption capacity was reduced from 2.72 to 1.7 mg/g [41]. Furthermore, the increase in removal percentage of copper (II) from 47% to 87% with the rising dose from 0.5 to 8 g/L was reported by Aydın et al., but the adsorption capacity was reduced from 9.6 to 1.1 mg/g [39].

Temperature: In the batch process, the temperature can affect the characteristics of adsorbents, the stability of adsorbate, and adsorbate-adsorbent interaction. With the rising temperature, the viscosity of the solution decreases, which helps in the transfer of pollutants from bulk solution to the surface of material [42]. The thermodynamic parameters help to estimate the characteristic of the batch adsorption process such as exothermic or endothermic, spontaneous or random nature and also signifies the favorability of temperature in the batch adsorption process. The spontaneity of adsorption is accounted by the negative values of Gibbs free energy (ΔG^0) and the enthalpy (ΔH^0) indicates the nature of the process whether it is exothermic or endothermic process.

According to Le chatelier's principle, the magnitude of adsorption decreases with increase in temperature. In Chemical adsorption, the removal first rises with increasing temperature then decreases (Fig. 5). Malana et al. [43] studied that with rising

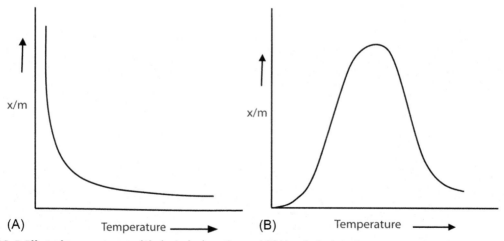

FIG. 5 Effect of temperature on (A) physical adsorption and (B) chemical adsorption.

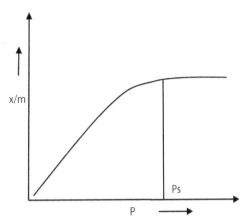

FIG. 6 Effect of pressure on batch adsorption.

temperature, the adsorption capacity of polymeric gels decreases, indicating the exothermic nature for Bromophenol blue, Malachite green, and, Methylene blue dye removal.

Pressure: In Adsorption Isotherm, with the increasing pressure, adsorption also increases up to a certain level, until the saturation of adsorbent occurs. After reaching the equilibrium level, no more uptake takes place no matter how high the pressure is applied. In Fig. 6, Ps is the saturation pressure; after that the graph becomes linear, as further adsorption of pollutants is not possible.

Hu et al. [44] studied the effect of adsorbate pressure (methylene blue concentration) and concluded that with the rising methylene blue concentration from 100 mg/L to 140 mg/L, the adsorption capacity increased from 137 to 139 mg/g, so it reached its equilibrium state.

Surface area: Adsorption is a surface phenomenon; therefore, it is proportional to specific surface area. Specific surface area can be described as that available portion for adsorption in total surface area. Large particles have less surface area than smaller particles, so their accessibility to particle pores is less. As the size is directly correlated with the pore accessibility, sorption by smaller particles is comparatively better as compared to the larger particles. However, this is achieved by breaking up larger particles into smaller ones opening some tiny scales, thereby increasing their surface area for better adsorption [45].

Coexisting ions: The presence of coexisting ions in solution affects the uptake capacity of the adsorbent by creating competition for the active sites of the material surface.

In general, wastewater consists of different types of components. Sometimes the mixture of compounds mutually enhances the adsorption process, and sometimes interferes with another. In mixed solution, the competition among adsorbates depends on the molecular size of the adsorbents, concentration of solutes, and relative affinity. Terangpi et al. [22] studied the effect of coexisting ions and concluded that nitrate and phosphate were capable of reduction of 28%–38% of uptake of Cr (VI) by aniline-formaldehyde condensate polymer. They also reported that at higher anionic concentration phosphate and nitrate showed more interference in removal capacity than chloride and sulfate. According to Islam and Patal [30], carbonate and phosphate ions showed more interference in nitrate removal than chloride and sulfate ions.

4 Mechanism in batch adsorption

In the batch process, pollutants come to the contact of the activated site of the adsorbent and adsorption takes place in four steps [46]. As a result, the removal of pollutants occurred.

4.1 Bulk solution transport

In bulk solution transport, the adsorbate movement occurs through the bulk liquid to the fixed film of surrounding liquid of adsorbent by dispersion and advection.

4.2 Film diffusion transport

Transit of pollutants to the entrance of pore of the adsorbent occurs via stagnant liquid film.

4.3 Pore transport

Adsorbates are adsorbed by the pore diffusion or surface diffusion, which occurs by molecular diffusion of pore liquid and diffusion along the adsorbent surface, respectively. Then adsorbates are attached to available adsorption sites. Both types of diffusion may occur individually or simultaneously.

In diffusion, across the stagnant liquid film, the flux is in direct correlation to the mass transfer coefficient multiplied by a driving force constituted by the marked difference between the particle external surface concentration and bulk concentration.

Pore diffusion: In this mechanism, the mass transfer of adsorbates occurs via pore liquid. In the case of one-species desorption, this mechanism is supported by the following equation of conservation. Fick's law in porous materials, is given by the following equation.

$$Jp = -Dp\frac{\partial CL}{\partial r},$$

where Jp = mass transfer rate in porous media, Dp = pore diffusivity coefficient, CL = concentration in liquid phase, and R = particle radial coordinate.

Therefore, it can be concluded that the mass transfer flux is directly proportional to the concentration gradient with respect to the particle radial coordinate and the (D_P). The particle porosity and tortuosity are the main characteristics to calculate the pore diffusivity coefficient.

The fraction of the solution contained in the solid particle is defined as porosity ε. Porosity defines the volume in adsorbent where adsorbate can disperse.

$$\varepsilon = \frac{Liquid\ \ volume(VL)}{Particle\ \ volume(Vp)}$$

Pores are specified by irregular-shaped geometries and the twisted diffusion path can be represented by the tortuosity factor τ:

$$\tau = \frac{Real\ \ pore\ \ length}{Ideal\ \ pore\ \ length}$$

The tortuosity factor τ increases with the increase in the irregularity of pores shape as the travel distance increases [47].

Regarding the ideal regular and linear shape of the pore, the coefficient of diffusion is corrected for the irregular geometry of the porous structure by the product of molecular diffusivity (D_m) and particle porosity divided by the tortuosity factor.

$$D_P = \frac{\varepsilon}{\tau} D_m$$

4.4 Surface diffusion

It is defined as the motion of adsorbate molecules at the adsorbent surface. The macroscopic conservation equation may be written as the phenomenon of surface diffusion in adsorption and can be expressed by Fick's law

$$Js = -Ds\frac{\partial Cs}{\partial r},$$

where Js = mass transfer rate on the solid surface, Ds = surface diffusivity, and Cs = concentration on the solid surface.

Generalizing the mass transfer of intraparticle, it is evident that both pore and surface diffusion operate concurrently. The pore solution concentration(c) and the adsorbed phase concentration(q) are supposed to be in equilibrium, thereby concluding the fact that the actual adsorption step occurs much faster than the mass transfer step. Among all of these steps, some steps are identified as slowest or rate-limiting steps. In physical adsorption, one of the diffusion transport steps acts as a rate-limiting step, whereas in the case of chemical adsorption, the adsorption step acts as the rate-limiting step. When the rate of adsorption is equal to the rate of desorption, then the equilibrium of the adsorption capacity of adsorbent occurs.

5 Adsorption isotherm

Isotherms are beneficial for explaining the theoretical uptake capacity of different adsorptive materials. Adsorption process usually used for the removal of pollutants from solution by using varieties of adsorbents. It continues until equilibrium occurred. The adsorption equilibrium is expressed by the quantity of pollutant adsorbed per unit mass of adsorbent.

$$q_e = \frac{C_0 - C_e}{m/V}$$

Here q_e: equilibrium uptake capacity of the material; C_0: initial concentration of the contaminated solution; C_e: equilibrium concentration; m: mass of adsorbent material; and V: volume of treated solution.

To represent different types of adsorption isotherms, variety of theoretical and empirical models have been proposed. Basically, any single equation that cannot give a satisfactory explanation of all mechanisms. Langmuir, Freundlich, Sips, and the Brunauer-Emmett-Teller (BET) models are some of the equations that find common use for describing single-species adsorption. An extended Langmuir equation is developed for a multi-component adsorption mechanism.

5.1 Langmuir isotherm

In 1916, Irving Langmuir published the isotherm model, which retained his name. The Langmuir isotherm is usually expressed as:

$$\frac{1}{q_{eq}} = \frac{1}{q_{max} . k_b . C_e} + \frac{1}{q_{max}},$$

where q_{qe} (mg/g) is the equilibrium amount of contaminant adsorbed in per gram of adsorbent, C_e (mg/L) is the equilibrium concentration of contaminant in solution, and q_{max} (mg/g) and k_b (l/mg) are temperature-dependent parameters describing the maximum uptake capacity of the solid surface and the energy constant, respectively.

The Langmuir is based on the following assumptions:

(i) Adsorbate molecules are adsorbed at a fixed number of equivalent active sites. All accessible sites have the same energy.
(ii) Adsorption is limited to monolayer; adsorbate molecules do not deposit to others.
(iii) The enthalpy of adsorption is the same for all molecules.
(iv) There are no interactions between molecules adsorbed on neighboring sites. And all the adsorption proceeds through the same mechanism. The nature of the isotherms is indicated by the value of the separation factor (K_R) value [48].

$$K_R = \frac{1}{1 + K_b C_0}$$

K_R value <1 means adsorption is favorable, and less than 1 signifies unfavorable conditions [49].

5.2 Freundlich adsorption isotherm

In 1909, Freundlich described an equation for representing the isothermal by quantifying the adsorbed gas by a unit mass of adsorbent at the present of pressure. This Freundlich Adsorption Isotherm equation is denoted as:

$$q_e = KC_e^{1/n}$$

Unlike the Langmuir model, the Freundlich equation does not have much limiting factor regarding types of adsorption or surface of adsorbent. Freundlich isotherm can deal with both physical and chemical adsorption and both homogeneous and heterogeneous surfaces [3]. Basically, Freundlich offers a simple equation instead of a theoretical model to interpret for the energetic heterogeneity of adsorption at different areas of the isotherm. The effect of temperature is problematic to justify through Freundlich isotherm because, with a rise in temperature, the solute concentration (C) would increase while the adsorbed mass (Q) would generally decrease. In most applications, however, the Freundlich equation is comparatively more expedient.

Freundlich Isotherm failed to predict the value of adsorption at higher concentrations, as it establishes the relationship of adsorption with lower concentration values of adsorbate. The value of qe increases with the increase in C_0 but as $n > 1$ it does not increase suddenly. Nowadays, Freundlich isotherm is widely used in multipollutant removal or in heterogeneous systems. The slope ranges are used to measure the heterogeneity of the surface of the adsorbent. The value closer to zero means more heterogeneous [50].

5.3 Sips isotherm

It is a combination of Langmuir and Freundlich isotherms, which acts as Freundlich isotherm at lower adsorbate concentration and Langmuir isotherm at higher adsorbate concentration by monolayer adsorption [50]. Batch process variables, such as pH, the concentration of pollutants, temperature, act as the controlling factor in this isotherm model [51].

$$\beta_s \ln(C_e) = - \ln\left(\frac{k_s}{q_e}\right) + \ln(a_s)$$

5.4 BET isotherm

Brunauer, Emmett, and Teller explained the multilayer physical adsorption at high pressure and low temperature, the thermal energy of gas decreases; therefore, more gaseous molecules could be available on per unit. Generally, Nitrogen gas is used as a standard adsorptive for analyzing the surface area except in the case of quantitative assessment of the microporous material (<0.7 nm), where the use of N_2 gas was found to be inappropriate. Therefore, in case of microporous materials, argon and carbon dioxide gases are used as alternative adsorptive molecules.

International Union of Pure and Applied Chemistry (IUPAC) has classified six categories of experimental gas adsorption isotherms (Fig. 7A). Based on internal pore width, pore can be classified as micropore (pore <2 nm); mesopore (pore 2–50 nm); macropore (pore >50 nm). The micropore is subdivided into ultramicropores (>0.7 nm) and supermicropores (0.7–2 nm) [53]. The gas adsorption technique only allows determining the volume of open pores, whereas close pores are not possible to determine. The type I is described by Langmuir adsorption type; it shows monolayer adsorption and indicates that micropores are present in adsorbent material. Type II shows physical adsorption, in which sorption progressively moves from submonolayer to multilayer on the surface of the material. The type III isotherm characterizes a comparatively weak gas-solid interaction on nonporous low-polarity solids and adsorption isotherm convex in the whole pressure range. In I, II, and III types the lines of the adsorption-desorption isotherm graphs coincide, so no hysteresis occurs. Type IV and V isotherms signify multilayer adsorption on mesoporous materials continues via capillary condensation. The appearance of hysteresis loops found due to capillary condensation-evaporation, which often does not occur at the same

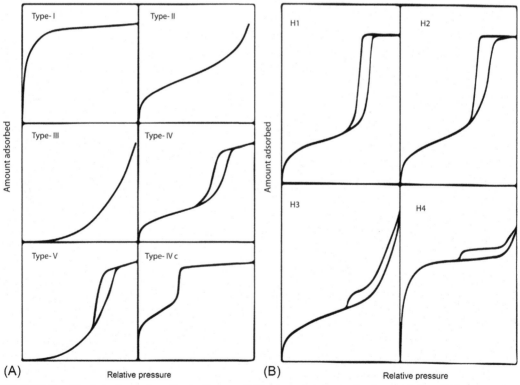

FIG. 7 (A) Classification of gas adsorption isotherm and (B) adsorption-desorption hysteresis loop [52].

pressure. Type IVc shows capillary reversible condensation-evaporation plot in mesoporous materials [54].

Hysteresis is attributed to cooperative effects due to the connectivity of the pore network, a combination of thermodynamic and kinetic effects porous materials. According to the IUPAC, hysteresis loops are classified into four types (Fig.7B) [52]. Type H1 is often associated with porous agglomerate materials with relatively uniform cylindrical pores geometry. H1 type loop on the adsorption isotherm also signifies facile pore connectivity and high pore size uniformity. Type H2 hysteresis loop featured steep and triangular in shape. It contains complex pore structure with narrow mouth. H3 types signify material forming slit-shaped pores and mainly comprised nonrigid aggregates of plate-like particles. Both microporous and mesoporous materials (having slit-shaped pores) can show hysteresis loop type H4, which features parallel and almost horizontal adsorption-desorption branches.

5.5 The extended Langmuir isotherm

This model is used to investigate the mechanism of binary and multispecies adsorption. The assumptions followed by this model are [55]:

  (i) a homogeneous surface for the energy of adsorption,
  (ii) there is no such interaction between adsorbate on the adsorbent surface, and
  (iii) every adsorption site is evenly accessible for all adsorbate.

$$q_{ei} = \frac{K_{L,i} C_{e,i}}{1 + \sum a_{L,i}^0 C_{e,i}}$$

For adsorbate 1

$$q_{e1} = \frac{K_{L,1} C_{e,1}}{1 + a_{L,1}^0 C_{e,1} + a_{L,2}^0 C_{e,2}}$$

For adsorbate 2

$$q_{e2} = \frac{K_{L,2} C_{e,2}}{1 + a_{L,1}^0 C_{e,1} + a_{L,2}^0 C_{e,2}}$$

Here, K_L and a_L are Langmuir isotherm constants.

6 Adsorption kinetics

The adsorption kinetics can assume the removal rate of the pollutant from the solution phase and gives important information for recognizing the mechanism of adsorption in the batch process [56]. There are two phases in adsorption kinetics: initially rapid removal phase followed by a slow phase toward equilibrium stage. The rates at which adsorbates are taken up by the adsorbent determine the adsorption kinetics, and they directly control the adsorption efficiency. A batch adsorption kinetic model was used to estimate the rate constants.

6.1 Pseudo-first-order equation

Different models are used to study the adsorption mechanism in the batch process. The reactive models for adsorption have been used generally for the kinetic data understanding of the batch process. In 1898, Lagergren proposed the equation of pseudo-first-order adsorption kinetics [57].

$$\frac{dq_t}{dt} = k_{s1}(q_1 - q_t)$$

Using following boundary conditions, i.e., $t = 0$ to $t = t$ and $q_t = 0$ to $q_t = q_t$.

$$\log(q_1 - q_t) = \log q_1 - \frac{K_{1t}}{2.303},$$

where k_{S1} = rate constant of first-order adsorption (1 min^{-1}), q_1 = amount of adsorbate adsorbed at equilibrium (mg/g), q_t = amount of adsorbate adsorbed on the surface of the adsorbent at any time t (mg/g).

Based on solution concentration, a reversible first-order reaction can be represented as

$$\ln\left[\frac{(C_0 - C_t)}{(C_0 - C_e)}\right] = (-k_{10} + k_{20})t$$

6.2 Pseudo-second-order equation

The second-order adsorption kinetics follows the equation of Ho and McKay [57]. In this model, chemisorption acts as a rate-limiting step that involves valence forces via interchange of electrons between two-phase (adsorbate-adsorbent) [58]. The equation is as follows.

$$\frac{dq_t}{dt} = k_2(q_e - q_t)^2,$$

where k_2 is the second-order rate coefficient. Applying the boundary conditions of $q_t = 0$ at $t = 0$ and $q_t = q_t$ at $t = t$ the equation is as follows:

$$q_t = \frac{t}{\left[\left(\frac{1}{k_2 q_e^2} + \frac{t}{q_e}\right)\right]}$$

The equation in the linear form is as follows:

$$\frac{t}{q_t} = \frac{1}{k_2 q_e^2} + \frac{1}{q_e}$$

Here k_2 is the rate constant(g mg^{-1} min^{-1}), acquired from the plots of t/q_t vs. t.

6.3 Elovich model

Elovich equation is useful to discuss the nature of chemical adsorption. According to the model, the energy of the solid surface is heterogeneous and follows second-order kinetics. This equation does not define the mechanism for interaction between adsorbent and pollutants. The Elovich equations [59] are generally expressed as

$$\frac{dq_t}{dt} = \alpha \exp(-\beta q_t),$$

where q_t = adsorption capacity at contact time t (mg/g), a = initial adsorption rate (mg/g min), and b = desorption constant (g/mg).

To simplify the Elovich equation, Chien and Clayton [60] assumed that

$$q_t = \beta \ln(\alpha\beta) + \beta \ln t,$$

where α and β are the Elovich coefficients, which represent the initial adsorption rate $(g\ mg^{-1}\ min^{-2})$ and the desorption coefficient $(mg/g\ min^{-1})$, respectively. The Elovich coefficients are obtained from the intercept plots of q_t vs. $\ln t$.

The linear form of this equation can be expressed as

$$qt = \frac{lna_e b_e}{b_e} + \frac{1}{b_e} \ln t,$$

where a_e = initial adsorption rate (mg/g min), and b_e = energy for activation of chemisorption $(g\ mg^{-1})$.

6.4 Intraparticle diffusion

The intraparticle diffusion model defines the relationship between the diffusion mechanism and the kinetic results. The diffusion in porous adsorbent, adsorbate molecules have been considered to propose an appropriate kinetic model for the process. The intraparticle diffusion, as reported earlier, has a pivotal role in controlling the rate of uptake of an adsorbate. In the model developed by Weber and Morris [61], the initial rate of intraparticle diffusion is calculated by linearization of equation

$$q_t = k_i t^{1/2} + C,$$

where C is the intercept and K_i is the rate constant intraparticle transport $(mg/g\ min^{1/2})$.

The intraparticle diffusion model can be determined through the plots the uptake and $t^{\frac{1}{2}}$ with zero intercept, which indicates it as a rate-limiting factor. If the plot does not go through the intercept, then intraparticle diffusion model, as well as other kinetics models, act as a rate-controlling factor [62].

It was noticed that the K_i values increased with solution temperature [62]. With increasing temperature, the pore diffusion in adsorbent materials also increased, and as an outcome, the intraparticle diffusion rate enhanced.

6.5 Kinetics of finite bath and infinite bath experiment

With the change in source type, the composition of wastewaters changes significantly. Therefore, the results of adsorption treatment and sorption rate characteristics also differ. Therefore, a definite applications procedure is needed [63]. Finite bath is a continuous-batch reactor system where the concentration of adsorbate decreases with proceeding batch reaction. It is a simple experimental adsorption study but mathematically challenging in interpretation. For avoiding these mathematical difficulties, the concept of an infinite bath was used by Snoeyink and Weber. It was assumed that in the spherical particle, the radial movement of adsorbate occurs, because of adsorption and diffusion mechanism [62].

$$D\left(\frac{\partial^2 C_f}{\partial r^2} + \frac{2}{r}\frac{C_f}{\partial r}\right) = \frac{\partial}{\partial t}\left(\epsilon C_f + (1-\epsilon)C_s\right),$$

where C_f = fluid-phase concentration, C_s = solid-phase concentration, ε = porosity, r = radial distance, and D = porosity diffusion coefficient.

Equation describes the concentration of fluid phase and solid phase after passing of time (t) and radial distance (r). In the batch process, adsorption always occurs rapidly, and nonlinear Langmuir model can describe the equilibrium.

$$C_s = \frac{C_s^* b C_f}{1 + b C_f}$$

b = Langmuir equilibrium energy constant; C_s^* = solid-phase concentration at monolayer coverage.

$$\frac{\partial C_s}{\partial t} = \frac{\partial C_s}{\partial C_f} \cdot \frac{\partial C_f}{\partial t}$$

Depending on the experimental condition in a continuous mixed bath reactor system, a boundary condition of either constant or change of concentration of adsorbate can be applied. Depending on the boundary condition, the equations are as follows:

Finite bath:

$$V\frac{\partial C_f}{\partial t} = -4\pi a^2 D\frac{\partial C_{f_i}}{\partial r}\ r \geq a, t \geq 0$$

$$D\frac{\partial C_{f_i}}{\partial r} = k_f\left(C_f - C_{f_i}\right) r = a, t \geq 0$$

Infinite bath:

$$C_f = C_{f_0}\ r \geq a, t \geq 0$$

$$D\frac{\partial C_{f_i}}{\partial r} = k_f\left(C_{f_0} - C_{f_i}\right) r = a, t \geq 0,$$

where V = volume per particle, a = particle radius, C_{fi} = interfacial fluid-phase concentration, K_f = liquid-phase mass transfer coefficient, and C_{f0} = initial fluid-phase concentration.

Here both the C_{fi} and K_f are important when film diffusion as external resistance is used. In an infinite bath system, the external concentration of solute is constant. To describe the rate of uptake, different complex models are used. For example, sometimes, an irreversible adsorption equilibrium model cannot be applied. The finite-bath experiment is an easier process, which analyzes decreased pollutant concentration with adsorption time. On another side, infinite-bath experiments are more challenging to initiate, particularly when mass transfer and nonlinear adsorption equilibria are involved as boundary effects. But also offers some reproducible results.

7 Thermodynamics

In batch process, thermodynamic consideration is useful to check its spontaneity. The change in Gibbs free energy (ΔG) indicates the spontaneity of a chemical reaction. The Gibbs free energy change is related to the equilibrium constant by the Van't Hoff equation:

$$\Delta G^0 = -RT \ln K_D,$$

where ΔG^0 is the free energy change (J/mol), R is the universal gas constant (8.314 J/mol K), and T is the absolute temperature (K). To determine the Gibbs free energy of the process, both energy and entropy factors are used. The following equations are used to calculate the thermodynamic parameters.

$$\Delta G^0 = \Delta H^0 - T\Delta S^0$$

After combining this equation [64],

$$\ln K_D = \frac{-\Delta G^0}{RT} = \frac{\Delta S^0}{R} - \frac{\Delta H^0}{R} - \frac{1}{T}$$

ΔH^0 = change in enthalpy (kJ/mol), ΔS^0 = change in entropy (kJ/mol K).

The free energy change (ΔG) must be negative for significant adsorption.

The estimation of the thermodynamic parameters, the spontaneity, feasibility nature of the adsorption process in terms of exothermic and endothermic is ascertained. The negative value of ΔG^0 confirms the feasibility and spontaneity of the adsorption process. The decrease in negative values of ΔG^0 with a simultaneous increase in temperature indicates that the adsorption process is more susceptible to higher temperatures [65]. Moreover, the negative values of ΔH^0 indicate the exothermic nature of the process. However, a positive value of ΔS^0 denotes the affinity of the adsorbent for the solute molecule and advocates some structural fluctuations in both adsorbate [66] (Table 2).

8 Desorption studies

Regeneration capacity of the adsorbent and recycling is another relevant aspect of the batch process. Recovery of the adsorbate is another important factor in adsorption process [66]. After estimating the amount of adsorption, the capacity for recycling the

Table 2   Isotherm models, kinetics model, and thermodynamic parameters for the adsorption of pollutants onto adsorbents.

Adsorbent	Adsorbate	Maximum adsorption capacity (mg/g)	Isotherm models	Kinetic models	Thermodynamic parameters	Reference
Modified activated carbon from tea leaf	Rhodamine B	750	Langmuir	Pseudo-second order	$\Delta G° = -3.75$ $\Delta H° = 50.2$ $\Delta S° = 0.178$	[67]
Mesoporous 2-line ferrihydrite	Fluoride	9.6	Freundlich	Pseudo-second order	$\Delta G° = -17.04$ $\Delta H° = -41.21$ $\Delta S° = -0.017$	[68]
Rice polish	Cadmium	9.7	Langmuir	–	$\Delta G° = -2.048$ $\Delta H° = -18.589$ $\Delta S° = -0.056$	[69]
KOH-activated carbon	MB	704	Langmuir	Pseudo-second order	$\Delta G° = -1.7$ $\Delta H° = 16.30$ $\Delta S° = 0.06$	[70]
Humic acid-modified expanded perlite	MB	78	Redlich-Peterson	Pseudo-second order	$\Delta G° = -9.20$ $\Delta H° = -24.83$ $\Delta S° = -0.0493$	[42]
Modified pineapple leaf powder	Methyl Orange dye	47.6	Langmuir	–	$\Delta G° = -3.746$ $\Delta H° = -8.7$ $\Delta S° = 0.016$	[71]
Biochar	Lead	4.2	Freundlich	Pseudo-second order	$\Delta G° = -14.6$ $\Delta H° = 41.1$ $\Delta S° = 0.187$	[72]
Shells of lentil	Copper	8.9	Langmuir	Pseudo-second order	$\Delta G° = -18.8$ $\Delta H° = 15.3$ $\Delta S° = 0.11$	[39]
Sawdust	Chromium	2.7	Langmuir	Pseudo-second order	$\Delta G° = -7.4$ $\Delta H° = -2.07$	[41]
Multiwall carbon nanotubes	Bismarck brown dye	18	Temkin	Pseudo-second order	$\Delta G° = -8.18$ $\Delta H° = 22.03$ $\Delta S° = 0.103$	[73]
Modified pea peel	Crystal violet	11	Sips	Pseudo-second order	$\Delta G° = -7.82$ $\Delta H° = 86.40$ $\Delta S° = 0.306$	[74]
polymeric gels	Bromophenol blue	2.9	Langmuir	Pseudo-second order	$\Delta G° = -3.987$ $\Delta H° = -82.366$ $\Delta S° = -0.2358$	[43]

Table 3 Regeneration studies of different adsorbent.

Adsorbent	Pollutant	Desorption reagent	% of Desorption	Adsorption-desorption cycle no.	Reference
Aniline-formaldehyde condensate polymer	Chromium (VI)	1 N HCl	86	4	[22]
Activated alumina	Cadmium (II)	0.2 M HCl	99	3	[15]
Cerium oxide nanoparticles	Chromium (VI)	0.1 M HCl	80	2	[75]
Amine-based polymer	Chromium (III)	0.5 M H_2SO_4	42	1	[25]
Aniline formaldehyde condensate polymer	Copper (II)	1 N H_2SO_4	97	1	[33]
Activated alumina	Lead	0.2 M HNO_3	98	2	[15]
Aniline formaldehyde condensate polymer	Chromium (VI)	1 N NaOH	1.1	4	[22]
Iron oxide nanoparticles	Erichrome Black- T	NaOH	86	3	[76]
Cow dung carbon	Fluoride	0.2 M NaOH	100	2	[77]
Alumina activated carbon composite	procion red MX-5B dye	Ethanol	92	3	[78]
Barley husk	Cibacron Red	50% methanol	52	1	[79]
Amine-based polymer	Chromium (III)	EDTA	5.1	1	[25]
Aniline formaldehyde condensate polymer	Copper (II)	0.2 N EDTA	100	1	[33]
Magnetic iron nanoadsorbent	Cadmium	HNO_3	98	5	[80]

adsorbent is studied [45]. The amount of desorbed adsorbate is calculated by measuring the concentration of liquid phase, after desorption study.

$$\text{Desorption}\% = \frac{\text{Desorption concentration}}{\text{Initial adsorbes concentration}} \times 100$$

Adsorbent is mixed and placed for shaking with different buffer solutions with different pH for a period of time (Table 3) [22,81]. Due to electrostatic attraction desorption occurred in solution with pH change. In case of dyes, different solvents like methanol, acetone, ethanol, H_2O_2 are used for desorption study [78].

9 Conclusion

The present chapter shows different factors affecting the adsorption capacity in the batch process; it helps to determine the favorable pH of the solution, adsorbent dose, concentration of the contaminant, favorable temperature for maximum utilization of batch process. A well-designed batch tank is required for efficient adsorption process for wastewater treatment. Use of isotherm model to analyze the type of interaction and layer formation between adsorbate-adsorbent was summarized, which could also help to determine the pore type, surface area, capillary condensation of the hysteresis loop. This chapter clearly explains the working mechanism of adsorption, which occurs at different steps during the batch adsorption process. It also discusses various theoretical modeling approaches and kinetic studies that help to understand the adsorption process.

Acknowledgments

The author would also like to acknowledge Mr. Abhishek N. Srivastava, research scholar IIT Delhi, and Mr. Sounak Bhattachariya, research scholar IIT Guwahati, for checking the draft.

Funding

The author wants to thank the Department of Science and Technology (DST)-INSPIRE, Government of India for providing fellowship no. DST/INSPIRE Fellowship/2016/ IF 160093.

References

[1] H. Patel, R.T. Vashi, Batch adsorption treatment of textile wastewater. in: Characterization and Treatment of Textile Wastewater, 2015, pp. 111–125, https://doi.org/10.1016/b978-0-12-802326-6.00004-6.

[2] R.N. Coimbra, C. Escapa, M. Otero, Adsorption separation of analgesic pharmaceuticals from ultra-pure and waste water: batch studies using a polymeric resin and an activated carbon. Polymers (Basel) 10 (2018), https://doi.org/10.3390/polym10090958.

[3] Z. Xu, J. Cai, B. Pan, Mathematically modeling fixed-bed adsorption in aqueous systems. J. Zhejiang Univ. Sci. A. 14 (2013) 155–176, https://doi.org/10.1631/jzus.a1300029.

[4] T.C. Nguyen, P. Loganathan, T.V. Nguyen, J. Kandasamy, R. Naidu, S. Vigneswaran, Adsorptive removal of five heavy metals from water using blast furnace slag and fly ash. Environ. Sci. Pollut. Res. 25 (2018) 20430–20438, https://doi.org/10.1007/s11356-017-9610-4.

[5] A. Arafat, S.A. Samad, et al., Textile dye removal from wastewater effluents using chitosan-ZnO nano-composite. J. Text. Sci. Eng. 5 (2015) 5–8, https://doi.org/10.4172/2165-8064.1000200.

[6] V. Kumar, N. Talreja, D. Deva, N. Sankararamakrishnan, A. Sharma, N. Verma, Development of bi-metal doped micro- and nano multi-functional polymeric adsorbents for the removal of fluoride and arsenic (V) from wastewater. Desalination 282 (2011) 27–38, https://doi.org/10.1016/j.desal.2011.05.013.

[7] I. Ali, Z.A. Alothman, M.M. Sanagi, Green synthesis of iron nano-impregnated adsorbent for fast removal of fluoride from water. J. Mol. Liq. 211 (2015) 457–465, https://doi.org/10.1016/j.molliq.2015.07.034.

[8] C. Xu, J. Li, F. He, Y. Cui, C. Huang, H. Jin, S. Hou, Al_2O_3-Fe_3O_4-expanded graphite nano-sandwich structure for fluoride removal from aqueous solution. RSC Adv. 6 (2016) 97376–97384, https://doi.org/10.1039/c6ra19390k.

[9] L. Cui, Y. Wang, L. Gao, L. Hu, L. Yan, Q. Wei, B. Du, EDTA functionalized magnetic graphene oxide for removal of Pb (II), Hg (II) and Cu (II) in water treatment: adsorption mechanism and separation property. Chem. Eng. J. 281 (2015) 1–10, https://doi.org/10.1016/j.cej.2015.06.043.

[10] S. Pacheco, M. Medina, F. Valencia, J. Tapia, Removal of inorganic mercury from polluted water using structured nanoparticles. J. Environ. Eng. 132 (2006) 342–349, https://doi.org/10.1061/(ASCE)0733-9372(2006)132.

[11] M.N. Haque, G.M. Morrison, G. Perrusqu, M. Gutierr, A.F. Aguilera, I. Cano-aguilera, J.L. Gardea-Torresdey, Characteristics of arsenic adsorption to sorghum biomass. J. Hazard. Mater. 145 (2007) 30–35, https://doi.org/10.1016/j.jhazmat.2006.10.080.

[12] V. Chandra, J. Park, Y. Chun, J.W. Lee, I. Hwang, K.S. Kim, Water-dispersible magnetite-reduced graphene oxide composite for arsenic removal, ACS Nano 4 (2010) 3979–3986.

[13] M. Llor, Removal of cadmium from aqueous solutions by adsorption onto orange waste. J. Hazard. Mater. 139 (2007) 122–131, https://doi.org/10.1016/j.jhazmat.2006.06.008.

[14] F.Y. Wang, H. Wang, J.W. Ma, Adsorption of cadmium (II) ions from aqueous solution by a new low-cost adsorbent—bamboo charcoal. J. Hazard. Mater. 177 (2010) 300–306, https://doi.org/10.1016/j.jhazmat.2009.12.032.

[15] T.K. Naiya, A.K. Bhattacharya, S.K. Das, Adsorption of Cd (II) and Pb (II) from aqueous solutions on activated alumina. J. Colloid and Interface Sci. 333 (2009) 14–26, https://doi.org/10.1016/j.jcis.2009.01.003.

[16] N.E. Ibisi, C.A. Asoluka, Use of agro-waste (*Musa paradisiaca* peels) as a sustainable biosorbent for toxic metal ions removal from contaminated water, Chem. Int. 4 (2018) 52–59.

[17] L. Dong, Z. Zhu, Y. Qiu, J. Zhao, Removal of lead from aqueous solution by hydroxyapatite/magnetite composite adsorbent. Chem. Eng. J. 165 (2010) 827–834, https://doi.org/10.1016/j.cej.2010.10.027.

[18] A. Afkhami, M. Saber-tehrani, H. Bagheri, Simultaneous removal of heavy-metal ions in wastewater samples using nano-alumina modified with 2,4-dinitrophenylhydrazine. J. Hazard. Mater. 181 (2010) 836–844, https://doi.org/10.1016/j.jhazmat.2010.05.089.

[19] D. Thakre, S. Jagtap, N. Sakhare, N. Labhsetwar, S. Meshram, S. Rayalu, Chitosan based mesoporous Ti-Al binary metal oxide supported beads for defluoridation of water. Chem. Eng. J. 158 (2010) 315–324, https://doi.org/10.1016/j.cej.2010.01.008.

[20] M. Leili, M. Fazlzadeh, A. Bhatnagar, Green synthesis of nano-zero-valent iron from Nettle and Thyme leaf extracts and their application for the removal of cephalexin antibiotic from aqueous solutions. Environ. Technol. 39 (2018) 1158–1172, https://doi.org/10.1080/09593330.2017.1323956.

[21] Y.C. Sharma, V. Srivastava, C.H. Weng, S.N. Upadhyay, Removal of Cr (VI) from wastewater by adsorption on iron nanoparticles. Can. J. Chem. Eng. 87 (2009) 921–929, https://doi.org/10.1002/cjce.20230.

[22] P. Terangpi, S. Chakraborty, M. Ray, Improved removal of hexavalent chromium from 10 mg/L solution by new micron sized polymer clusters of aniline formaldehyde condensate. Chem. Eng. J. 350 (2018) 599–607, https://doi.org/10.1016/j.cej.2018.05.171.

[23] K. Gong, Q. Hu, L. Yao, M. Li, D. Sun, Q. Shao, B. Qiu, Z. Guo, Q. Hu, L. Yao, M. Li, D. Sun, Q. Shao, B. Qiu, Z. Guo, Ultrasonic pretreated sludge derived stable magnetic active carbon for Cr(VI) removal from wastewater. ACS Sustain. Chem. Eng. 6 (2018) 7283–7291, https://doi.org/10.1021/acssuschemeng.7b04421.

[24] M. Fazlzadeh, K. Rahmani, A. Zarei, H. Abdoallahzadeh, F. Nasiri, R. Khosravi, A novel green synthesis of zero valent iron nanoparticles (NZVI) using three plant extracts and their efficient application for removal of Cr(VI) from aqueous solutions. Adv. Powder Technol. 28 (2017) 122–130, https://doi.org/10.1016/j.apt.2016.09.003.

[25] P.A. Kumar, M. Ray, S. Chakraborty, Adsorption behaviour of trivalent chromium on amine-based polymer aniline formaldehyde condensate. Chem. Eng. J. 149 (2009) 340–347, https://doi.org/10.1016/j.cej.2008.11.030.

[26] Y.C. Sharma, S.N. Upadhyay, Removal of a cationic dye from wastewaters by adsorption on activated carbon developed from coconut coir, Energy Fuels 23 (2009) 2983–2988.

[27] A. Tor, Y. Cengeloglu, Removal of congo red from aqueous solution by adsorption onto acid activated red mud. J. Hazard. Mater. 138 (2006) 409–415, https://doi.org/10.1016/j.jhazmat.2006.04.063.

[28] B. Ramaraju, P. Manoj, K. Reddy, C. Subrahmanyam, Low cost adsorbents from agricultural waste for removal of dyes. Environ. Progr. Sustain. Energy 4 (2014), https://doi.org/10.1002/ep.11742.

[29] T. Santhi, S. Manonmani, V.S. Vasantha, Y.T. Chang, A new alternative adsorbent for the removal of cationic dyes from aqueous solution. Arab. J. Chem. 9 (2016) S466–S474, https://doi.org/10.1016/j.arabjc.2011.06.004.

[30] M. Islam, R. Patel, Synthesis and physicochemical characterization of Zn/Al chloride layered double hydroxide and evaluation of its nitrate removal efficiency. Desalination 256 (2010) 120–128, https://doi.org/10.1016/j.desal.2010.02.003.

[31] C.P. Devatha, A.K. Thalla, S.Y. Katte, Green synthesis of iron nanoparticles using different leaf extracts for treatment of domestic waste water. J. Clean. Prod. 139 (2016) 1425–1435, https://doi.org/10.1016/j.jclepro.2016.09.019.

[32] A.H. Khan, S. Ahmad, A. Ahmad, Role of sawdust in the removal of copper (II) from industrial wastes, Water Res. 32 (1998) 3085–3091.

[33] G.P. Kumar, P.A. Kumar, S. Chakraborty, M. Ray, Uptake and desorption of copper ion using functionalized polymer coated silica gel in aqueous environment. Sep. Purif. Technol. 57 (2007) 47–56, https://doi.org/10.1016/j.seppur.2007.03.003.

[34] M. Ghaedi, S. Hajjati, Z. Mahmudi, I. Tyagi, S. Agarwal, A. Maity, V.K. Gupta, Modeling of competitive ultrasonic assisted removal of the dyes – methylene blue and Safranin-O using Fe_3O_4 nanoparticles. Chem. Eng. J. 268 (2015) 28–37, https://doi.org/10.1016/j.cej.2014.12.090.

[35] W.J. Weber, Adsorption processes. Pure Appl. Chem. 37 (2008) 375–392, https://doi.org/10.1351/pac197437030375.

[36] B. Dhal, H.N. Thatoi, N.N. Das, B.D. Pandey, Chemical and microbial remediation of hexavalent chromium from contaminated soil and mining/metallurgical solid waste: a review. J. Hazard. Mater. 250–251 (2013) 272–291, https://doi.org/10.1016/j.jhazmat.2013.01.048.

[37] B. Markiewicz, I. Komorowicz, A. Sajnóg, M. Belter, D. Barałkiewicz, Chromium and its speciation in water samples by HPLC/ICP-MS - technique establishing metrological traceability: a review since 2000. Talanta 132 (2015) 814–828, https://doi.org/10.1016/j.talanta.2014.10.002.

[38] R. Rakhunde, L. Deshpande, H.D. Juneja, Chemical speciation of chromium in water: a review. Crit. Rev. Environ. Sci. Technol. 42 (2012) 776–810, https://doi.org/10.1080/10643389.2010.534029.

[39] H. Aydın, Y. Bulut, Removal of copper (II) from aqueous solution by adsorption onto low-cost adsorbents. J. Environ. Manage. 87 (2008) 37–45, https://doi.org/10.1016/j.jenvman.2007.01.005.

[40] N. Fiol, I. Villaescusa, Determination of sorbent point zero charge: usefulness in sorption studies. Environ. Chem. Lett. 7 (2009) 79–84, https://doi.org/10.1007/s10311-008-0139-0.

[41] S.S. Baral, S.N. Das, P. Rath, Hexavalent chromium removal from aqueous solution by adsorption on treated sawdust. Biochem. Eng. J. 31 (2006) 216–222, https://doi.org/10.1016/j.bej.2006.08.003.

[42] W.J. Luo, Q. Gao, X.L. Wu, C.G. Zhou, Removal of cationic dye (methylene blue) from aqueous solution by humic acid-modified expanded perlite: experiment and theory. Sep. Sci. Technol. 49 (2014) 2400–2411, https://doi.org/10.1080/01496395.2014.920395.

[43] M.A. Malana, S. Ijaz, M.N. Ashiq, Removal of various dyes from aqueous media onto polymeric gels by adsorption process: their kinetics and thermodynamics. Desalination 263 (2010) 249–257, https://doi.org/10.1016/j.desal.2010.06.066.

[44] T. Hu, Q. Liu, T. Gao, K. Dong, G. Wei, J. Yao, Facile preparation of tannic acid-poly(vinyl alcohol)/sodium alginate hydrogel beads for methylene blue removal from simulated solution. ACS Omega 3 (2018) 7523–7531, https://doi.org/10.1021/acsomega.8b00577.

[45] V.K. Gupta, R. Jain, M. Shrivastava, A. Nayak, Equilibrium and thermodynamic studies on the adsorption of the dye tartrazine onto waste "coconut husks" carbon and activated carbon. J. Chem. Eng. Data 55 (2010) 5083–5090, https://doi.org/10.1021/je100649h.

[46] Metcalf and Eddy, Wastewater Engineerig; Treatment and Reuse, McGraw-Hill, Boston, 2003.

[47] V. Russo, M. Trifuoggi, M. Di Serio, R. Tesser, Fluid-solid adsorption in batch and continuous processing: a review and insights into Modeling. Chem. Eng. Technol. 40 (2017) 799–820, https://doi.org/10.1002/ceat.201600582.

[48] Y.S. Ho, Removal of copper ions from aqueous solution by tree fern. Water Res. 37 (2003) 2323–2330, https://doi.org/10.1016/S0043-1354(03)00002-2.

[49] F.M. Mohammed, E.P.L. Roberts, A.K. Campen, N.W. Brown, Wastewater treatment by multi-stage batch adsorption and electrochemical regeneration. J. Electrochem. Sci. Eng. 2 (2012) 223–236, https://doi.org/10.5599/jese.2012.0019.

[50] K.Y. Foo, B.H. Hameed, Insights into the modeling of adsorption isotherm systems. Chem. Eng. J. 156 (2010) 2–10, https://doi.org/10.1016/j.cej.2009.09.013.

[51] R. Sips, On the structure of a catalyst surface. J. Chem. Phys. 16 (1948) 490–495, https://doi.org/10.1063/1.1746922.

[52] M. Kruk, M. Jaroniec, Gas adsorption characterization of ordered organic-inorganic nanocomposite materials. Chem. Mater. 13 (2001) 3169–3183, https://doi.org/10.1021/cm0101069.

[53] M. Thommes, Physical adsorption characterization of nanoporous materials. Chem.-Ing.-Tech. 82 (2010) 1059–1073, https://doi.org/10.1002/cite.201000064.

[54] R. Ryoo, I.S. Park, S. Jun, C.W. Lee, M. Kruk, M. Jaroniec, Synthesis of ordered and disordered silicas with uniform pores on the border between micropore and mesopore regions using short double-chain surfactants. J. Am. Chem. Soc. 123 (2001) 1650–1657, https://doi.org/10.1021/ja0038326.

[55] K.K.H. Choy, J.F. Porter, G. Mckay, Langmuir isotherm models applied to the multicomponent sorption of acid dyes from effluent onto activated carbon. J. Chem. Eng. Data 45 (2000) 575–584, https://doi.org/10.1021/je9902894.

[56] B. Cheng, Y. Le, W. Cai, J. Yu, Synthesis of hierarchical Ni(OH)$_2$ and NiO nanosheets and their adsorption kinetics and isotherms to Congo red in water. J. Hazard. Mater. 185 (2011) 889–897, https://doi.org/10.1016/j.jhazmat.2010.09.104.

[57] Y.S. Ho, G. Mckay, Pseudo-second order model for sorption processes, Process Biochem. 34 (1999) 451–465.

[58] D. Robati, Pseudo-second-order kinetic equations for modeling adsorption systems for removal of lead ions using multi-walled carbon nanotube. J. Nanostruct. Chem. 3 (2013) 55, https://doi.org/10.1186/2193-8865-3-55.

[59] H. Moussout, H. Ahlafi, M. Aazza, H. Maghat, Critical of linear and nonlinear equations of pseudo-first order and pseudo-second order kinetic models. Karbala Int. J. Mod. Sci. 4 (2018) 244–254, https://doi.org/10.1016/j.kijoms.2018.04.001.

[60] S.H. Chien, W.R. Clayton, Application of Elovich equation to the kinetics of phosphate release and sorption in Soils1. Soil Sci. Soc. Am. J. 44 (2010) 265, https://doi.org/10.2136/sssaj1980.03615995004400020013x.

[61] B. Tanhaei, A. Ayati, M. Lahtinen, M. Sillanpää, Preparation and characterization of a novel chitosan/Al$_2$O$_3$/magnetite nanoparticles composite adsorbent for kinetic, thermodynamic and isotherm studies of methyl orange adsorption. Chem. Eng. J. 259 (2015) 1–10, https://doi.org/10.1016/j.cej.2014.07.109.

[62] S.M. Yakout, E. Elsherif, Batch kinetics, isotherm and thermodynamic studies of adsorption of strontium from aqueous solutions onto low cost rice-straw based carbons, Carbon Sci. Technol. 3 (2010) 144–153. http://www.applied-science-innovations.com.

[63] F.A. Digiano, W.J. Weber, Sorption Kinetics in Infinite-Bath Experiments, Wiley, 1973.

[64] V.C. Srivastava, I.D. Mall, I.M. Mishra, Adsorption thermodynamics and isosteric heat of adsorption of toxic metal ions onto bagasse fly ash (BFA) and rice husk ash (RHA). Chem. Eng. J. 132 (2007) 267–278, https://doi.org/10.1016/j.cej.2007.01.007.

[65] A.B. Zaki, M.Y. El-Sheikh, J. Evans, S.A. El-Safty, Kinetics and mechanism of the sorption of some aromatic amines onto Amberlite IRA-904 anion-exchange resin. J. Colloid Interface Sci. 221 (2000) 58–63, https://doi.org/10.1006/jcis.1999.6553.

[66] V.K. Gupta, Equilibrium uptake, sorption dynamics, process development, and column operations for the removal of copper and nickel from aqueous solution and wastewater using activated slag, a low-cost adsorbent. Ind. Eng. Chem. Res. 37 (1998) 192–202, https://doi.org/10.1021/ie9703898.

[67] M. Goswami, P. Phukan, Enhanced adsorption of cationic dyes using sulfonic acid modi fi ed activated carbon. J. Environ. Chem. Eng. 5 (2017) 3508–3517, https://doi.org/10.1016/j.jece.2017.07.016.

[68] B.-S. Zhu, L.-T. Kong, J.-H. Liu, Y. Jia, Z. Jin, B. Sun, T. Luo, A facile precipitation synthesis of mesoporous 2-line ferrihydrite with good fluoride removal properties. RSC Adv. 5 (2015) 84389–84397, https://doi.org/10.1039/C5RA15619J.

[69] K.K. Singh, R. Rastogi, S.H. Hasan, Removal of cadmium from wastewater using agricultural waste 'rice polish'. J. Hazard. Mater. 121 (2005) 51–58, https://doi.org/10.1016/j.jhazmat.2004.11.002.

[70] K.C. Bedin, A.C. Martins, A.L. Cazetta, O. Pezoti, V.C. Almeida, KOH-activated carbon prepared from sucrose spherical carbon: adsorption equilibrium, kinetic and thermodynamic studies for methylene blue removal. Chem. Eng. J. 286 (2016) 476–484, https://doi.org/10.1016/j.cej.2015.10.099.

[71] A.A. Kamaru, N.S. Sani, N. Ahmad, N. Nik, Raw and surfactant-modified pineapple leaf as adsorbent for removal of methylene blue and methyl orange from aqueous solution. Desalin. Water Treat. 57 (2015) 18836–18850, https://doi.org/10.1080/19443994.2015.1095122.

[72] Z. Liu, F. Zhang, Removal of lead from water using biochars prepared from hydrothermal liquefaction of biomass. J. Hazard. Mater. 167 (2009) 933–939, https://doi.org/10.1016/j.jhazmat.2009.01.085.

[73] A.M. Kamil, F.H. Abdalrazak, Adsorption of bismarck brown r dye onto multiwall carbon nanotubes. J. Environ. Anal. Chem. 1 (2016) 1–6, https://doi.org/10.4172/jreac.1000104.

[74] T.A. Khan, R. Rahman, E.A. Khan, Decolorization of bismarck brown R and crystal violet in liquid phase using modified pea peels: non-linear isotherm and kinetics modeling. Model. Earth Syst. Environ. 2 (2016) 1–11, https://doi.org/10.1007/s40808-016-0195-6.

[75] S. Recillas, J. Colón, E. Casals, E. González, V. Puntes, A. Sánchez, X. Font, Chromium VI adsorption on cerium oxide nanoparticles and morphology changes during the process. J. Hazard. Mater. 184 (2010) 425–431, https://doi.org/10.1016/j.jhazmat.2010.08.052.

[76] B. Saha, S. Das, J. Saikia, G. Das, Preferential and enhanced adsorption of different dyes on iron oxide nanoparticles: a comparative study. J. Phys. Chem. C 115 (2011) 8024–8033, https://doi.org/10.1021/jp109258f.

[77] S. Rajkumar, S. Murugesh, V. Sivasankar, A. Darchen, T.A.M. Msagati, T. Chaabane, Low-cost fluoride adsorbents prepared from a renewable biowaste: syntheses, characterization and modeling studies. Arab. J. Chem. 12 (2015) 3004–3017, https://doi.org/10.1016/j.arabjc.2015.06.028.

[78] F. Fatma, P.L. Hariani, F. Riyanti, W. Sepriani, Desorption and re-adsorption of procion red MX-5B dye on alumina-activated carbon composite. Indones. J. Chem. 18 (2018) 222–228, https://doi.org/10.22146/ijc.23927.

[79] T. Robinson, B. Chandran, P. Nigam, Studies on desorption of individual textile dyes and a synthetic dye effluent from dye-adsorbed agricultural residues using solvents, Bioresour. Technol. 84 (2002) 299–301.

[80] V.K. Gupta, A. Nayak, Cadmium removal and recovery from aqueous solutions by novel adsorbents prepared from orange peel and Fe_2O_3 nanoparticles. Chem. Eng. J. 180 (2012) 81–90, https://doi.org/10.1016/j.cej.2011.11.006.

[81] H. Lu, G. Yi, S. Zhao, D. Chen, L.H. Guo, J. Cheng, Synthesis and characterization of multi-functional nanoparticles possessing magnetic, up-conversion fluorescence and bio-affinity properties. J. Mater. Chem. 14 (2004) 1336–1341, https://doi.org/10.1039/b315103d.

2

Removal of heavy metals from industrial effluents by using biochar

Manash Gope and Rajnarayan Saha

DEPARTMENT OF CHEMISTRY, NATIONAL INSTITUTE OF TECHNOLOGY DURGAPUR (NITD), DURGAPUR, WEST BENGAL, INDIA

1 Introduction

Heavy metals pose a threat to human health due their nonbiodegradability and toxic nature with availability in all natural and man-made environments. Heavy metals are discharged into environment due to various man-made activities, such as disposal of municipal and industrial waste released due to different activities, heating of houses, emissions from factories, heavy traffic, etc. [1]. They are mainly released into the environment from different point sources such as battery manufacturing industries, mining industries, metal plating works, steel industries, paper industries, cement industries, etc. [2]. Untreated effluent from industries are directly released into river water creating various diseases in fish, which are no longer safe for consumption. Recent studies reported that some developing countries discharge more than 80% and 70% of sewage effluent and industrial waste, respectively, into surface water without any treatment [3]. Heavy metals are very harmful to living organism as well as humans in very low concentration [4]. Heavy metals have carcinogenic and non-carcinogenic health effects on humans and have various negative effects on plant physiology. They are bioaccumulated in the food chain because they are not biodegradable and cause various effects to human health [5–7]. Table 1 abridges the health effects of heavy metals. Every day, industries generate a huge amount of waste, and this becomes a burden to us and creates various environmental problems. So, the effluents and wastewater released from these industries need to be treated as the overgrowth of human population and parallel outbreak of industrialization leads to more demand for safe water. Treated and recycled wastewater can be used as alternative sources, which decreases the worldwide demand for safe water [12]. The event of raised concentration of heavy metals in wastewater poses contests in remediation of the environment. Table 1 also abridges a few heavy metals positioned by their poisonousness by the Comprehensive Environmental Response Compensation and Liability Act (CERCLA) [8]. Likewise, it also records the present maximum allowable contaminant levels (MCLs) given by the Environmental Protection Agency (EPA) for these metals [9].

Intelligent Environmental Data Monitoring for Pollution Management. https://doi.org/10.1016/B978-0-12-819671-7.00002-6

Table 1 The maximum contaminant level (MCL) standards for the most hazardous heavy metals [8–11].

Heavy metals	Toxicities	MCL (mg/L)	MACL by EPA (mg/L)	Rank (CERCLA)
As	Dermal appearances, visceral malignancies, vascular ailment	0.05	0.010	1
Cd	Kidney injury, renal damage, causes human cancer	0.01	0.005	7
Cr	Headache, diarrhea, nausea, unsettled stomach, causes cancer	0.05	0.1	17
Cu	Hepatic impairment, Wilson disease, sleeplessness	0.25	1.3	125
Hg	Rheumatoid arthritis, and renal disease, circulatory system damage, and nervous system impairment	0.00003	0.002	3
Ni	Skin problem, nausea, chronic asthma, coughing, causes cancer	0.20	–	57
Zn	Misery, tiredness, neurological symptoms, and dehydration	0.80	0.015	2
Pb	Brain damage, Renal diseases, cardiovascular system, and nervous system damage	0.006	5.0	75

Note: Maximum allowable contaminant level was derived from the 2011 Compensation and Liability Act list of priority unsafe substances and EPA's drinking water standards [8].

There are various processes to deal with heavy metal concentrations in wastewater. Out of these processes, sorption has been recognized as an outstanding technique for exclusion of heavy metals from polluted water. Present review work has focused on different processes with their respective advantages and disadvantages of heavy metal removal from polluted water. Adsorption process using biochar, which is more effective and cheaper compared with other conventional processes, is discussed in detail in this chapter. Moreover, the variety of feedstocks (raw materials) and their removal efficiency is also discussed in the present chapter.

2 Industrial effluents and heavy metal pollution

A number of elements are present in industrial effluents, out of which heavy metals (Cd, Cr, Cu, Ni, As, Pb, Hg, and Zn) are a special group having density of 5 g/cc. Industrial wastewater carries a substantial quantity of toxic heavy metals that produce a problematic condition for safe utilization of farming soil [13]. Table 2 summarizes the sources of heavy metals.

3 Conventional processes of removal heavy metals from effluent

Many conventional technologies have been practiced worldwide for heavy metal removal from wastewater. Every single technology has some advantages and disadvantages [11]. Table 3 shows the advantage and disadvantages of each heavy metal removal technique.

Table 2   Sources of heavy metals [11].

Heavy metals	Sources of contaminations
Pb	Smoking, vehicle emissions, burning of coal, mining, antiknock agents, paints, pesticides, lead-acid batteries, ceramics, plastic, in alloys, sheets, pipes, or tubing.
Cr	Blowdown of cooling tower, Metal plating and coating operation, production of steels, textiles and leather tanning, timber treatment.
Cd	Ni-Cd batteries, pigments, anti-corrosive metal coatings, plastic stabilizers, amalgams, neutron absorbers in nuclear reactors, steel industries, cooling tower, electroplating.
As	Pesticides, fungicides, mining, production, metallurgy, and protection (preserve) of timber.
Hg	Deposition of minerals, fossil fuel or ores, pest killer, batteries, paper manufacture.
Cu	Pesticides manufacturing, mining, piping of metals, chemical industries.
Zn	Wood pulp making, ground and newspaper production, steel works with galvanizing lines, Zinc alloys (bronze, brass), anti-corrosion coating, batteries, refineries.
Ni	Battery manufacturing, Ni/Cd batteries, amalgams, zinc base molding, printing, coating and electroplating, silver refineries.

Table 3   Heavy metal removal techniques, and their advantages and disadvantages [10,11].

Method	Advantages	Drawbacks
Ion exchange	<ul><li>Great treatment capacity</li><li>High exclusion efficiency</li><li>Quick kinetics</li></ul>	<ul><li>Very costly owing to artificial resins</li><li>Revival of the resins leads to significant secondary pollution</li><li>Creation of waste products</li></ul>
Chemical precipitation	<ul><li>Easy operation</li><li>Small investment cost</li></ul>	<ul><li>Sludge formation</li><li>Additional cost for sludge dumping</li><li>Useless for lower concentrations of heavy metals</li></ul>
Membrane filtration	<ul><li>Great segregation selectivity</li><li>Little space requirement</li><li>Pressure necessity less</li></ul>	<ul><li>Membrane fouling leads to high operational cost</li><li>Complicated process</li><li>Low permeate flux</li></ul>
Electrodialysis	<ul><li>High separation selectivity</li></ul>	<ul><li>Operational cost very high because of energy consumption and membrane fouling</li></ul>
Photocatalysis	<ul><li>Elimination of metals as well as organic pollutant concurrently</li><li>Nontoxic byproducts</li></ul>	<ul><li>Restricted applications</li><li>Extensive duration time</li></ul>
Coagulation and flocculation	<ul><li>Destroys turbidity along with heavy metal exclusion</li><li>Formed sludge with decent sludge settling and de-watering characteristics</li></ul>	<ul><li>Greater generation of sludge volume</li><li>It can't treat the wastewater in terms of heavy metals, and must be trailed by other treatment practices</li></ul>
Flotation	<ul><li>Great metal choosiness</li><li>High removal efficacy</li><li>Excess overflow rates</li><li>Short detention stage</li><li>Cost-effective</li></ul>	<ul><li>High preliminary investment cost</li><li>Maintenance and operation costs are very high</li></ul>
Adsorption	<ul><li>Easy operating conditions</li><li>More metal binding capabilities</li><li>Wide range of pH</li><li>Run under very little cost</li></ul>	<ul><li>Selectivity is less</li></ul>

3.1 Chemical precipitation

In this method, precipitates of carbonates, hydroxide, and sulfide (CO^{3-}, OH^-, and S^{2-}, respectively), which are insoluble, have been formed by the reaction between chemicals and metal ions, which can be segregated from water through sedimentation or filtration. This strategy is best suitable for eviction of bivalent heavy metals, such as Cu (2^+), Cd (2^+), and Zn (2^+) [11]. Out of insoluble precipitates, sulfide precipitation removes a huge amount of pollutant [14]. This technique is more affordable but usually nominated for treatment of polluted water comprising high concentrations of heavy metal ions. Chemical precipitation is not suitable for low metal ion concentrations.

3.2 Ion exchange

Ion exchange is an alternative practice that eradicates metal ions from polluted water. An ion exchanger is a resin that accomplishes swapping cations or anions from adjacent materials. Resins are either natural or synthetic with the capability to exchange ions with the wastewater metal [11]. Artificial resins are generally favored due to their complete elimination abilities of heavy metal from the wastewater [15]. Although ion exchange is a well-recognized and effective procedure of metal removal, it has some disadvantages. Ion exchange is nonselective and enormously dependent on the pH of the solution; the medium gets easily fouled by different materials in the wastewater; and the resin requires retrieval on a routine basis to uphold effective exclusion of the pollutants that can raise the expenditure of the entire unit process [10,11]. An obvious drawback is corrosion to the electrodes, and frequent replacement of the electrodes is required [16].

3.3 Membrane filtration

Membrane filtration has gotten extensive consideration as it can expel heavy metals. Membrane filtration is a partition procedure for heavy metals driven by pressure, which is based on size exclusion. Electrodialysis (ED), nanofiltration, ultrafiltration, and reverse osmosis are examples of this process for actual exclusion of heavy metals from wastewater [10]. However, this process isn't monetarily feasible due to higher maintenance and operational costs.

3.4 Photocatalysis process

Photocatalysis is a renowned advanced oxidation processes (AOPs) that uses nontoxic semiconductors instead of chemical compounds [17]. The photocatalysis process became widespread as an efficient technique because of its simple design, low-cost operation, and high removal efficiency [18]. This process is commanding in abolishing pollutants such as heavy metals from wastewater. In spite of the expansion of the photocatalysis processes, it has some restrictions like producing unwanted byproducts, fault in absorbing visible light, and recombination of electron or whole [19].

3.5 Electrodialysis

ED is an innovative liquid hybrid membrane process used in the elimination of heavy metals from wastewater of diverse industries [17]. ED has ion-discriminating exchange membranes (IEMs) that do not transfer cations and anions at the same time. There are two different membranes: cation exchange membranes and anion exchange membranes [20]. Electrical potential or concentration slope is the reason for ion transportation through these membranes [17]. ED depresses the scaling and fouling and has higher retrieval rates. But on the other side, it has some disadvantages such as a requirement of more plumbing and electrical controls.

3.6 Electrochemical treatments

Electrochemical technologies are cost-effective, more efficient, and more compact. Three key electrochemical treatment processes are electrocoagulation (EC), electroflotation (EF), and electrodeposition. Wastewater treatment through electrochemical technique has received less consideration so far due to the requirement of large investments of money and expensive electricity supply. EC is a simple and creative skill but never acknowledged as a dependable method due to its poor organized reactor design and problems of electrode trustworthiness [17]. On other side, EF became widespread because of its flexibility, easiness in strategy and process, environmental compatibility, low running expenses, and small and compressed units [21]. But the limitation of EF is that it can only be used for lower concentration solutions (50 ppm).

3.7 Adsorption

Adsorption is a physicochemical technique that depends on mass exchange between the fluid phase (wastewater) and the solid phase (adsorbent). These adsorbents give huge surface areas and high permeable structures. The additional purpose of this system is that the adsorbents will be restored by desorption. Adsorbents can be retrieved from horticultural waste, industrial byproducts, or natural materials. Until this point of time, a variety of adsorbents have been explored. Required qualities include a huge explicit surface region, suitable pore and surface structure, simple to produce and recover, as well as great mechanical characteristics. The most utilized adsorbents incorporate carbon materials, activated carbon (AC), carbon nanotubes (CNT), sawdust molecular sieves, silica gel, normal clay, or other rising biomaterials [22–24]. To fulfil the guidelines, different exclusion systems (e.g., ion exchange, precipitation, EC, actuated carbon adsorption, and pressed bed filtration) have been used in wastewater treatment facilities. The majority of these strategies are overpriced, and there is a requirement to grow more cheap innovations with minimal cost adsorbents. Biochar can satisfy these prerequisites as an adsorbent in wastewater treatment process.

4 Biochar: The adsorbent

Among numerous treatment techniques, adsorption is a fast and worldwide method for eliminating heavy metals with high efficiency and low cost. Numerous sorbents (natural and artificial materials) have been developed to date [25]. Ali [26] confirmed that carbon-based adsorbent was evidenced to be the most profitable in the elimination of heavy metal from polluted water. Biochar is not as carbonized (additional hydrogen and oxygen persist in its structure) compared with AC [27,28] and has newly recognized large consideration due to its several applications and benefits [29]. Biochar is black carbon derived from pyrolysis of carbon-rich biomass in anaerobic condition or with limited supply of oxygen. Other than heavy metal removal from wastewater, biochar also plays a significant role in carbon sequestration, soil fertility enhancement [30], bioenergy production [31], and environmental remediation [28]. Biochar formation is a way to minimize solid waste, and the cost of biochar formation is not so high. A huge amount of solid waste biomass is produced daily, which is used to form biochar, which is a porous and solid derivative of pyrolysis [32,33]. Pyrolysis is a process in which thermal changes occur in organic material in an anaerobic condition and temperature maintained between 300°C and 900°C [34]. Due to low cost and other advantages of biochar, it is an effective and widely accepted process for metal removal, but it can be more effective after some modifications including oxidization of the surface of the biochar, exploration, and functionalization. These modifications are used to increase the sorption of biochar through increasing the surface or sorption area, and the addition of a functional group in the surface of the biochar increases the sorption of pollutants [35]. Trapping of heavy metals in biochar is a type of transformation of heavy metals from bioavailable to relatively stable form [36]. Production of biochar reduces the amount of waste material as well as reduces the cost of waste management; it also helps remove heavy metals from environmental media, and it also produces renewable bioenergy by anaerobic digestion and pyrolysis [2]. Adsorption of heavy metals into biochar matrix is pH and temperature dependent. Research interest in the subtraction of heavy metal from water increased due to the availability and cost-effectiveness of biochar. Biochar is produced at <700°C temperature, in anaerobic conditions, and is widely used for various purposes (adopter, capacitors, adsorbents, etc.) due to its stability, obtainability, and lower price [37]. Pyrolysis that occurs in high temperatures can remove more heavy metals from the media than pyrolysis occurring at low temperature [32]. Various types of catalysts are present that can increase the activity of biochar; nanoparticles are one that can increase the adsorption capacity of biochar [38,39]. Numerous current publications have provided indications of biochar's outstanding ability to arrest organic [40,41] and inorganic pollutants [42,43] in soil and water systems.

There are various treatment processes present, but adsorption is the rapid and widespread technology in the field of heavy metal removal as well as biochar of various types depending on the carbon, oxygen, and hydrogen content, and their absorbent capacity are also different [44]. Various commercial adsorbents are available, but their cost is high. However, biochar is a low-cost material with high efficiency [45]. Several natural and synthetic sorbents are present, but among them carbonaceous adsorbents are more profitable to remove organic and inorganic pollutants.

Adsorption of metals into biochar is the most effective process for metal removal from the sludge of acid mine drainage [46]. Various research revealed that material generated from a blast furnace is known as slag, and this slag is used as a good sorbent for Zn and Cd from wastewater [47,48]. Bagasse fly ash, which is a byproduct of sugarcane industries, is actively used to remove of Cu, Zn, Cd, Ni, and Cr from wastewater [12]. *Aspergillus versicolor* is a type of fungus that is a very good absorbent of Hg [49]. *Oedogonium* sp is a type of algae, and this alga is a very good adsorbent of Zn and Cr from polluted water [25]. Biochar is an effective absorbent due to (1) negatively charged carbon surface easily traps positively charged metal ions and makes them stable and inactive; (2) ionic alteration between positively charged protons of biochar with negatively charged cations of metals; (3) absorptive contact between delocalized π electron of carbon [47]. Acceptance of biochar use for metal removal is increasing each day, and it is widely accepted by the researchers due to its efficiency, low cost of operation, and easy way to manufacture.

Temperature during the pyrolysis process has great influence on heavy metal adsorption through biochar [33]. The efficiency of biochar decreases with increasing temperature. If pyrolysis occurs in high temperature, the ash content of char become high, porosity of char increases, and the functional group present in surface also decreases in number. On the other hand, some metals are adsorbed by biochar produced in high temperature (800–900°C) [33]. Sometimes researchers mixed in some synthetic material to modify and increase the efficiency of the biochar. In a comparative study, it was observed that sulfurized biochar was more efficient than normal wood biochar in the case of mercury adsorption [50]. Preparation of magnetic biochar is a new concept for heavy metal adsorption. Son et al. [51] prepared magnetic biochar for heavy metal removal from a polluted solution. In this process, normal biochar was prepared from macro-algal biomass, and then the biochar was deep in $FeCl_3$ solution and was agitated through magnetic stirrer for 30 min. Biochar is produced through pyrolysis, and it is a high carbon-containing material, produced from degradation of organic matter in an anaerobic condition when temperature is maintained. In the present scenario, burning of agricultural waste is very harmful for the environment due to the production of huge amounts of carbon dioxide as well as other environmental pollution. But if we use these huge amounts of agricultural biomass to form biochar, it will be beneficial for the country and will be a pollution-reduction step. The economy of our country is agriculture-based, and this agriculture produces tons of waste every year. Use of these low-cost agricultural wastes for biochar production will decrease the cost of biochar [2]. Low-cost biochar is used as a cost-beneficial remover of heavy metal contaminant from various solutions and is developing into a wastewater treatment skill [2,52–55]. Biochars made up of agronomic residues, animal discards, and woody constituents were also tested for their efficiency for heavy metals like copper, nickel, and cadmium sorption [2,55,56]. Inyang et al. [2] also reported that biochar prepared from sugarcane bagasse through anaerobic condition is a more active sorbent of Pb than the biochar prepared from undigested bagasse, and its activity is higher than the marketable AC. Biochar can be produced by pyrolysis of sewage sludge [46], macro-algae [32,51], agricultural waste [2,34,37,39,44,57], dairy products [2,12], etc.

5 Preparation of biochar

Biochars are made from a variety of biomasses that have diverse chemical and physical properties. Wide-ranging feedstock biomasses have been utilized in the making of biochar [58]. Organic waste [59], agricultural waste [60–62], kitchen waste, sewage sludge [46], bioenergy crops [52], and forest residues [63] are used in production of biochar. A variety of properties such as volatile constituent, water content, and particle size in feedstocks affects the characteristics of obtained biochar. Properties of biomass that are more lignin and less cellulose lead to higher yield of biochar [60]. More lignin content in the biomass feedstock leads to higher porosity in biochar. Biochars produced at higher temperatures, which results in a lower oxygen-to-carbon ratio, are anticipated to be π-donors, whereas biochars obtained at lower temperatures are likely to be π-acceptors [64].

Production of biochar has been achieved through pyrolysis (slow, intermediate, fast), gasification, hydrothermal carbonization, or flash carbonization [65] (Fig. 1). Biochar used to be produced in earthen kilns in modern retorts. Pyrolysis, gasification, and combustion procedure are carried out in in the earthen kiln layer, whereas pyrolysis and combustion

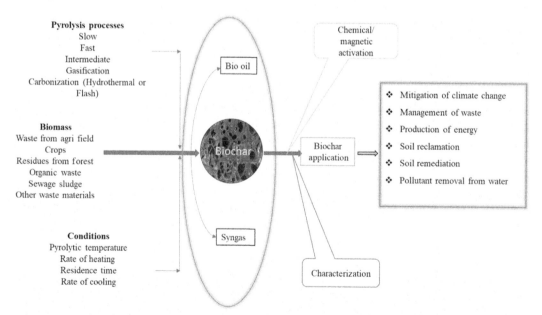

FIG. 1 Diagram of production and applications of biochar with affecting factors. *Source: O.D. Nartey, B. Zhao, Biochar preparation, characterization, and adsorptive capacity and its effect on bioavailability of contaminants: an overview, Adv. Mater. Sci. Eng. 2014 (2014) 715398, https://doi.org/10.1155/2014/715398.*

processes are materially parted by a metal barrier in modern retorts [66]. Pyrolysis process, whether it is slow or fast, can be determined by the temperature in the furnace, pressure, and the heating rate. Slow pyrolysis can be defined as a nonstop process, where feedstock biomass is moved into an oxygen-free externally heated kiln or furnace. On the other hand, fast pyrolysis is subjected to very quick heat transfer or rapid heating rate ($100–1000°C/s$) [67]. The properties of the biochar product are heavily dependent on pyrolysis type, size of biomass, and residence time in furnace [66]. For the gasification process, the biomass feedstock is oxidized up to some level in the chamber at about 800°C temperature at atmospheric or raised pressure [68]. The prime product in gasification pyrolysis is gas with little or no biochar, as liquids are formed. The aqueous carbonization of biomass feedstock is begun by utilizing high pyrolytic temperature ($200–250°C$) to a biomass in a suspension fluid under high environmental weight for a few hours. It produces solids, fluid, and gaseous items [69] in a process called wet pyrolysis [70]. Because oxygen is disconnected to the reactor with the biomass fluid suspension, this clarification is promptly acknowledged. Then again, with flash carbonization of the biomass, a glare fire is lit at a raised pressure (at around 1–3 MPa) underneath a stuffed bed biomass. The fire rises through the carbonization bed against the descending progression of air provided to the procedure.

Crude feedstocks were dried at 60–103°C for 24 to 72 h relying on different analysts [2,50,57,71,72]. Dried feedstocks were converted into biochar through slow pyrolysis at 300–700°C for 1 to 6 h in an N_2 environment in a furnace [2,50,57,71,72]. Very high temperature is not preferable for biochar formation because a high temperature increases the ash content of char [57]. The obtained biochar materials were then ground to fine particles and used for characterization and further applications.

6 Properties of biochar

The surface area is very important, as it can control the efficiency of the biochar in removal of heavy metals from wastewater. The surface area depends on the pyrolysis condition, raw feedstock, etc. Brunauer Emmett Teller (BET) method is used to calculate the surface area and pore volume. The surface area ($m^2\,g^{-1}$) of biochar increased many times from its feedstock. Different researchers reported different enhancement of biochar surface areas. Kim et al. [73] stated enhancement of surface area of pine cone biochar from 2.9 to 175 $m^2\,g^{-1}$; Gao et al. [57] reported 6.95 to 41.4 $m^2\,g^{-1}$ for rice straw biochar and 6.08 to 34.8 $m^2\,g^{-1}$ for sewage; and Devi and Saroha [36] reported 4.8 to 67 $m^2\,g^{-1}$ for paper mill sludge biochar.

The Fourier transform infrared (FTIR) spectrum shows some sharp peaks, which defined some functional groups associated with biochars [36]. The FTIR window is recorded in the range of 4000–500 cm^{-1} [7]. C=C stretch indicates the aromaticity of the biochar [73]. The peaks at 3405–3443 cm^{-1} signify —OH stretching vibration [34], and the trough observed at 3753 cm^{-1} is signifies the occurrence of amine and —OH groups [36]. The broad band around 1400 cm^{-1} and thin band around 860 cm^{-1} are accredited to the existence of CaO and $CaCO_3$ in the biochar, respectively [74]. The biochar handled several functional groups, such as O—H, C—H, C=O and C—N, C=N, phenols, aromatic ring, aromatic groups, and C—Cl at 3300 cm^{-1}, 2930 cm^{-1}, 1650 cm^{-1} 1440 cm^{-1}, 1375 cm^{-1}, 1035 cm^{-1}, 780 cm^{-1}, and 500–550 cm^{-1}, respectively [43,45,50].

Field emission scanning electron microscope (FESEM) images of biochar and raw feedstock clearly show that biochar contains a porous structure, providing effective adsorption sites compared with raw feedstock (Fig. 2).

Biochar has some generalized properties such as carbon content, hydrogen, nitrogen, oxygen, sulfur, ash content, pH, and cation exchange capacity (CEC) [3]. As the pyrolysis temperature increases, the carbon (C) content also raises with the temperature, whereas hydrogen (H), sulfur (S), and oxygen (O) content decline with the rise in pyrolysis temperature [3]. Increase or decrease in nitrogen (N) content depends on the pyrolysis condition and used feedstock. Yield (%) of biochar decreases with the increase in pyrolytic temperature (Table 4). The pH of biochars fluctuates broadly based on the source of the feedstock. The pyrolysis temperature is a crucial aspect affecting the pH. For example, the pH of the biochar from oak wood obtained at 200°C is less (pH = 4.6), but the biochar was neutral and alkaline (pH 6.9 and 9.5, respectively) at 400°C and 600°C [3,75]. The conclusion of Zhang et al. [76] is that CEC of wood biochar declines with rising pyrolysis temperature [77]. As well as the previously mentioned properties, one of the other properties of biochar is exclusion of heavy metals from contaminated water.

7 Removal of heavy metals by biochar

Biochar plays a vital role in removal of heavy metals from polluted water through several mechanisms such as physical sorption, precipitation, complexation, ion exchange, and electrostatic interaction (chemisorption). Biochars have high sorption ability for heavy metals like AC because of their surface [78]. The high surface area and pores in biochars have a high attraction for metals due to the physical sorption of metallic ions onto the char surface and retention within the pores [79]. Most of the biochars have negatively charged surfaces that adsorb metals with a positive charge through electrostatic attractions. Various complexes have been formed between metals with specific ligands and functional groups present in biochars [80] or precipitates of their solid mineral phases [56,81].

The particular job of every procedure in substantial metal adsorption differs impressively, relying upon the target metal, ionic condition of the solution, and the carbon adsorbent [75,82]. One of the significant factors is pH of the solution, as it impacts metal speciation as well as the surface charge of adsorbent. Overall chemisorption (ionic exchange, surface complexation, precipitation) adopts a progressively larger job in the expulsion of substantial metals from fluid solution than physical sorption, electrostatic interaction, and physical adsorption [83]. Similarly, it can be assumed that all processes can be at work in a specific wastewater.

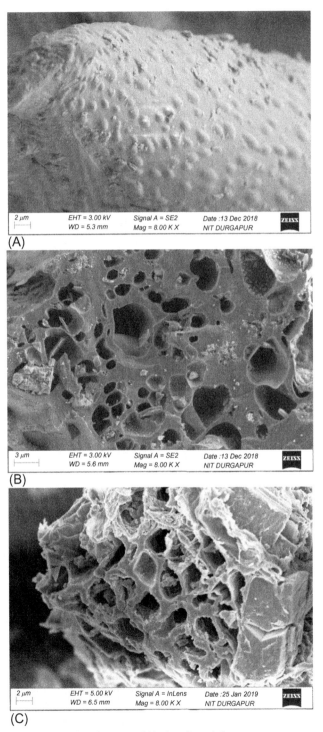

FIG. 2 FESEM images of a feedstock (A) and prepared biochar (B and C).

Table 4   Biomass feedstock results of various kinds of pyrolysis.

Process	Liquid or bio oil (%)	Solid or biochar (%)	Gas or syngas (%)
Fast pyrolysis (high heating rate)	73–75	10–12	12–13
Intermediate pyrolysis (moderate temperature, moderate heating rate)	50	20–25	20–25
Slow pyrolysis (slow heating rate)	30	35	35
Gasification	5	8–10	85
Hydrothermal carbonization	–	–	–
Flash carbonization	–	50	50

Source: O.D. Nartey, B. Zhao, Biochar preparation, characterization, and adsorptive capacity and its effect on bioavailability of contaminants: an overview, Adv. Mater. Sci. Eng. 2014 (2014) 715398, https://doi.org/10.1155/2014/715398.

The capability of biochar depends on the procedure of char formation, temperature, used material, etc. Researchers are working all over the world on biochar preparation in various ways to increase the efficiency of char for heavy metal removal capacity from various mediums. The base of the working procedure of biochar is ion exchange as biochar traps various heavy metals in its surface. Ionic exchange between heavy metals and carboxyl and hydroxyl groups is one of the chief processes for heavy metal adsorption by carbon adsorbents like biochar [84]. The effectiveness of the ion-exchange method in adsorbing substantial metals onto biochar surfaces depends to a great extent on the particle size of the metal and functional groups present on the adsorbent surface [84]. CEC is a significant pointer of heavy metal adsorption when ion exchange is the leading scheme. The efficiency of a biochar is proportional with the negative ion present on the surface of the biochar. Gao et al. [57] from China worked on cadmium ion removal through biochar using rice straw and sewage sludge. Rice straw and sewage sludge both are treated as waste material. Sometimes rice straw is used as animal food, but a large portion of this is treated as waste, and farmers burn it, which creates environmental pollution. According to Agrafioti et al. [85], micropores of char are filled with ash, which decreases the surface area and reduces efficiency of char. Then the biochar was washed with HCl followed by distilled water to remove the ions; this stepwise washing is helpful to make pH constant. Removal of ions is an important step because it helps to prevent interference during adsorption of heavy metal ions. In rice straw biochar, adsorption capacity by precipitation was at maximum 700°C temperature (57%). In sewage sludge biochar, adsorption capacity by precipitation mechanism was at maximum 700°C temperature (62%). The efficiency of the two biochars were different mainly due to the bonding action of Cd ion with $-\pi$ electron coordinator. This result was also supported by Wang et al. [34] who worked on biochar prepared from bamboo, corn straw, and pig manure. CEC in rice straw biochar played an important role during cadmium ion removal in 300°C temperature, but the efficiency of CEC decreased with rising temperature.

Inyang et al. [2] worked on the elimination process of heavy metals from aqueous solution by biochars. In this work, biochar was prepared from anaerobically digested dairy products and sugar beet residue, which is an agricultural waste. Digested residues were dried at 80°C temperature before pyrolysis. Dry residues were pyrolyzed at 600°C temperature, and resultant biochar was washed through deionized water several times to remove ion interferences from the char. Surface charge, surface area, and pore volume were higher in the biochar produced from the animal waste, which increased the ability of metal removal from aqueous solution. Pb^{2+} (99%) and Cu^{2+} (98%) removal efficiency was very high, but Cd^{2+} (57%) and Ni^{2+} (26%) removal efficiency was observed to be low in the biochar produced from animal waste. Heavy metal removal ability was Cd > Ni > Pb > Cu and Pb > Cu > Cd > Ni for biochar produced from sugar beet residue and dairy waste residue, respectively. The Pb ion adsorption ability of a biochar from a solution depends on the high electronegativity constant of the external sides of the char [86]. Very high heavy metal removal capacity of dairy products advises that anaerobic digestion might be useful to produce excellent biochar for heavy metal adsorption. Uchimiya et al. [55] showed that the success of biochar to trap heavy metal ions hinge on the heavy metal contamination type. Adsorption ability of the two prepared biochars were similar to commercial ACs. These two anaerobically manufactured biochars can be used as a substitute for AC. Using prepared biochar instead of AC is economically beneficial and eco-friendly as well because these reduce sludge [87].

Lu et al. [46] from China worked on Pb ion removal from acid mine drainage by using biochar. The water released from the mine area contains very high acid, which is responsible for the degradation of soil and water as well as the ecological misbalance of that area. Sewage sludge was used for biochar production in this study. Temperature was maintained at 110–120°C to dry up the sludge, which was pyrolyzed at 555°C temperature. Sewage sludge contains little amount of heavy metals itself, but those heavy metals successfully became stable during pyrolysis, and the risk of metal escape is insignificant. Capability of metal adsorption of sewage sludge biochar might be lower than the activated alumina, thermos-modified silica in higher pH, whereas in acidic pH, the efficiency of sewage sludge biochar increases. A wide range of pH (2.0–5.0) has been selected for the whole mechanism. Alkaline earth materials like Ca^{2+} and Mg^{2+} are replaced by the Pb from the surface of char. These alkaline earth materials are bound with various functional group like R-O-Me or R-COO-Me, etc., of char. Surface complexation mainly occurs between free hydroxyl or carboxyl functional groups and heavy metal ions. Metallic conversion occurs between lead ion and sodium or calcium or magnesium ion. There is no such exact beginning procedure for use of biochar, and due to its low-cost operation process, it can be easily applicable for the removal of lead ions from acid mine drainage. Pb sorption capacity of biochar increased with increasing pH value indicating alkaline conditions of biochar are favorable for heavy metal adsorption. Through surface complexation, the amount of Pb ion bound with the free carboxyl group is proportional with pH. Removing protons is a process occurring in carboxyl and hydroxyl groups through which pH increases the surface charge of biochar and makes it capable of adsorbing more Pb^{2+}. Hydrolyzed Pb^{2+} are prepared for more adsorption into biochar than free Pb^{2+} because electrical repulsion between biochar surface and free Pb^{2+} is less [88,89]. The carboxyl and hydroxyl group present on the surface of the biochar acts as a positive ion gainer and easily binds the Pb ion and forms a complex on the surface of the char. In alkaline, pH-free Pb ions are hydrolyzed and form hydrolyzed Pb ions. The initial cost of biochar formation is very low because the raw material used during biochar formation is waste or discarded materials, and due to low cost, this sorption procedure is easily acceptable. Sewage sludge biochar has a good neutralization effect on the acidic solution.

During insertion of nanoparticles [38,75,90] chemical stimulation [91] is used to enhance the heavy metal adsorption capacity of biochar. Stimulation of biochar by hydroxides [91], steam or nitrogen run [92], heat up, and magnetic alteration [93] have been reported in various studies. A paper from China by Luo et al. [39] worked on the modification of biochar prepared from agricultural waste to remove Cd from water. Corncobs were used for char preparation at 350°C, 450°C, and 550°C temperature. In this paper, the prepared biochar was modified to increase the char efficiency for heavy metal adoption. The char was modified with a solution of NaOH, urea, and distilled water (in a temperature between −5°C and −100°C). Acrylonitrile is the final chemical to mix with the previous mixture. Increased pyrolysis temperature increases pore size of char. Through modification of biochar, C—N group is grafted into the char at low pyrolysis temperature. With increasing temperature, the pore size of the char is increased. Maximum surface area observed in the modified char was previously pyrolyzed in 350°C temperature. In this paper, acrylonitrile was used to modify the biochar. Due to this modification, biochar receives cyanogen (C—N), which provides more metal binding positions, expands the surface area, and increases chemical activities. The Langmuir adsorption isotherm models revealed that the adsorption of Cd was an exchange of ions between char and heavy metals.

Ni et al. [45] worked on the removal of Cd and Pb through biochar from aqueous solution. In this paper, synthetic Pb and Cd solution was prepared. In the preliminary stage, the sludge was digested anaerobically in an anaerobic mesophilic digester. At 80°C temperature, the previously arranged sludge was dried, and after drying the sludge was pyrolyzed at 600°C temperature. Sludge itself is a nutrient-rich material, and char prepared from this sludge contains various types of inorganic substances. The prepared biochar was cleaned with water for removal of ions. Pb has a greater adsorption rate in the prepared biochar than the Cd. The heavy metals were adsorbed in the surface of char when the metal concentration was low, but with increasing metal concentration, the metals entered into the inner part of the char and filled all the adsorption places of the char. In 1.0 mmol/L metal concentration, the adsorption capacity of Pb was 0.61 mmol/g and 0.44 mmol/g for Cd. Langmuir, Freundlich, and Dubinin-Radushkevich are the three single-factor models used to evaluate the heavy metal adsorption of char. Competitive Langmuir, noncompetitive Langmuir, and Langmuir-Freundlich models are the three binary factor isotherm models used to estimate metal adsorption activity between Pb and Cd. Carbon content is very high in anaerobically digested sludge than other digestion processes. Reports show that char prepared in pH 7.0 gives rise to a high amount of surface negative charge in char, and these negative charges actively reduce the positive charge of heavy metal ions. O—H, C—H, C=O and C—N, C=N, phenols, aromatic ring, aromatic groups, and C—Cl are the functional groups present in the biochar prepared from the anaerobically digested sludge. It was reported that the Pb adsorption capacity of biochar prepared from digested whole sugar beet (0.20 mmol/g), pinewood (0.02 mmol/g), rice husk (0.01 mmol/g), and sesame straw (0.49 mmol/g) were less than the biochar prepared from anaerobically digested sludge (0.61 mmol/g). It was also reported in this paper that the Cd adsorption capacity of biochar prepared from dairy manure (0.28 mmol/g), pine (0.01 mmol/g), oak (0.05 mmol/g), and municipal sewage sludge (0.36 mmol/g) were less than the biochar prepared from anaerobically digested sludge (0.61 mmol/g).

Another work on biochar formation from marine macro-algae *Saccharina japonica*, *Sargassum fusiforme*, and pinewood saw dust to remove heavy metals from aqueous solution was presented by Poo et al. [32]. A temperature gradient of 250°C, 400°C, 500°C, 600°C, and 700°C was used to form biochar during pyrolysis. Waste marine macro-algal biomass was used to prepare biochar for removal of heavy metals from aqueous solutions. *Saccharina japonica* and *Sargassum fusiforme*, common names Kelp and Hijikia, respectively, were used to form the biochar. Cu, Cd, and Zn were the selected heavy metals, which were removed by the algal biochar. Biochar is very efficient in the case of heavy metal removal from wastewater, but the extraction process of heavy metals from the char is not easy. This problem can be overwhelmed through the preparation of biochar with magnetic properties. Different oxides of iron are used during pyrolysis to magnetize the char. Biochar made up of marine algae is a better absorber of heavy metal over traditional pinewood biochar. Due to magnetization, the surface pores of the char are occupied by iron oxides and decrease heavy metal adsorption capability. The oxygen content of biochar formed from *Saccharina japonica* (27.4%) and *Sargassum fusiforme* (23.4%) are higher than pinewood (9.37%), but carbon content is less in *Saccharina japonica* and *Sargassum fusiforme* than pine wood, so the O/C ratio is high in algal biochar. Magnetization of char increases carbon, nitrogen, and sulfur content, but oxygen content is decreased slightly. O—H, C=O, and C—O functional groups are present in the surface of the algal biochar, and oxygen-containing functional groups increase the heavy metal adsorption rate of char. Sometimes —OH group also binds a large number of metals. Oxygen is strong electronegative element, which is why hydrophilic oxygen releases hydrogen [65]. *Saccharina japonica* and *Sargassum fusiforme* biochar removes maximum amount of Cu than other metals from all samples because Cu has a very good capability to bind with oxygen-containing functional group. Cd removal by magnetized *Saccharina japonica* biochar ($23.1 \, \mathrm{m^2 \, g^{-1}}$) and *Sargassum fusiforme* biochar ($19.4 \, \mathrm{m^2 \, g^{-1}}$) was greater than magnetized oak bark biochar ($7.40 \, \mathrm{m^2 \, g^{-1}}$) and magnetized oak wood biochar ($2.87 \, \mathrm{m^2 \, g^{-1}}$). Cu removal by magnetic nut shield is $50.6 \, \mathrm{m^2 \, g^{-1}}$, by *Saccharina japonica* biochar is $55.8 \, \mathrm{m^2 \, g^{-1}}$, by *Sargassum fusiforme* biochar is $47.7 \, \mathrm{m^2 \, g^{-1}}$, and by pectin iron oxide magnetic nanocomposite is $48.9 \, \mathrm{m^2 \, g^{-1}}$. Zn removal by *Saccharina japonica* biochar, *Sargassum fusiforme* biochar, and magnetic empty fruit branch biochar are $22.2 \, \mathrm{m^2 \, g^{-1}}$, $19.1 \, \mathrm{m^2 \, g^{-1}}$, and $4.10 \, \mathrm{m^2 \, g^{-1}}$, respectively. The potentiality of pinewood saw dust to remove heavy metals increases with temperature. At 250°C temperature, pinewood saw dust biochar removes 103%, 3.99%, and 6.24% of Cu^{2+}, Cd^{2+}, and Zn^{2+}, respectively, but at 700°C temperature removes 81.25%, 46.36%, and 46.59% of Cu^{2+}, Cd^{2+}, and Zn^{2+} respectively. Biochar from *Saccharina japonica* can remove 70.79%, 69.34%, and 63.45% of Cu^{2+}, Cd^{2+}, and Zn^{2+}, respectively. Biochar from *Saccharina japonica* can remove 98% of all heavy metals at 400°C temperature. *Sargassum fusiforme* can remove 97% of heavy metals at a temperature between 400°C and 600°C, but at 700°C temperature, the efficiency decreases to 87.6%; 88% Cd and Zn were removed from solution at 700°C temperature, and maximum removal was observed at 600°C temperature (90%). The author showed that the pH increases with temperature, and this increased pH decreased the metal removal efficiency. *S. japonica*, *S. fusiforme*, and pinewood sawdust at temperature 500°C, 600°C, and 700°C, respectively, adsorbed maximum amount of Cu (98.6, 94.1, and 8.9 mg g^{-1}). *S. japonica*, *S. fusiforme*, and pinewood sawdust at a temperature of 700°C adsorbed the maximum amount of Cd (60.7, 37.2, and 5.0 mg g^{-1}). *S. japonica*, *S. fusiforme*, and pinewood sawdust at a temperature of 700°C adsorbed the maximum amount of Zn (84.3, 43.0, and 5.2 mg g^{-1}). Metal adsorbance depends on ionic radius, atomic weight, and electronegativity [105]. Electronegativity of copper was higher than Cd and Zn, which is why copper was easily absorbed by the char than Cd and Zn.

The biochars from solid waste and sewage sludge adsorbed more than 99% of the Cr (III), which is probably associated with electrostatic interactions among adsorbent's negative surface charge and Cr (III) cations. Sewage sludge biochar and solid waste biochar removes Cr (VI) by 89% and 44%, respectively. Cr (VI) sorption by biochar can be done via these probable techniques such as (a) anionic adsorption, (b) adsorption-coupled reduction, (c) anionic and cationic adsorption, and (d) reduction and cationic adsorption. If pH is on the lower side (acidic), the Cr (VI) reduced to Cr (III), and adsorption process takes place easily [72]. Agrafioti et al. [72] stated that the reason of less elimination of Cr (VI) in their study was due to the pH values (pH 7.0–9.5).

Complexation of functional groups, adsorption through ion exchange [106], or coprecipitation with mineral components [98] can be the reason of Hg (II) sorption onto the biochar surface. Low ash-containing biochar possibly rejects the coprecipitation mechanism [49]. Biochar surface and arsenic speciation are the two regulatory factors that control the sorption of Arsenic (As) onto adsorbents like biochar. Table 5 shows the pyrolysis temperature and used feedstock to remove different heavy metals.

8 Conclusion

Among all the heavy metal removal techniques, adsorption via biochars is a cost-effective and highly efficient one. This review reveals a huge variety of feedstock used in production of biochar that may remove heavy metals from wastewater. Efficiency of produced biochar depends on the pyrolytic condition and used feedstock. Biochars were obtained from an extensive variety of biomass feedstocks such as sewage, solid waste, forest residues, and agricultural wastes at various pyrolysis conditions. The attributes of biochars were recognized by physical and substance techniques, which uncover the fundamental structure

Table 5   Heavy metal removal and used biochar feedstock

Metal	Biochar feedstock	% of Removal	Temperature (°C)	References
Cd	Rice straw	–	300	[57]
	Sewage sludge	–	500	
			700	
			800	
Cd, Cu, Ni, Pb	Dairy waste	Cd (57%), Pb (99%), Cu(98%), Ni (26%)	600	[2]
	Sugar beet	>97%		
Cd, Cr, Cu, Pb, Ni, Zn, Hg	Sewage sludge	–	550	[46]
Cd	Raw corncobs (agromaterials)	–	350	[39]
			450	
			550	
Cd (II), Pb (II)	Sludge	–	600	[45]
Cd, Cr, Cu, Pb, Zn	Sesame straw biochar	–	700	[44]
Hg	Wood biochar	–	600	[50]
	Mixture of wood and sulfur element			
Cu, Cd, Zn	Macro-algae (*Saccharina japonica*)	>98% (at 400°C)	250	[32]
			400	
	Macro-algae (*Sargassum fusiforme*)	>86% (at 500°C)	500	
			600	
	Pine sawdust	>81% (Cu), 46% (Cd), 47% (Zn)	700	
Cd, Cu, Zn	*Saccharina japonica* (Kelp)	–	500	[51]
	Sargassum fusiforme (Hijikia)			
	Magnetic biochar (*Saccharina japonica* + $FeCl_3$ solution)			
	Magnetic biochar (*Sargassum fusiforme* + $FeCl_3$ solution)			
Cd, Pb	Peanut shell	–	400	[37]
Cd, Cu, Pb	Biomass	–	600	[80]
Zn, Cu, Cr, Ni, Cd, Mn	Textile dying sludge	–	300	[34]
			400	
			500	
			600	
			700	
Cd, Cu, Zn	Dairy manure	–	200	[47]
			350	
Cd, Cu, Pb	Bamboo	–	600	[35]
	Sugarcane bagasse			
	Hickory wood			
	Peanut hull			
Cr	Oak wood	–	400–450	[94]
Cr	Oak bark	–	400–450	[94]
Chromium	Coconut coir	–	250–600	[95]

Continued

Table 5   Heavy metal removal and used biochar feedstock—cont'd

Metal	Biochar feedstock	% of Removal	Temperature (°C)	References
Pb, Cu, Zn, and Cd	Dairy manure	2%–40.9%	350	[12]
Pb, Cu, Zn, and Cd	Rice husk	38.4%–100%		
Copper and zinc	Corn straw	–	600	[96]
Cu	Orange waste	–	300, 600	[97]
Cu	Compost	–	300, 600	[97]
Pb	Sugarcane bagasse	–	600	[81]
Hg	Soybean stalk	75%–87%	700	[98]
Pb, Cu	Waste tire rubber ash	–	500	[99]
As, Cd, Pb	Wood and bark chars	–	400, 450	[54]
Cd, Cu, Ni, Pb	Cottonseed hull char	–	350	[100]
As, Pb	Pinewood	–	600	[101]
Cr	Rice hull	–	400	[102]
As	Municipal solid wastes	–	400, 500, 600	[103]
Cr	Cotton fiber	–	850	[104]

and properties of biochar. Because of its enormous surface zone, surface charge, and functional groups, biochar has incredible potential to adsorb substantial metal and natural contaminants. Several kinds of interactions/exchanges including ionic exchange, electrostatic attraction, physical adsorption, and chemical complexation and/or precipitation are largely responsible for restricting water pollutants onto the biochar. Use of biochar should diminish the bioavailability, harmfulness, and mobility of natural and inorganic contaminations. This can possibly be valuable for immobilization of pollutants/impurities with high concentrations/quantity.

Acknowledgment

The first author is grateful for financial support given by N-PDF scheme (Project No. PDF/2017/000530) under DST-SERB.

References

[1] Z.S. Szolnoki, A. Farsang, I. Puskás, Cumulative impacts of human activities on urban garden soils: origin and accumulation of metals, Environ. Pollut. 177 (2013) 106–115.

[2] M. Inyang, B. Gao, Y. Yao, Y. Xue, A.R. Zimmerman, P. Pullammanappallil, X. Cao, Removal of heavy metals from aqueous solution by biochars derived from anaerobically digested biomass, Bioresour. Technol. 110 (2012) 50–56.

[3] S.M. Shaheen, N.K. Niazi, N.E. Hassan, I. Bibi, H. Wang, D.C. Tsang, et al., Wood-based biochar for the removal of potentially toxic elements in water and wastewater: a critical review, Int. Mater. Rev. 64 (4) (2018) 216–247.

[4] F. Fu, Q. Wang, Removal of heavy metal ions from wastewaters: a review, J. Environ. Manag. 92 (3) (2011) 407–418.

[5] X. Gong, D. Huang, Y. Liu, G. Zeng, R. Wang, J. Wan, C. Zhang, M. Cheng, X. Qin, W. Xue, Stabilized nanoscale zerovalent iron mediated cadmium accumulation and oxidative damage of *Boehmeria nivea* (L.) Gaudich cultivated in cadmium contaminated sediments, Environ. Sci. Technol. 51 (19) (2017) 11308–11316.

[6] W. Xue, D. Huang, G. Zeng, J. Wan, C. Zhang, R. Xu, M. Cheng, R. Deng, Nanoscale zero-valent iron coated with rhamnolipid as an effective stabilizer for immobilization of Cd and Pb in river sediments, J. Hazard. Mater. 341 (2018) 381–389.

[7] D. Huang, C. Liu, C. Zhang, R. Deng, R. Wang, W. Xue, et al., Cr (VI) removal from aqueous solution using biochar modified with Mg/Al-layered double hydroxide intercalated with ethylenediaminetetraacetic acid, Bioresour. Technol. 276 (2019) 127–132.

[8] N.K. Srivastava, C.B. Majumder, Novel biofiltration methods for the treatment of heavy metals from industrial wastewater, J. Hazard. Mater. 151 (1) (2008) 1–8.

[9] M.I. Inyang, B. Gao, Y. Yao, Y. Xue, A.R. Zimmerman, A. Mosa, P. Pullammanappallil, X. Cao, A review of biochar as a low-cost adsorbent for aqueous heavy metal removal, Crit. Rev. Environ. Sci. Technol. 46 (4) (2016) 406–433.

[10] M.A. Barakat, New trends in removing heavy metals from industrial wastewater, Arab. J. Chem. 4 (4) (2011) 361–377.

[11] M. Bilal, J.A. Shah, T. Ashfaq, S.M.H. Gardazi, A.A. Tahir, A. Pervez, H. Haroon, Q. Mahmood, Waste biomass adsorbents for copper removal from industrial wastewater—a review, J. Hazard. Mater. 263 (2013) 322–333.

[12] X. Xu, X. Cao, L. Zhao, Comparison of rice husk-and dairy manure-derived biochars for simultaneously removing heavy metals from aqueous solutions: role of mineral components in biochars, Chemosphere 92 (8) (2013) 955–961.

[13] J.U. Ahmad, M.A. Goni, Heavy metal contamination in water, soil, and vegetables of the industrial areas in Dhaka, Bangladesh, Environ. Monit. Assess. 166 (1–4) (2010) 347–357.

[14] A.H.M. Veeken, S. De Vries, A. Van der Mark, W.H. Rulkens, Selective precipitation of heavy metals as controlled by a sulfide-selective electrode, Sep. Sci. Technol. 38 (1) (2003) 1–19.

[15] B. Alyüz, S. Veli, Kinetics and equilibrium studies for the removal of nickel and zinc from aqueous solutions by ion exchange resins, J. Hazard. Mater. 167 (1–3) (2009) 482–488.

[16] T.A. Kurniawan, G.Y. Chan, W.H. Lo, S. Babel, Physico–chemical treatment techniques for wastewater laden with heavy metals, Chem. Eng. J. 118 (1–2) (2006) 83–98.

[17] A. Azimi, A. Azari, M. Rezakazemi, M. Ansarpour, Removal of heavy metals from industrial wastewaters: a review, ChemBioEng Rev. 4 (1) (2017) 37–59.

[18] L. Jiang, Y. Wang, C. Feng, Application of photocatalytic technology in environmental safety, Procedia Eng. 45 (2012) 993–997.

[19] R. Molinari, A. Caruso, T. Poerio, Direct benzene conversion to phenol in a hybrid photocatalytic membrane reactor, Catal. Today 144 (1–2) (2009) 81–86.

[20] A. Basile, A. Cassano, N.K. Rastogi (Eds.), Advances in Membrane Technologies for Water Treatment: Materials, Processes and Applications, Elsevier, 2015.

[21] S. Zodi, O. Potier, F. Lapicque, J.P. Leclerc, Treatment of the textile wastewaters by electrocoagulation: effect of operating parameters on the sludge settling characteristics, Sep. Purif. Technol. 69 (1) (2009) 29–36.

[22] Y. Zhu, W. Fan, T. Zhou, X. Li, Removal of chelated heavy metals from aqueous solution: a review of current methods and mechanisms, Sci. Total Environ. 678 (2019) 253–266.

[23] J. Cai, M. Lei, Q. Zhang, J.R. He, T. Chen, S. Liu, et al., Electrospun composite nanofiber mats of cellulose@ organically modified montmorillonite for heavy metal ion removal: design, characterization, evaluation of absorption performance, Compos. A: Appl. Sci. Manuf. 92 (2017) 10–16.

[24] U. Habiba, A.M. Afifi, A. Salleh, B.C. Ang, Chitosan/(polyvinyl alcohol)/zeolite electro spun composite nanofibrous membrane for adsorption of Cr6+, Fe^{3+} and Ni^{2+}, J. Hazard. Mater. 322 (2017) 182–194.

[25] V.K. Gupta, A. Rastogi, Biosorption of hexavalent chromium by raw and acid-treated green alga *Oedogonium hatei* from aqueous solutions, J. Hazard. Mater. 163 (1) (2009) 396–402.

[26] M.U. Ali, G. Liu, B. Yousaf, Q. Abbas, H. Ullah, M.A.M. Munir, B. Fu, Pollution characteristics and human health risks of potentially (eco) toxic elements (PTEs) in road dust from metropolitan area of Hefei, China, Chemosphere 181 (2017) 111–121.

[27] D. Mohan, R. Sharma, V.K. Singh, P. Steele, C.U. Pittman Jr., Fluoride removal from water using biochar, a green waste, low-cost adsorbent: equilibrium uptake and sorption dynamics modeling, Ind. Eng. Chem. Res. 51 (2) (2012) 900–914.

[28] D. Mohan, A. Sarswat, Y.S. Ok, C.U. Pittman Jr., Organic and inorganic contaminants removal from water with biochar, a renewable, low cost and sustainable adsorbent – a critical review, Bioresour. Technol. 160 (2014) 191–202.

[29] J.H. Park, Y.S. Ok, S.H. Kim, J.S. Cho, J.S. Heo, R.D. Delaune, D.C. Seo, Competitive adsorption of heavy metals onto sesame straw biochar in aqueous solutions, Chemosphere 142 (2016) 77–83.

[30] D. Laird, P. Fleming, B. Wang, R. Horton, D. Karlen, Biochar impact on nutrient leaching from a Midwestern agricultural soil, Geoderma 158 (3–4) (2010) 436–442.

[31] J.L. Field, C.M. Keske, G.L. Birch, M.W. DeFoort, M.F. Cotrufo, Distributed biochar and bioenergy coproduction: a regionally specific case study of environmental benefits and economic impacts, GCB Bioenergy 5 (2) (2013) 177–191.

[32] K.M. Poo, E.B. Son, J.S. Chang, X. Ren, Y.J. Choi, K.J. Chae, Biochars derived from wasted marine macro-algae (*Saccharina japonica* and *Sargassum fusiforme*) and their potential for heavy metal removal in aqueous solution, J. Environ. Manag. 206 (2018) 364–372.

[33] T. Chen, Y. Zhang, H. Wang, W. Lu, Z. Zhou, Y. Zhang, L. Ren, Influence of pyrolysis temperature on characteristics and heavy metal adsorptive performance of biochar derived from municipal sewage sludge, Bioresour. Technol. 164 (2014) 47–54.

[34] X. Wang, C. Li, Z. Li, G. Yu, Y. Wang, Effect of pyrolysis temperature on characteristics, chemical speciation and risk evaluation of heavy metals in biochar derived from textile dyeing sludge, Ecotoxicol. Environ. Saf. 168 (2019) 45–52.

[35] Y. Zhou, B. Gao, A.R. Zimmerman, J. Fang, Y. Sun, X. Cao, Sorption of heavy metals on chitosan-modified biochars and its biological effects, Chem. Eng. J. 231 (2013) 512–518.

[36] P. Devi, A.K. Saroha, Synthesis of the magnetic biochar composites for use as an adsorbent for the removal of pentachlorophenol from the effluent, Bioresour. Technol. 169 (2014) 525–531.

[37] S. Wan, J. Wu, S. Zhou, R. Wang, B. Gao, F. He, Enhanced lead and cadmium removal using biochar-supported hydrated manganese oxide (HMO) nanoparticles: behavior and mechanism, Sci. Total Environ. 616 (2017) 1298–1306.

[38] J. Wu, Y. Yi, Y. Li, Z. Fang, E.P. Tsang, Excellently reactive Ni/Fe bimetallic catalyst supported by biochar for the remediation of decabromodiphenyl contaminated soil: reactivity, mechanism, pathways and reducing secondary risks, J. Hazard. Mater. 320 (2016) 341–349.

[39] M. Luo, H. Lin, B. Li, Y. Dong, Y. He, L. Wang, A novel modification of lignin on corncob-based biochar to enhance removal of cadmium from water, Bioresour. Technol. 259 (2018) 312–318.

[40] M. Inyang, B. Gao, A. Zimmerman, M. Zhang, H. Chen, Synthesis, characterization, and dye sorption ability of carbon nanotube–biochar nanocomposites, Chem. Eng. J. 236 (2014) 39–46.

[41] A.U. Rajapaksha, M. Vithanage, M. Zhang, M. Ahmad, D. Mohan, S.X. Chang, Y.S. Ok, Pyrolysis condition affected sulfamethazine sorption by tea waste biochars, Bioresour. Technol. 166 (2014) 303–308.

[42] J. Meng, X. Feng, Z. Dai, X. Liu, J. Wu, J. Xu, Adsorption characteristics of Cu (II) from aqueous solution onto biochar derived from swine manure, Environ. Sci. Pollut. Res. 21 (11) (2014) 7035–7046.

[43] Y. Yao, B. Gao, M. Inyang, A.R. Zimmerman, X. Cao, P. Pullammanappallil, L. Yang, Biochar derived from anaerobically digested sugar beet tailings: characterization and phosphate removal potential, Bioresour. Technol. 102 (10) (2011) 6273–6278.

[44] J.H. Park, Y.S. Ok, S.H. Kim, J.S. Cho, J.S. Heo, R.D. Delaune, D.C. Seo, Competitive adsorption of heavy metals onto sesame straw biochar in aqueous solutions, Chemosphere 142 (2015) 77–83.

[45] B.J. Ni, Q.S. Huang, C. Wang, T.Y. Ni, J. Sun, W. Wei, Competitive adsorption of heavy metals in aqueous solution onto biochar derived from anaerobically digested sludge, Chemosphere 219 (2019) 351–357.

[46] H. Lu, W. Zhang, Y. Yang, X. Huang, S. Wang, R. Qiu, Relative distribution of Pb^{2+} sorption mechanisms by sludge-derived biochar, Water Res. 46 (3) (2012) 854–862.

[47] X. Xu, X. Cao, L. Zhao, H. Wang, H. Yu, B. Gao, Removal of Cu, Zn, and Cd from aqueous solutions by the dairy manure-derived biochar, Environ. Sci. Pollut. Res. 20 (1) (2013) 358–368.

[48] V.K. Gupta, I. Ali, V.K. Saini, Defluoridation of wastewaters using waste carbon slurry, Water Res. 41 (15) (2007) 3307–3316.

[49] X. Xu, A. Schierz, N. Xu, X. Cao, Comparison of the characteristics and mechanisms of Hg (II) sorption by biochars and activated carbon, J. Colloid Interface Sci. 463 (2016) 55–60.

[50] J.H. Park, J.J. Wang, B. Zhou, J.E. Mikhael, R.D. DeLaune, Removing mercury from aqueous solution using sulfurized biochar and associated mechanisms, Environ. Pollut. 244 (2019) 627–635.

[51] E.B. Son, K.M. Poo, J.S. Chang, K.J. Chae, Heavy metal removal from aqueous solutions using engineered magnetic biochars derived from waste marine macro-algal biomass, Sci. Total Environ. 615 (2018) 161–168.

[52] L. Beesley, M. Marmiroli, The immobilisation and retention of soluble arsenic, cadmium and zinc by biochar, Environ. Pollut. 159 (2) (2011) 474–480.

[53] Z. Liu, F.S. Zhang, Removal of lead from water using biochars prepared from hydrothermal liquefaction of biomass, J. Hazard. Mater. 167 (1–3) (2009) 933–939.

[54] D. Mohan, C.U. Pittman Jr., M. Bricka, F. Smith, B. Yancey, J. Mohammad, et al., Sorption of arsenic, cadmium, and lead by chars produced from fast pyrolysis of wood and bark during bio-oil production, J. Colloid Interface Sci. 310 (1) (2007) 57–73.

[55] M. Uchimiya, I.M. Lima, K. Thomas Klaswong, S. Chang, L.H. Wartelle, J.E. Rodgers, Immobilization of heavy metal ions (CuII, CdII, NiII, and PbII) by broiler litter-derived biochars in water and soil, J. Agric. Food Chem. 58 (9) (2010) 5538–5544.

[56] X. Cao, L. Ma, B. Gao, W. Harris, Dairy-manure derived biochar effectively sorbs lead and atrazine, Environ. Sci. Technol. 43 (9) (2009) 3285–3291.

[57] L.Y. Gao, J.H. Deng, G.F. Huang, K. Li, K.Z. Cai, Y. Liu, F. Huang, Relative distribution of Cd2 + adsorption mechanisms on biochars derived from rice straw and sewage sludge, Bioresour. Technol. 272 (2019) 114–122.

[58] D. Angin, Effect of pyrolysis temperature and heating rate on biochar obtained from pyrolysis of safflower seed press cake, Bioresour. Technol. 128 (2013) 593–597.

[59] W. Song, M. Guo, Quality variations of poultry litter biochar generated at different pyrolysis temperatures, J. Anal. Appl. Pyrolysis 94 (2012) 138–145.

[60] O.D. Nartey, B. Zhao, Biochar preparation, characterization, and adsorptive capacity and its effect on bioavailability of contaminants: an overview. Adv. Mater. Sci. Eng. 2014 (2014), 715398. https://doi.org/10.1155/2014/715398.

[61] X. Zhao, W. Ouyang, F. Hao, C. Lin, F. Wang, S. Han, X. Geng, Properties comparison of biochars from corn straw with different pretreatment and sorption behaviour of atrazine, Bioresour. Technol. 147 (2013) 338–344.

[62] T.T. Wang, J. Cheng, X.J. Liu, W. Jiang, C.L. Zhang, X.Y. Yu, Effect of biochar amendment on the bio-availability of pesticide chlorantraniliprole in soil to earthworm, Ecotoxicol. Environ. Saf. 83 (2012) 96–101.

[63] T. Xu, L. Lou, L. Luo, R. Cao, D. Duan, Y. Chen, Effect of bamboo biochar on pentachlorophenol leachability and bioavailability in agricultural soil, Sci. Total Environ. 414 (2012) 727–731.

[64] L. Yang, M. Jin, C. Tong, S. Xie, Study of dynamic sorption and desorption of polycyclic aromatic hydrocarbons in silty-clay soil, J. Hazard. Mater. 244 (2013) 77–85.

[65] M. Zhang, B. Gao, S. Varnoosfaderani, A. Hebard, Y. Yao, M. Inyang, Preparation and characterization of a novel magnetic biochar for arsenic removal, Bioresour. Technol. 130 (2013) 457–462.

[66] V. Asensio, F.A. Vega, M.L. Andrade, E.F. Covelo, Tree vegetation and waste amendments to improve the physical condition of copper mine soils, Chemosphere 90 (2) (2013) 603–610.

[67] S. Meyer, B. Glaser, P. Quicker, Technical, economical, and climate-related aspects of biochar production technologies: a literature review, Environ. Sci. Technol. 45 (22) (2011) 9473–9483.

[68] D.P. Oliver, Y.F. Pan, J.S. Anderson, T.F. Lin, R.S. Kookana, G.B. Douglas, L.A. Wendling, Sorption of pesticides by a mineral sand mining by-product, neutralised used acid (NUA), Sci. Total Environ. 442 (2013) 255–262.

[69] A. Funke, F. Ziegler, Hydrothermal carbonization of biomass: a summary and discussion of chemical mechanisms for process engineering, Biofuels Bioprod. Biorefin. 4 (2) (2010) 160–177.

[70] J.A. Libra, K.S. Ro, C. Kammann, A. Funke, N.D. Berge, Y. Neubauer, et al., Hydrothermal carbonization of biomass residuals: a comparative review of the chemistry, processes and applications of wet and dry pyrolysis, Biofuels 2 (1) (2011) 71–106.

[71] Z. Fan, Q. Zhang, B. Gao, M. Li, C. Liu, Y. Qiu, Removal of hexavalent chromium by biochar supported nZVI composite: batch and fixed-bed column evaluations, mechanisms, and secondary contamination prevention, Chemosphere 217 (2019) 85–94.

[72] E. Agrafioti, D. Kalderis, E. Diamadopoulos, Arsenic and chromium removal from water using biochars derived from rice husk, organic solid wastes and sewage sludge, J. Environ. Manag. 133 (2014) 309–314.

[73] K.H. Kim, J.Y. Kim, T.S. Cho, J.W. Choi, Influence of pyrolysis temperature on physicochemical properties of biochar obtained from the fast pyrolysis of pitch pine (*Pinus rigida*), Bioresour. Technol. 118 (2012) 158–162.

[74] A. Méndez, J. Paz-Ferreiro, F. Araujo, G. Gasco, Biochar from pyrolysis of deinking paper sludge and its use in the treatment of a nickel polluted soil, J. Anal. Appl. Pyrolysis 107 (2014) 46–52.

[75] H. Li, X. Dong, E.B. da Silva, L.M. de Oliveira, Y. Chen, L.Q. Ma, Mechanisms of metal sorption by biochars: biochar characteristics and modifications, Chemosphere 178 (2017) 466–478.

[76] H. Zhang, R.P. Voroney, G.W. Price, Effects of temperature and processing conditions on biochar chemical properties and their influence on soil C and N transformations, Soil Biol. Biochem. 83 (2015) 19–28.

[77] J.M. Novak, K.B. Cantrell, D.W. Watts, Compositional and thermal evaluation of lignocellulosic and poultry litter chars via high and low temperature pyrolysis, BioEnergy Res. 6 (1) (2013) 114–130.

[78] G.N. Kasozi, A.R. Zimmerman, P. Nkedi-Kizza, B. Gao, Catechol and humic acid sorption onto a range of laboratory-produced black carbons (biochars), Environ. Sci. Technol. 44 (16) (2010) 6189–6195.

[79] S. Kumar, V.A. Loganathan, R.B. Gupta, M.O. Barnett, An assessment of U (VI) removal from groundwater using biochar produced from hydrothermal carbonization, J. Environ. Manag. 92 (10) (2011) 2504–2512.

[80] M.C. Wang, G.D. Sheng, Y.P. Qiu, A novel manganese-oxide/biochar composite for efficient removal of lead (II) from aqueous solutions, Int. J. Environ. Sci. Technol. 12 (5) (2015) 1719–1726.

[81] M. Inyang, B. Gao, W. Ding, P. Pullammanappallil, A.R. Zimmerman, X. Cao, Enhanced lead sorption by biochar derived from anaerobically digested sugarcane bagasse, Sep. Sci. Technol. 46 (12) (2011) 1950–1956.

[82] M.I. Shariful, T. Sepehr, M. Mehrali, B.C. Ang, M.A. Amalina, Adsorption capability of heavy metals by chitosan/poly (ethylene oxide)/activated carbon electrospun nanofibrous membrane, J. Appl. Polym. Sci. 135 (7) (2018) 45851.

[83] V.K. Gupta, O. Moradi, I. Tyagi, S. Agarwal, H. Sadegh, R. Shahryari-Ghoshekandi, et al., Study on the removal of heavy metal ions from industry waste by carbon nanotubes: effect of the surface modification: a review, Crit. Rev. Environ. Sci. Technol. 46 (2) (2016) 93–118.

[84] X. Yang, Y. Wan, Y. Zheng, F. He, Z. Yu, J. Huang, et al., Surface functional groups of carbon-based adsorbents and their roles in the removal of heavy metals from aqueous solutions: a critical review, Chem. Eng. J. 366 (2019) 608–621.

[85] E. Agrafioti, G. Bouras, D. Kalderis, E. Diamadompulos, Biochar production by sewage sludge pyrolysis, J. Anal. Appl. Pyrol. 101 (2013) 72–78.

[86] T. Shi, S. Jia, Y. Chen, Y. Wen, C. Du, H. Guo, Z. Wang, Adsorption of Pb (II), Cr (III), Cu (II), Cd (II) and Ni (II) onto a vanadium mine tailing from aqueous solution, J. Hazard. Mater. 169 (1–3) (2009) 838–846.

[87] A. Callegari, A.G. Capodaglio, Properties and beneficial uses of (bio) chars, with special attention to products from sewage sludge pyrolysis, Resources 7 (1) (2018) 20.

[88] P. Huang, D.W. Fuerstenau, The effect of the adsorption of lead and cadmium ions on the interfacial behavior of quartz and talc, Colloids Surf. A Physicochem. Eng. Asp. 177 (2–3) (2001) 147–156.

[89] Y. Yin, C.A. Impellitteri, S.J. You, H.E. Allen, The importance of organic matter distribution and extract soil: solution ratio on the desorption of heavy metals from soils, Sci. Total Environ. 287 (1–2) (2002) 107–119.

[90] J.R. Kim, E. Kan, Heterogeneous photocatalytic degradation of sulfamethoxazole in water using a biochar-supported TiO_2 photocatalyst, J. Environ. Manag. 180 (2016) 94–101.

[91] L. Trakal, V. Veselská, I. Šafařík, M. Vítková, S. Cíhalová, M. Komárek, Lead and cadmium sorption mechanisms on magnetically modified biochars, Bioresour. Technol. 203 (2016) 318–324.

[92] X. Zhang, S. Zhang, H. Yang, Y. Feng, Y. Chen, X. Wang, H. Chen, Nitrogen enriched biochar modified by high temperature CO_2–ammonia treatment: characterization and adsorption of CO_2, Chem. Eng. J. 257 (2014) 20–27.

[93] X. Kong, Y. Liu, J. Pi, W. Li, Q. Liao, J. Shang, Low-cost magnetic herbal biochar: characterization and application for antibiotic removal, Environ. Sci. Pollut. Res. 24 (2017) 6679–6687.

[94] D. Mohan, S. Rajput, V.K. Singh, P.H. Steele, C.U. Pittman Jr., Modeling and evaluation of chromium remediation from water using low cost bio-char, a green adsorbent, J. Hazard. Mater. 188 (1–3) (2011) 319–333.

[95] Y.S. Shen, S.L. Wang, Y.M. Tzou, Y.Y. Yan, W.H. Kuan, Removal of hexavalent Cr by coconut coir and derived chars–the effect of surface functionality, Bioresour. Technol. 104 (2012) 165–172.

[96] X. Chen, G. Chen, L. Chen, Y. Chen, J. Lehmann, M.B. McBride, A.G. Hay, Adsorption of copper and zinc by biochars produced from pyrolysis of hardwood and corn straw in aqueous solution, Bioresour. Technol. 102 (2011) 8877–8884.

[97] F.M. Pellera, A. Giannis, D. Kalderis, K. Anastasiadou, R. Stegmann, J.Y. Wang, E. Gidarakos, Adsorption of Cu (II) ions from aqueous solutions on biochars prepared from agricultural by-products, J. Environ. Manag. 96 (1) (2012) 35–42.

[98] H. Kong, J. He, Y. Gao, H. Wu, X. Zhu, Cosorption of phenanthrene and mercury (II) from aqueous solution by soybean stalk-based biochar, J. Agric. Food Chem. 59 (22) (2011) 12116–12123.

[99] H.Z. Mousavi, A. Hosseynifar, V. Jahed, S.A.M. Dehghani, Removal of lead from aqueous solution using waste tire rubber ash as an adsorbent, Braz. J. Chem. Eng. 27 (1) (2010) 79–87.

[100] M. Uchimiya, S. Chang, K.T. Klasson, Screening biochars for heavy metal retention in soil: role of oxygen functional groups, J. Hazard. Mater. 190 (1–3) (2011) 432–441.

[101] S. Wang, B. Gao, Y. Li, A. Mosa, A.R. Zimmerman, L.Q. Ma, et al., Manganese oxide-modified biochars: preparation, characterization, and sorption of arsenate and lead, Bioresour. Technol. 181 (2015) 13–17.

[102] Y. Han, X. Cao, X. Ouyang, S.P. Sohi, J. Chen, Adsorption kinetics of magnetic biochar derived from peanut hull on removal of Cr (VI) from aqueous solution: effects of production conditions and particle size, Chemosphere 145 (2016) 336–341.

[103] H. Jin, S. Capareda, Z. Chang, J. Gao, Y. Xu, J. Zhang, Biochar pyrolytically produced from municipal solid wastes for aqueous As (V) removal: adsorption property and its improvement with KOH activation, Bioresour. Technol. 169 (2014) 622–629.

[104] J. Zhu, H. Gu, J. Guo, M. Chen, H. Wei, Z. Luo, et al., Mesoporous magnetic carbon nanocomposite fabrics for highly efficient Cr (VI) removal, J. Mater. Chem. A 2 (7) (2014) 2256–2265.

[105] M. Mohapatra, S. Anand, Synthesis and applications of nano-structure iron oxides/hydroxides-a review, Int. J. Eng. Sci. Technol. 2 (2010) 127–146.

[106] M. Kılıç, M.E. Keskin, S. Mazlum, N. Mazlum, Hg (II) and Pb (II) adsorption on activated sludge biomass: effective biosorption mechanism, Int. J. Miner. Process. 87 (1–2) (2008) 1–8.

3

Nanoparticles: A new tool for control of mosquito larvae

Arghadip Mondal, Priyanka Debnath, and Naba Kumar Mondal

ENVIRONMENTAL CHEMISTRY LABORATORY, DEPARTMENT OF ENVIRONMENTAL SCIENCE, THE UNIVERSITY OF BURDWAN, BARDHAMAN, WEST BENGAL, INDIA

1 Introduction

Mosquitoes are disease-causing vectors that are major causative agents of human filariasis, malaria, and many other viral diseases like dengue, Japanese encephalitis, Zika, and West Nile virus. According to the World Health Organization report, India, Indonesia, Bangladesh, and Nigeria are most affected. At present, many biological processes like mosquito larvae-eating fish, application of bit granules to larval habitat, plant-based ovicides, larvicides, and many physical processes like mosquito netting and mosquito-killing electrical traps, and the most common chemical controls such as dichlorodiphenyltrichloroethane, dieldrin, organophosphorus, and fenitrothion are used for the control of mosquitoes. But none of these processes can control mosquitoes to a satisfactory level.

Therefore, an alternative method is required that should give the desired result as well as being safer for the environment. Numerous recent literature highlighted that mosquito larvae can be controlled by nanoparticles (NPs). Physics Nobel laureate Richard Phillips Feynman first introduced nanotechnology in his famous lecture entitled, "There's plenty of room at the bottom" [1]. Prof. Norio Taniguchi invented the term "nanotechnology" in 1974 [2]. Typically, NPs are defined as agglomerated atoms and molecules within a 1–100 nm size range. NPs are classified depending on their dimensionality, uniformity, morphology, agglomeration, and composition [3,4]. Previous literature shows biologically synthesized silver, gold, zinc, and copper NPs are widely used for mosquito control. But among them, which NP is most suitable as a larvicidal agent as well as eco-friendly in nature has not yet been discovered. With this as our background, we will thoroughly discuss biological synthesis procedures, characterization of NPs, and application of NPs toward mosquito larvae in this chapter.

2 Nanoparticle synthesis

Synthesis of dimensionally controlled particles in large quantities and studying their properties to investigate novel applications is a preferred activity for researchers in

Intelligent Environmental Data Monitoring for Pollution Management. https://doi.org/10.1016/B978-0-12-819671-7.00003-8

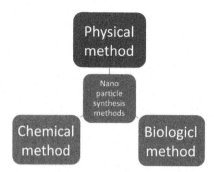

FIG. 1 Three major synthesis types of nanoparticles.

nanomaterials. Commercially large quantities of NPs are synthesized through physical, chemical, and biological (Fig. 1) procedures in a short time period.

Physical and chemical synthesized NP applications are limited because of the use of heavy toxic compounds through microemulsion, aqueous solution, and photoreduction of salt solution [5]. Through an evaporation-condensation method, various NPs have been previously synthesized like lead sulfide (PbS), cadmium sulfide (CdS), and fullerene (a type of carbon nanostructure). Chemically synthesized NPs are mostly produced through chemical production by an inorganic and organic reducing agent such as ascorbate, tollens reagent, elemental hydrogen, polyol process, sodium citrate, N-dimethylformamide (DMF), and sodium borohydride ($NaBH_4$) [6]. But in the biological synthesis process, we do not require any external power or extra chemicals. In this chapter, detailed focus will be on the ability of biomolecules for synthesis of NPs.

2.1 Biological synthesis of nanoparticles

NPs are in demand in overall medicine, biomedical devices, and engineering sector, as well as several environmentally important areas, such as environmental remediation and renewable energy. Many NP-impregnated materials are also now commercially available [7]. With this background, the development of eco-friendly and safe methods for biofabricated production is now of great interest due to facile production and huge applications in various fields. Biologically synthesized NPs show better control size and shape, so this route of synthesis is more acceptable [8,9]. In this chapter, we have focused mainly on two types of biomaterials: plant- and micro-organism-mediated materials for NP synthesis (Fig. 2). Detailed discussion of these two types of biomaterials are as follows.

2.1.1 Plant-mediated nanoparticle synthesis
In the last few years, biogenic synthesized nanoresearch has been rather comprehensive due to the availability of raw materials. Biogenic NPs can be synthesized by using different parts of the plant such as the leaf, root, flower, bark, fruit, seeds, and stem (Fig. 3).

These plant extracts contain some biomolecules that are mainly responsible for the rapid reduction of metal salts along with stabilization of the synthesized nanomaterial [10].

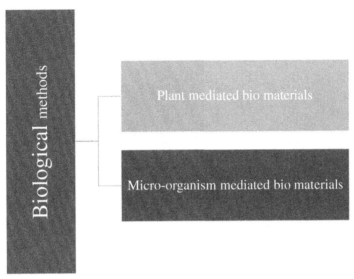

FIG. 2 Two major biological components for nanoparticle synthesis.

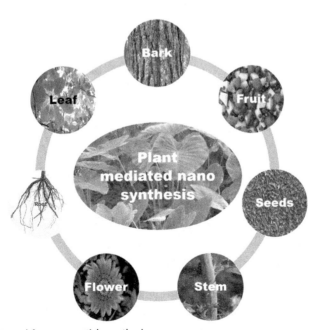

FIG. 3 Major plant part used for nanoparticle synthesis.

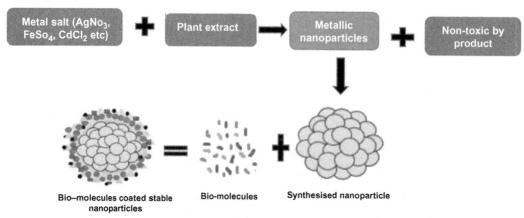

FIG. 4 Basic principle of biologically synthesized nanoparticles.

Plant extracts contain many compounds like enzymes, vitamins, organic acids, proteins, polysaccharides, and lipids. Moreover, many phytochemicals such as flavonoids, terpenoids, alkaloids, proteins, phenols, and carbohydrates, rather than amine, hydroxyl, and methoxide are present in the plant, and these functional groups act as both reducing and capping agents.

By following the steps (Fig. 4), plant extract-mediated NPs were synthesized: (1) Plant extract was prepared by using various plant parts or the whole plant. (2) A metal salt solution was separately prepared. (3) Finally, plant extracts and metal salt solution were mixed at the desirable conditions.

Temperature, pH, and concentration of metal salt solution play important roles in the synthesis and stability of NPs, although the exact synthesis mechanisms of NPs are not completely understood yet. Different salts are being used for synthesis of different heavy metal NPs. Table 1 shows the different types of NPs that are synthesized using different plant parts.

2.1.2 Micro-organism-mediated nanoparticle synthesis

Besides plants, another group of micro-organisms like fungi, yeast, bacteria, and algae are used for biological synthesis (Fig. 5). Micro-organisms are also preferred for NPs synthesis, because of their easy cultivation, fast growth rate, and their ability to grow at ambient conditions.

Owing to their adaptability to a metal toxic environment, micro-organisms show an internal power to biologically synthesize NPs of inorganic metal by following reduction mechanism via intracellular (which involves conveying the ions inside the microbial cell to form NPs within the sight of enzymes) and extracellular (which involves confining the ions to the cell surface thus reducing the ions in the presence of enzymes) routes. Naturally, micro-organisms capture metal ions from the environment and turn them into their elemental form through their own enzymatic activities [27]. Table 2 shows the different types of NPs that were synthesized from different micro-organisms.

Table 1   Checklist of plant-mediated synthesized nanoparticles.

Plant name	Plant parts	Synthesized nanoparticles	Reference
Azadirachta indica	Leaf	Silver	[11]
Citrus limon	Juice	Silver	[7]
Couroupita guianensis	Fruit	Iron oxide	[12]
Jatropha curcas	Latex	Silver	[13]
Tagetes sp.	Flower	Silver	[14]
Azadirachta indica	Leaf	Titanium oxide	[15]
Coriandrum sativum	Leaf	Silver	[16]
Camellia sinensis	Leaf	Silver and gold	[17]
Boerhaavia diffusa	Leaf	Silver	[18]
Calliandra haematocephala	Leaf	Silver	[19]
Coriandrum sativum	Leaf	Gold	[20]
Olea europaea	Leaf	Gold	[21]
Ocimum tenuiflorum	Leaves and peel	Silver	[22]
Aloe vera	Leaf	Gold and silver	[23]
Nyctanthes arbortristis	Flower	Gold	[24]
Azadirachta indica	Leaf	Silver	[25]
Desmodium gangeticum	Leaf	Silver	[26]

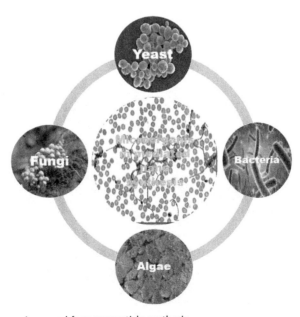

FIG. 5 Different micro-organisms used for nanoparticle synthesis.

Table 2 Checklist of micro-organism-mediated synthesized nanoparticles.

Micro-organism	Type of micro-organism	Synthesized nanoparticles	References
Alternaria alternate	Fungi	Silver	[14]
Aspergillus flavus	Fungi	Silver	[28]
Thermomonospora sp.	Bacteria	Gold	[29]
Sargassumwightii Greville	Algae	Gold	[30]
Galaxaura elongate	Algae	Gold	[31]
Stoechospermum marginatum	Algae	Gold	[32]
Penicillium sp.	Fungi		
E. coli and	Bacteria	Silver	[33]
S. aureus	Bacteria		
Bacillus sp.	Bacteria	Titanium oxide	[34]
Bifurcaria bifurcate	Algae		
Chlamydomonas reinhardtii	Algae		
Chlorella vulgaris	Algae		
Ecklonia cava	Algae	Iron	[35]
Oscillatoria willei	Bacteria		
Pithophora oedogonia	Algae		
Klebsiella pneumonia	Bacteria	Silver	[36]
Pseudomonas aeruginosa	Bacteria	Copper	
Acidithiobacillus sp.	Bacteria	Cadmium	[37]
Bacillus amyloliquefaciens	Bacteria	Titanium dioxide	[38]
Lactobacilli sp.	Bacteria	Silver and titanium	[39]
Yarrowialipolytica NCIM 3589	Yeast	Silver	[40]
		Gold	
Yarrowialipolytica NCYC 789	Yeast	Silver	[41]

3 Nanoparticle characterizations

It is important to fully characterize and understand the structure of synthesized NPs. Over the years, many methods like UV-visible spectroscopy, Fourier transformed infrared spectroscopy (FTIR), dynamic light scattering (DLS), X-ray diffraction (XRD), Transmission electron microscopy (TEM), Scanning electron microscopy (SEM), and Energy-dispersive X-ray spectroscopy (EDX) have been developed for this purpose. Detailed explanations of these characterization methods are given in the following sections.

3.1 UV-visible spectrophotometry

After synthesis of NPs, UV-visible spectrophotometry (UV-vis spectroscopy) is a first conformational test. This is one of the most simple, feasible, quick, and nonhazardous characterization methods. This instrument involves measuring the amount of UV or visible radiation absorbed by a substance in a liquid nanosolution. The fundamental principle of this instrument governs the quantitative spectrophotometric analysis by following Lambert-Beer's law that states that the absorption of light is directly proportional to both

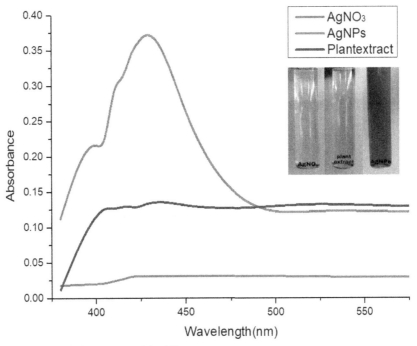

FIG. 6 UV-visible spectra of silver nanoparticles [6].

concentration and the path length of the absorbing medium in the light path. It is expressed as:

$$T = I/I_0 \tag{1}$$

where I is the intensity of the transmitted light, I_0 is the intensity of incident light, and T is the transmittance. In nanocharacterization, intensity of the peak is denoted as synthesis rate of NP, and right or left shifting of the peak is related to the particle size. Additionally, different NPs produce different significant peaks at different wavelengths. Mondal et al. [6] proposed that synthesized silver NPs from *Colocasia esculenta* showed a sharp peak at around 430 nm wavelength (Fig. 6). This figure also suggests that plant extracts and silver salt does not show any significant peak.

So, this peak is significant for silver NPs. This preliminary result confirms the synthesis of silver NPs.

3.2 Fourier transform infrared spectroscopy

Functional compounds and chemical composition of synthesized NPs was detected through FTIR. In FTIR, the electromagnetic spectral range is 4000–600 cm^{-1}. Each absorption peak is produced because of the interaction between the functional group and chemical bonds of NPs. Many biomolecules are present in biologically synthesized NPs, which

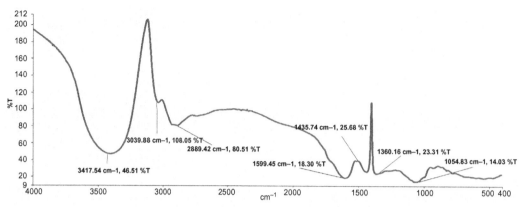

FIG. 7 FTIR spectra of silver nanoparticles [6].

are solely responsible not only for reduction of the metal salt but also as capping and stabilizing agents for the newly formed NPs. These biomolecules are detected through the FTIR study. Mondal et al. [6] synthesized silver NPs using *Colocasia esculenta* stem extract as a reducing and capping agent. The FTIR spectrum of the synthesized NPs is shown in Fig. 7. In this figure, IR peaks are observed around at 1054.83, 1360.16, 1435.74, 1599.45, 2889.42, 3039.88 and 3417.54 cm^{-1}.

Each peak denotes different chemical compounds present in the biosynthesized NPs. Fig. 7 denotes that —C—O stretching vibrations of alcoholic groups that represent 1054.83 cm^{-1} located at a peak, and a 1360.16 cm^{-1} located peak arises due to —C—O—C stretching modes. The peak emerged at 3039.88 and 3417 cm^{-1} are attributed to hydroxyl functional groups in polyphenol and phenol, respectively. The stretching vibration at 1435.74 cm^{-1} is due to C—N or O—H vibration, whereas the peak around 1599.45 cm^{-1} is denoted as the carboxyl group. The peak at 2889.42 cm^{-1} shows that the Ag NPs are bound to a protein via a carboxylic group and —C—C— aromatic ring, respectively.

3.3 Transmission electron microscopy

TEM is a very essential tool for revealing the NP's shape, size, and distribution pattern. TEM analysis uses a high energy beam of electrons shoring through a very thin specimen layer on a copper grid, resulting in an interaction between the specimen's atoms and the electronic beam. The basic principle of TEM is the same as a light microscope, but it uses electrons instead of light. Debnath et al. [42] reported that synthesized gold NPs have ~17–24 nm average size, which was calculated from TEM analysis. On the other hand, TEM was also used to determine the shape of synthesized NPs. Here, it is observed that the biosynthesized Au NPs have spherical, triangular, and rod shapes (Fig. 8).

From this TEM image, it can also be concluded that synthesized NPs are well distributed.

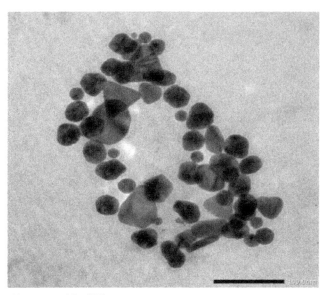

FIG. 8 TEM image of gold nanoparticles [42].

3.4 Scanning electron microscopy

Like TEM, SEM is another microscopic technique for analysis of NP's surface morphology at different magnification ranges. SEM micrograph provides details about the surface morphology of NPs. The main principle of SEM imaging is that an electron beam passes through a fine probe, and the whole sample is scanned. Then the sample emits electromagnetic radiation due to interaction between the sample and electrons. Medda et al. [43] synthesized silver NPs from *Aloe vera* extract. The SEM image (Fig. 9) of this experiment shows that the synthesized Ag NPs were spherical, triangular, rectangular, and cubical in shape.

3.5 Energy-dispersive X-ray spectroscopy

EDX is an analytical tool used for chemical or elemental analysis of NPs. It relies on the interaction of an X-ray excitation source and NPs. Its characterization capability is based on the fundamental principle that different elements have different peaks due to their unique atomic structures [44]. In nanotechnology, EDX is a very strong and user-friendly tool for qualitative and quantitative elemental profile analysis. Debnath et al. [42] highlighted in their research paper that synthesized silver NPs produced a significant peak in EDX. Fig. 10 shows a broad Ag peak due to metallic silver NPs and another small gold peak because, at the time of EDX sample preparation, metallic gold was used as a sample coating agent.

FIG. 9 SEM image of silver nanoparticles [43].

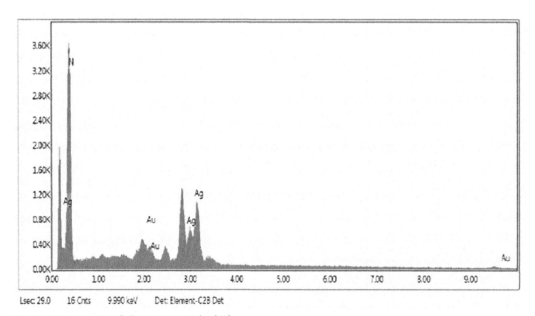

FIG. 10 EDX spectrum of silver nanoparticles [42].

3.6 X-ray diffraction

To determine a particle's crystal structure, surface lattice parameters, synthesized particle average size, and particle phase formation nature, we can use XRD analysis. The radiation source used in an X-ray beam interacts with the specimen. The fundamental principle of this experiment is Bragg's law:

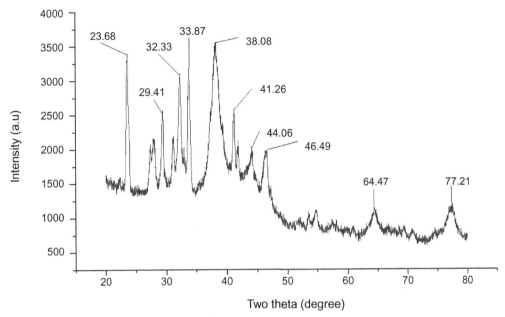

FIG. 11 XRD spectrum of silver nanoparticles [6].

$$n\lambda = 2d(\sin\theta) \qquad (2)$$

where n = integer, λ = the wavelength of the incident wave, d = the interplanar spacing of the crystal, and θ = the angle of incidence. This is a simple and faster nondestructive method. Mondal et al. [6] characterized their synthesized silver NPs through XRD. From this spectrograph (Fig. 11), 10 different peaks were observed at 23.68 to 77.21, which corresponded to crystal morphology.

The mean silver NP size was 16.16 nm, which was calculated from the XRD data followed by Debye-Scherrer methods through Bragg's law.

3.7 Dynamic light scattering

The accurate size of the particle can also be measured through DLS. As a function of time when light is scattered by molecules in Brownian motion, that intensity measurement is the main principle of this technique. The NP's size can be measuring by the change in the intensity of light scattered from the sample. The particle's distribution, the average size, and the polydispersity index of the molecules in the solution are measured through this technique [45]. Mondal et al. [6] characterized their synthesized silver NPs through a DLS particle size analyzer.

From Fig. 12, it can be reported that the particle's charge was −25.1. Previous literature suggested that lower (−25.1 mV) particles charge are facile the nanoparticles synthesis process.

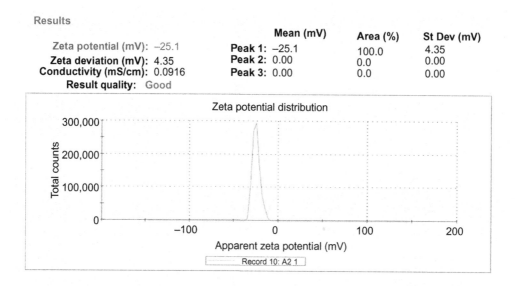

FIG. 12 Zeta potential report of synthesized silver nanoparticles [6].

4 Application

Green synthesis is more useful compared with chemical- and physical-based synthesis of NPs, which have many applications in the field of medical and biological science. In the recent past, many physicochemical methods have been commercially used for synthesis of different types of metal NPs. These synthesized NPs have been used in different application fields such as nanolabeling, drug delivery, antimicrobial, antifungal, larvicidal, and cancer treatment. In this chapter, detailed focus will be on mosquito larvicidal activity of biologically synthesized NPs.

4.1 Larvicidal activity of nanoparticles

With the progress of modern technology, many researchers around the world have focused on the biological synthesis of silver NPs and its application as a mosquitocidal agent. Green synthesized Ag NPs have been synthesized from many natural products like *Azadirecta indica* [11], *Tagets* sp. [14], *Calliandra haematocephala* [19], *Ocimum tenuiflorum* [22], *Aspergillus flavus* [28], *Panicillium* sp. [33], *Klebsiella pneumonia* [36], and many other biological agents. Recently, Raja et al. [19] proposed that synthesized silver NPs have larvicidal properties, and the LC_{50} values are 31.87 mg/L and 26.78 mg/L against *Aedes aegypti* during 24 and 48 h, respectively. Similarly, Found et al. [46] synthesized *Cassia fistula*-mediated silver NPs. This NP shows good larvicidal activity against *Aedes albopictus* and *Culex pipiens*. They reported 16.5 and 8.0 mg/L Ag NPs were required for 50%

Table 3   Green synthesized silver nanoparticles and its activity toward mosquito larvicidal activity.

SI No.	Biological agent	Family	Size of NP	Target species	LC$_{50}$/LC$_{90}$	Reference
1.	*Cassia fistula*	Fabaceae	Ag NPs (16.8 nm)	*Aedes albopictus*	8.3 mg/L (LC$_{50}$)	[46]
				Culex pipiens pallens	9.7 mg/L (LC$_{50}$)	
2.	*Halodule uninervis*	Cymodoceaceae	Ag NPs (30–35 nm)	*Aedes aegypti*	12.455 mg/L	[47]
3.	*Hedychium coronarium*	Zingiberaceae	Ag NPs (9.54–49 nm)	*Aedes aegypti*	72.618 mg/L (LC$_{50}$)	[48]
4.	*Turbinaria ornata*	Sargassaceae	Ag NPs (2–32 nm)	*Aedes aegypti*	0.738 mg/L (LC$_{50}$)	[49]
				Anopheles stephensi	1.134 mg/L (LC$_{50}$)	
				Culex quinquefasciatus	1.494 mg/L (LC$_{50}$)	
5.	*Sargassum polycystum*	Sargassaceae	Ag NPs (20–88 nm)	*Aedes aegypti*	0.3 mg/L (LC$_{50}$)	[50]
				Anopheles stephensi	3.7 mg/L (LC$_{50}$)	
				Culex quinquefasciatus	0.57 mg/L (LC$_{50}$)	
				Culex tritaeniorhynchus	4.8 mg/L (LC$_{50}$)	

mortality of *Aedes albopictus* at 24 and 48 h, respectively. But in the case of *Culex pipiens*, 9.7 and 3.2 mg/L Ag NPs were required for 50% mortality at 24 and 48 h, respectively. Mahyoub et al. [47] also synthesized silver NPs from *Halodule uninervis*. Although these synthesized NPs show moderate efficacy as a larvicidal agent, the LC$_{50}$ value was 12.455 mg/L against *Aedes aegypti* at 24 h. Additionally, Kalimuthu et al. [48] reported *Hedychium coronarium*-mediated Ag NP activity toward *Aedes aegypti* was not at a satisfactory level compared with other synthesized Ag NPs. During 24 h, LC$_{50}$ value was recorded at 72.618 mg/L against *Aedes aegypti*. However, Deepak et al. [49] synthesized silver NPs that showed tremendous results as a mosquito larvicide. Required silver NPs for 50% mortality was recorded as 0.73, 1.134, and 1.494 mg/L against *Aedes* sp., *Anopheles* sp., and *Culex* sp., respectively. Table 3 shows the larvicidal activity of synthesized silver NPs from different biological sources.

On the other hand, developing new reliable eco-friendly processes for copper NP synthesis is growing day by day. Among the key advantages, larvicidal efficacy of biologically synthesized copper NP from various sources such as *Tridax procumbens* [51], *Artocarpus heterophyllus* [52], *Tagetes* sp. [53], *Helianthus* sp. [54], and *Aegle marmelos* [55] are already investigated and reported. Recently, Selvan et al. [51] reported mosquito larvicidal activity of CuO NP against dengue, Zika, and chikungunya causing vector *Aedes aegypti* having a LC_{50} value of 4.209 mg/L. Similarly, Sharon et al. [52] investigated the efficacy of CuO NP against dengue vector *Aedes aegypti* and reported the LC_{50} value of 5.08 mg/L. Previously, Mondal and Hajra [53] successfully synthesized copper NPs using petal extracts of *Tagetes* sp. and *Helianthus* sp., and the LC_{50} value was recorded as 34.4 mg/L as a larvicidal agent against *Culex quinquifasciatus*. Angajala et al. [55] synthesized copper NPs from *Aegle marmelos* extract and evaluated the mosquito larvicidal efficacy on *Anopheles stephensi*, *Aedes aegypti*, and *Culex quinquefasciatus*. The LC_{50} value was recorded as 57.21, 60.81, and 74.6 mg/L, respectively. Synthesis of copper NPs has been also investigated utilizing many ubiquitous fungal species. Very recently, Kalaimurugan et al. [56] demonstrated the extracellular synthesis of copper NPs using the whole-cell biomass of *Fusarium proliferatum* (YNS2). Moreover, the larvicidal activity of synthesized NP was evaluated on *Anopheles stephensi*, *Aedes aegypti*, and *Culex quinquefasciatus* where the LC_{50} value was recorded as 39.25 mg/L, 81.34 mg/L, and 21.84 mg/L, respectively. Table 4 represents the mosquito larvicidal efficacy of copper NPs.

Another semiconductor, zinc NPs have been well established as a larvicidal agent against mosquito vectors. Zinc is an efficient semiconductor and can be synthesized biologically using *Plectranthus amboinicus* [57], *Lobelia leschenaultiana* [58], *Myristica fragrans* [59], *Scadoxus multiflorus* [60], *Macroalga Sargassum* [61], and many other different biological agents. Zinc or zinc oxide NPs have mosquito larvicidal efficacy, which is represented in Table 5. Synthesis of zinc oxide NPs using *Plectranthus amboinicus* and its efficacy as a larvicidal agent against *Anopheles stephensi*, *Culex quinquefasciatus* and *C. tritaeniorhynchus* was reported by Vijayakumar et al. [57].

Table 4   Green synthesized copper nanoparticles and its activity toward mosquito larvicidal activity.

SI No.	Biological agent	Family	Size of NP	Target species	LC_{50}/LC_{90}	Reference
1.	*Tridax procumbens*	Asteraceae	CuO NPs (16 nm)	*Aedes aegypti*	4.209 mg/L (LC_{50})	[51]
2.	*Artocarpus heterophyllus*	Moraceae	CuO NPs (132 nm)	*Aedes aegypti*	5.08 mg/L (LC_{50})	[52]
3.	*Tagetes* sp.; *Helianthus* sp.	Asteraceae	Cu NPs (5–20 nm)	*Culex quinquifasciatus*	34.4 mg/L (LC_{50})	[53]
4.	*Aegle marmelos*	Rutaceae	Cu NPs (50–100 nm)	*Anopheles stephensi*, *Aedes aegypti*, and *Culex quinquefasciatus*	57.21, 60.81, and 74.6 mg/L (LC_{50})	[55]
5.	*Fusarium proliferatum* (YNS2)	Nectriaceae	Cu NPs (10–50 nm)	*Anopheles stephensi*, *Aedes aegypti*, and *Culex quinquefasciatus*	39.25 mg/L, 81.34 mg/L, and 21.84 mg/L (LC_{50})	[56]

Table 5   Green synthesized zinc nanoparticles and its activity toward mosquito larvicidal activity.

Sl No.	Biological agent	Family	Size of NP	Target species	LC_{50}/LC_{90}	Reference
1.	*Plectranthus amboinicus*	Lamiaceae	ZnO NPs (20–50 nm)	*Anopheles stephensi, Culex quinquefasciatus,* and *Culex tritaeniorhynchus*	3.11 mg/L (LC_{50}), 4.6 mg/L (LC_{90}), 3.1 mg/L (LC_{50}), 4.5 mg/L (LC_{90}) and 4.2 mg/l (LC_{50}), 5.7 (LC_{90}), respectively	[57]
2.	*Lobelia leschenaultiana*	Campanulaceae	ZnO NPs (10–30 nm)	*Aedes aegypti*	1.57 mg/L	[56]
3.	*Myristica fragrans*	Myristicaceae	ZnO NPs (100 nm)	*Aedes aegypti*	3.4 mg/L (LC_{50}), 14.63 mg/L (LC_{90})	[59]
4.	*Scadoxus multiflorus*	Amaryllidaceae	ZnO NPs (31 nm)	*Aedes aegypti*	34.04 mg/L (LC_{50})	[60]
5.	*Macroalga Sargassum*	Sargassaceae	ZnO NPs (100 nm)	*Anopheles stephensi*	4.330 mg/L (LC_{50})	[61]

The LC_{50} and LC_{90} values were recorded as 3.11 mg/L (LC_{50}) and 4.6 mg/L (LC_{90}) for *Anopheles stephensi*. But in the case of *Culex quinquefasciatus* and *Culex tritaeniorhynchus*, the LC_{50} and LC_{90} values were recorded as 3.1 mg/L, 4.5 mg/L, and 4.2 mg/L, 5.7 mg/L, respectively. Recently, Banumathi et al. [58] effectively used ZnO NPs as a larvicidal agent, and they reported tremendous results; 1.57 mg/L ZnO NP is required for 50% of mortality of *Aedes aegypti.* Similarly, Ashokan et al. [59] also used biosynthesized ZnO NPs on *Aedes aegypti* larvae, and the LC_{50} was recorded as 3.4 mg/L, whereas the biosynthesized NP shows moderate efficacy for 90% of mortality, i.e., 14.63 mg/L. Another investigation was done by Al-Dhabi and Arasu [60], which also confirms the larvicidal efficacy of ZnO NPs against *Aedes aegypti* showing 50% mortality at 38.04 mg/L. These synthesized NPs showed poor results compared with other synthesized zinc NPs. Moreover, Murugan et al. [61] selected *Anopheles stephensi* as a target organism to examine the larvicidal activity of ZnO NPs, and 50% mortality at 4.330 mg/L was reported. Along with silver, copper, and zinc NPs, biosynthesized gold NPs can effectively be utilized as a mosquito larvicidal agent. Many researchers have successfully used gold NPs as a mosquito larvicide (Table 6). Aqueous extract of a seaweed (*Turbinaria ornata* (Turner)) [54], flower extract of *Couroupita guianensis* [62], and the leaf extract of *Hybanthus enneaspermus* [64] have shown potential to act both as reducing and capping agents for the synthesis of gold NPs.

Some fungal species, i.e., *Chrysosporium tropicum* [63] and *Aspergillus niger* [65] have also shown tremendous ability to reduce Au (III) ions to form gold NPs Au (0). Recently, Deepak et al. [54] investigated the larvicidal efficiency of biosynthesized Au NPs, and it shows moderate efficacy as a larvicidal agent. The LC_{50} and LC_{90} value were recorded

Table 6 Green synthesized gold nanoparticles and its activity toward mosquito larvicidal activity.

Sl No.	Biological agent	Family	Size of NP	Target species	LC$_{50}$/LC$_{90}$	Reference
1.	*Turbinaria ornata*	Sargassaceae	Au NPs (20–30 nm)	*Anopheles stephensi*	12.79 and 78.70 mg/L	[54]
2.	*Couroupita guianensis*	Lecythidaceae	Au NPs (29.2–43.8 nm)	*Anopheles stephensi*	17.36 mg/L (larva I), 19.79 mg/L (larva II), 21.69 mg/L (larva III), 24.57 mg/L (larva IV) (LC$_{50}$)	[62]
3.	*Chrysosporium tropicum*	Onygenaceae	Au NPs (20–50 nm)	*Aedes aegypti*	12 mg/L (LC$_{50}$) and 36.30 mg/L (LC$_{90}$)	[63]
4.	*Hybanthus enneaspermus*	Violaceae	Au NPs (19–27 nm)	*Anopheles stephensi* and *Culex tritaeniorhynchus*	28.15 and 26.05 mg/L, respectively	[64]
5.	*Aspergillus niger*	Trichocomaceae	Au NPs (10–30 nm)	*Anopheles stephensi* and *Aedes aegypti*	1.65 mg/L (LC$_{50}$), 24 mg/L (LC$_{90}$) and 24 mg/L (LC$_{50}$), 3.15 mg/L (LC$_{90}$), respectively	[65]

as 12.79 and 78.70 mg/L, respectively, against *Anopheles stephensi*. However, Subramaniam et al. [62] reported Au NPs were acting as a larvicidal agent against *Anopheles stephensi* larvae and the LC$_{50}$ was 17.36 mg/L (larva I), 19.79 mg/L (larva II), 21.69 mg/L (larva III), and 24.57 mg/L (larva IV), whereas Soni and Prakash [63] used *Chrysosporium tropicum*, a fungal species, for Au NP synthesis where it showed a moderate effect against *Aedes aegypti*. The LC$_{50}$ and LC$_{90}$ values were reported as 12 mg/L and 36.30 mg/L, respectively. Yoganandham et al. [64] also utilized plant extract of *Hybanthus enneaspermus* to synthesize Au NPs, which shows 50% mortality at 28.15 mg/L against *Anopheles stephensi*. But in the case of *Culex tritaeniorhynchus*, LC$_{50}$ was 26.05 mg/L.

4.2 Toxicological effect of nanoparticles on nontarget organism

This NP concept arises from an environmental safety background, because commercially available mosquito-controlling agents are harmful to the environment. From the previous literature, we showed that larvicidal activity of silver, gold, copper, and zinc was comparatively good than another biological agent. Govindarajan et al. [66] proposed that *Barlenia cristata*-mediated silver NPs was safer to the nontarget organism. They found that toxicity of synthesized NPs was much less. The LC$_{50}$ value was recorded at 633.26, 684.25, and

866.92 mg/L against *Diplonychus indicus, Anisops bouvieri,* and *Gambusia affinis,* respectively. They also suggested that swimming activity of three nontarget organisms didn't change throughout the 7 days of observation. Murugan et al. [67] also showed that silver NPs boosted the predation rate of water bugs toward mosquito larvae. Bai et al. [68] proposed from their experiment that zinc oxide NP's effect on zebra fish is very low; 50 to 100 mg/L zinc oxide NPs show less than 10% mortality toward zebra fish. An experiment by Farkash et al. [69] found that gold NPs don't have any impact on cytogenetics of *Oncorhynchus myhiss.* Hejazy et al. [70] said that copper NPs are very useful, but their toxicity effects have not been thoroughly studied compared with the other three NPs.

5 Research gap

At present, mosquito larvae are controlled by various chemical reagents, but those chemicals are harmful to the environment; for this reason, this new nanotechnology was simultaneously developed. Nowadays, many researchers are synthesizing various NPs from biological agents. The biological agent here acts as both a reducing and capping substrate. But which factor or secondary metabolites of this biological agent play the main role behind this synthesis is still not properly understood. Detection of these respective biomolecules will make the synthesis technique easier and quicker.

At the same time, this nanotechnology is the center of attraction of world researchers due to its higher reactivity. Many research papers have been published so far, which has established that NPs can effectively kill mosquito larvae, but till now, the mechanism of this interaction between NPs and mosquito larvae is not clear. So this is one of the most important areas that needs to be explored. Moreover, the efficacy of NPs as a larvicidal agent on natural environment as well as the impact of this nanoagent on nature has yet to be investigated.

6 Conclusion

This chapter mainly focused on metallic NP synthesis from cheap, easily available biological agents and detailed discussion about characterization techniques like TEM, SEM, XRD, EDX, UV-vis spectroscopy, and DLS. Those synthesized NPs act as an excellent mosquito larvicidal agent at laboratory conditions and controlled natural conditions without any detrimental effect on the nontarget organism and aquatic body. This chapter also suggests that silver and zinc/zinc oxide NPs showed higher larvicidal efficacy than gold and copper NPs. But there are few areas like mode of interaction between NPs and mosquito larvae, biological secondary metabolites activity toward nanosynthesis, and synthesized NP activity on the natural environment that are still not clearly understood. Therefore, it can be concluded that green synthesized NPs could be an effective tool toward control of mosquito larvae without affecting nontarget organisms, but further research is needed for the total scenario.

Acknowledgments

The authors acknowledge their sincere thanks to the funding agency, WBDST Memo No: 126 [Sanc.]/ST/P/ S&T/15G-10/2015 for providing the necessary funds for conducting the present research. The authors also extend their sincere gratitude to all the faculty members and staff, Department of Environmental Science, The University of Burdwan, for their moral support and valuable suggestions in preparation of this manuscript.

References

[1] M.M. Kholoud, A. El-Nour, A. Eftaiha, A. Al-Warthan, A.A. Reda, Synthesis and applications of silver nanoparticles, Arab. J. Chem. 3 (2010) 135–140.

[2] H.A. Salam, R.P.M. Kamaraj, P. Jagadeeswaran, S. Gunalan, R. Sivaraj, Plants: green route for nanoparticle synthesis, Int. Res. J. Biol. Sci. 1 (5) (2012) 85–90.

[3] G. Rajakumar, A.A. Rahuman, Larvicidal activity of synthesized silver nanoparticles using *Eclipta prostrata* leaf extract against filariasis and malaria vectors, Acta Trop. 118 (3) (2011) 196–203.

[4] S. Sudrik, N. Chaki, V. Chavan, S. Chavan, Silver nanocluster redox-couple-promoted non classical electron transfer: an efficient electrochemical Wolff rearrangement of α-diazoketones, Eur. J. Chem. 12 (2006) 859–864.

[5] T. Sinha, M. Ahmaruzzaman, A. Bhattacharjee, A simple approach for the synthesis of silver nanoparticles and their application as a catalyst for the photodegradation of methyl violet 6B dye under solar irradiation, J. Environ. Chem. Eng. 2 (4) (2014) 2269–2279.

[6] A. Mondal, A. Hajra, W.A. Shaikh, S. Chakraborty, N.K. Mondal, Synthesis of silver nanoparticle with *Colocasia esculenta* (L.) stem and its larvicidal activity against *Culex quinquefasciatus* and *Chironomus* sp, Asian Pac. J. Trop. Biomed. 9 (2019) 510–517.

[7] X. Li, H. Xu, Z. Chen, G. Chen, Biosynthesis of nanoparticles by micro-organisms and their applications, J. Nanomater. (2011) 1–16.

[8] M. Popescu, A. Velea, A. Lorinczi, Biogenic production of nanoparticles, Dig. J. Nanomater. Biostruct. 5 (2010) 1035–1040.

[9] S. Iravani, Green synthesis of metal nanoparticles using plants, Green Chem. 13 (10) (2011) 2638–2650.

[10] G. Benelli, Plant-mediated biosynthesis of nanoparticles as an emerging tool against mosquitoes of medical and veterinary importance: a review, Parasitol. Res. 115 (1) (2016) 23–34.

[11] Y. Mohammad, S. Jaspreet, M.K. Tripathi, P. Singh, R. Shrivastava, Green synthesis of silver nanoparticles using leaf extract of common arrowhead houseplant and its anticandidal activity, Pharmacogn. Mag. 13 (4) (2018) 840–844.

[12] G. Sathishkumar, V. Logeshwaran, S. Sarathbabu, P.K. Jha, M. Jeyaraj, C. Rajkuberan, S. Sivaramakrishnan, Green synthesis of magnetic Fe_3O_4 nanoparticles using *Couroupita guianensis* Aubl. fruit extract for their antibacterial and cytotoxicity activities. Artif. Cells Nanomed. Biotechnol. 46 (3) (2017) 589–598, https://doi.org/10.1080/21691401.2017.1332635.

[13] H. Padalia, P. Moteriya, S. Chanda, Green synthesis of silver nanoparticles from marigold flower and its synergistic antimicrobial potential. Arab. J. Chem. 8 (5) (2015) 732–741, https://doi.org/10.1016/j.arabjc.2014.11.015.

[14] M. Gajbhiye, J. Kesharwani, A. Ingle, A. Gade, M. Rai, Fungus-mediated synthesis of silver nanoparticles and their activity against pathogenic fungi in combination with fluconazole. Nanomedicine 5 (4) (2009) 382–386, https://doi.org/10.1016/j.nano.2009.06.005.

[15] R. Sankar, K. Rizwana, K.S. Shivashangari, V. Ravikumar, Ultra-rapid photocatalytic activity of *Azadirachta indica* engineered colloidal titanium dioxide nanoparticles. Appl. Nanosci. 5 (6) (2014) 731–736, https://doi.org/10.1007/s13204-014-0369-3.

[16] R. Sathyavathi, M.B. Krishna, S.V. Rao, R. Saritha, D.N. Rao, Biosynthesis of silver nanoparticles using *Coriandrum Sativum* leaf extract and their application in nonlinear optics. Adv. Sci. Lett. 3 (2) (2010) 138–143, https://doi.org/10.1166/asl.2010.1099.

[17] N.A. Begum, S. Mondal, S. Basu, R.A. Laskar, D. Mandal, Biogenic synthesis of Au and Ag nanoparticles using aqueous solutions of black tea leaf extracts. Colloids Surf. B: Biointerfaces 71 (1) (2009) 113–118, https://doi.org/10.1016/j.colsurfb.2009.01.012.

[18] P.P.N.K. Vijay, S.V.N. Pammi, P. Kollu, K.V.V. Satyanarayana, U. Shameem, Green synthesis and characterization of silver nanoparticles using *Boerhaavia diffusa* plant extract and their antibacterial activity. Ind. Crop. Prod. 52 (2014) 562–566, https://doi.org/10.1016/j.indcrop.2013.10.050.

[19] H. Raja, V. Ramesh, V. Thivaharan, Green biosynthesis of silver nanoparticles using *Calliandra haematocephala* leaf extract, their antibacterial activity and hydrogen peroxide sensing capability. Arab. J. Chem. 10 (2) (2017) 253–261, https://doi.org/10.1016/j.arabjc.2015.06.023.

[20] K.B. Narayanan, N. Sakthivel, Coriander leaf mediated biosynthesis of gold nanoparticles. Mater. Lett. 62 (30) (2008) 4588–4590, https://doi.org/10.1016/j.matlet.2008.08.044.

[21] M.M.H. Khalil, E.H. Ismail, F. El-Magdoub, Biosynthesis of Au nanoparticles using olive leaf extract. Arab. J. Chem. 5 (4) (2012) 431–437, https://doi.org/10.1016/j.arabjc.2010.11.011.

[22] P. Logeswari, S. Silambarasan, J. Abraham, Synthesis of silver nanoparticles using plants extract and analysis of their antimicrobial property. J. Saudi Chem. Soc. 19 (3) (2015) 311–317, https://doi.org/10.1016/j.jscs.2012.04.007.

[23] S.P. Chandran, M. Chaudhary, R. Pasricha, A. Ahmad, M. Sastry, Synthesis of gold nanotriangles and silver nanoparticles using *Aloe vera* plant extract. Biotechnol. Prog. 22 (2) (2006) 577–583, https://doi.org/10.1021/bp0501423.

[24] R.K. Das, N. Gogoi, U. Bora, Green synthesis of gold nanoparticles using *Nyctanthes arbor tristis* flower extract. Bioprocess Biosyst. Eng. 34 (5) (2011) 615–619, https://doi.org/10.1007/s00449-010-0510-y.

[25] S. Ahmed, A.M. Saifullah, B.L. Swami, S. Ikram, Green synthesis of silver nanoparticles using *Azadirachta indica* aqueous leaf extract. J. Radiat. Res. Appl. Sci. 9 (1) (2016) 1–7, https://doi.org/10.1016/j.jrras.2015.06.006.

[26] H. Bar, D.K. Bhui, G.P. Sahoo, P. Sarkar, S.P. De, A. Misra, Green synthesis of silver nanoparticles using latex of Jatrophacurcas. Colloids Surf. A Physicochem. Eng. Asp. 339 (1–3) (2009) 134–139, https://doi.org/10.1016/j.colsurfa.2009.02.008.

[27] X. Li, H. Xu, Z.S. Chen, G. Chen, Biosynthesis of nanoparticles by micro-organisms and their applications. J. Nanomater. 2011 (2011), 270974. https://doi.org/10.1155/2011/270974.

[28] N. Vigneshwaran, N.M. Ashtaputre, P.V. Varadarajan, R.P. Nachane, K.M. Paralikar, R.H. Balasubramanya, Biological synthesis of silver nanoparticles using the fungus *Aspergillus flavus*. Mater. Lett. 61 (6) (2007) 1413–1418, https://doi.org/10.1016/j.matlet.2006.07.042.

[29] A. Ahmad, S. Senapati, M.I. Khan, R. Kumar, M. Sastry, Extracellular biosynthesis of monodisperse gold nanoparticles by a novel extremophilic actinomycete, *Thermomonospora* sp. Langmuir 19 (8) (2003) 3550–3553, https://doi.org/10.1021/la0267721.

[30] G. Singaravelu, J.S. Arockiamary, V.G. Kumar, K. Govindaraju, A novel extracellular synthesis of monodisperse gold nanoparticles using marine alga, *Sargassum wightii Greville*. Colloids Surf. B: Biointerfaces 57 (1) (2007) 97–101, https://doi.org/10.1016/j.colsurfb.2007.01.010.

[31] N. Abdel-Raouf, N.M. Al-Enazi, I.B.M. Ibraheem, Green biosynthesis of gold nanoparticles using *Galaxaura elongata* and characterization of their antibacterial activity. Arab. J. Chem. 10 (2010) 3029–3039, https://doi.org/10.1016/j.arabjc.2013.11.044.

[32] A.R.F. Arockiya, C. Parthiban, K.V. Ganesh, P. Anantharaman, Biosynthesis of antibacterial gold nanoparticles using brown alga, *Stoechospermum marginatum* (kützing). Spectro Chim. Acta A: Mol. Biomol. Spectrosc. 99 (2012) 166–173, https://doi.org/10.1016/j.saa.2012.08.081.

[33] D. Singh, V. Rathod, S. Ninganagouda, J. Hiremath, A.K. Singh, J. Mathew, Optimization and characterization of silver nanoparticle by endophytic fungi *Penicillium sp.* isolated from *Curcuma longa* (turmeric) and application studies against MDR *E. coli* and *S. aureus*. Adv. Appl. Bioinform. Chem. 2014 (2014), 408021. https://doi.org/10.1155/2014/408021.

[34] S.P. Suriyaraj, T. Vijayaraghavan, P. Biji, R. Selvakumar, Adsorption of fluoride from aqueous solution using different phases of microbially synthesized TiO_2 nanoparticles. J. Environ. Chem. Eng. 2 (1) (2014) 444–454, https://doi.org/10.1016/j.jece.2014.01.013.

[35] K.S. Siddiqi, A. Husen, Fabrication of metal and metal oxide nanoparticles by algae and their toxic effects. Nanoscale Res. Lett. 11 (1) (2016) 363, https://doi.org/10.1186/s11671-016-1580-9.

[36] B.A. Providence, A.A. Chinyere, A.A. Ayi, O.O. Charles, T.A. Elijah, H.L. Ayomide, Green synthesis of silver monometallic and copper-silver bimetallic nanoparticles using *Kigelia africana* fruit extract and evaluation of their antimicrobial activities. Int. J. Phys. Sci. 13 (3) (2018) 24–32, https://doi.org/10.5897/ijps2017.4689.

[37] G. Ulloa, B. Collao, M. Araneda, B. Escobar, S. Álvarez, D. Bravo, J.M. Pérez-Donoso, Use of acidophilic bacteria of the genus *Acidithio bacillus* to biosynthesize CdS fluorescent nanoparticles (quantum dots) with high tolerance to acidic pH. Enzym. Microb. Technol. 95 (2016) 217–224, https://doi.org/10.1016/j.enzmictec.2016.09.005.

[38] R. Khan, M.H. Fulekar, Biosynthesis of titanium dioxide nanoparticles using *Bacillus amyloliquefaciens* culture and enhancement of its photocatalytic activity for the degradation of a sulfonated textile dye Reactive Red 31. J. Colloid Interface Sci. 475 (2016) 184–191, https://doi.org/10.1016/j.jcis.2016.05.001.

[39] A.K. Jha, K. Prasad, Biosynthesis of metal and oxide nanoparticles using Lactobacilli from yoghurt and probiotic spore tablets. Biotechnol. J. 5 (3) (2010) 285–291, https://doi.org/10.1002/biot.200900221.

[40] M. Agnihotri, S. Joshi, A.R. Kumar, S. Zinjarde, S. Kulkarni, Biosynthesis of gold nanoparticles by the tropical marine yeast *Yarrowia lipolytica* NCIM 3589. Mater. Lett. 63 (15) (2009) 1231–1234, https://doi.org/10.1016/j.matlet.2009.02.042.

[41] M. Apte, D. Sambre, S. Gaikawad, S. Joshi, A. Bankar, A. Kumar, S. Zinjarde, Psychrotrophic yeast *Yarrowia lipolytica* NCYC 789 mediates the synthesis of antimicrobial silver nanoparticles via cell-associated melanin. AMB Express 3 (1) (2013) 32, https://doi.org/10.1186/2191-0855-3-32.

[42] P. Debnath, A. Mondal, A. Hajra, C. Das, N.K. Mondal, Cytogenetic effects of silver and gold nanoparticles on *Allium cepa* roots. J. Genet. Eng. Biotechnol. 16 (2018) 519–526, https://doi.org/10.1016/j.jgeb.2018.07.007.

[43] S. Medda, A. Hajra, U. Dey, P. Bose, N.K. Mondal, Biosynthesis of silver nanoparticles from *Aloe vera* leaf extract and antifungal activity against *Rhizopus* sp. and *Aspergillus* sp. Appl. Nanosci. 5 (7) (2014) 875–880, https://doi.org/10.1007/s13204-014-0387-1.

[44] F.K. Alsammarraie, W. Wang, P. Zhou, A. Mustapha, M. Lin, Green synthesis of silver nanoparticles using turmeric extracts and investigation of their antibacterial activities. Colloids Surf. B: Biointerfaces 171 (2018) 398–405, https://doi.org/10.1016/j.colsurfb.2018.07.059.

[45] P. Elia, R. Zach, S. Hazan, S. Kolusheva, Y. Porat, Green synthesis of gold nanoparticles using plant extracts as reducing agents, Int. J. Nanomedicine (2014) 4007–4021.

[46] H. Fouad, L. Hongjie, D. Hosni, J. Wei, G. Abbas, H. Ga'al, M. Jianchu, Controlling *Aedes albopictus* and *Culex pipiens* pallens using silver nanoparticles synthesized from aqueous extract of *Cassia fistula* fruit pulp and its mode of action. Artif. Cells Nanomed. Biotechnol. 46 (3) (2017) 558–567, https://doi.org/10.1080/21691401.2017.1329739.

[47] J.A. Mahyoub, A.T. Aziz, C. Panneerselvam, K. Murugan, M. Roni, S. Trivedi, G. Benelli, Seagrasses as sources of mosquito nano-larvicides? Toxicity and uptake of *Halodule uninervis*-biofabricated silver nanoparticles in dengue and zika virus vector *Aedes aegypti*. J. Clust. Sci. 28 (1) (2016) 565–580, https://doi.org/10.1007/s10876-016-1127-3.

[48] K. Kalimuthu, C. Panneerselvam, C. Chou, L.C. Tseng, K. Murugan, K.H. Tsai, G. Benelli, Control of dengue and Zika virus vector *Aedes aegypti* using the predatory copepod Megacyclops formosanus: synergy with *Hedychium coronarium*-synthesized silver nanoparticles and related histological changes in targeted mosquitoes. Process Saf. Environ. Prot. 109 (2017) 82–96, https://doi.org/10.1016/j.psep.2017.03.027.

[49] P. Deepak, R. Sowmiya, R. Ramkumar, G. Balasubramani, D. Aiswarya, P. Perumal, Structural characterization and evaluation of mosquito-larvicidal property of silver nanoparticles synthesized from the seaweed, *Turbinaria ornata* (Turner) J. Agardh 1848. Artif. Cells Nanomed. Biotechnol. 45 (5) (2016) 990–998, https://doi.org/10.1080/21691401.2016.1198365.

[50] S. Vinoth, S.G. Shankar, P. Gurusaravanan, B. Janani, J.K. Devi, Anti-larvicidal activity of silver nanoparticles synthesized from *Sargassum polycystum* against mosquito vectors. J. Clust. Sci. 30 (2018) https://doi.org/10.1007/s10876-018-1473-4.

[51] S.M. Selvan, K.V. Anand, K. Govindaraju, S. Tamilselvan, V.G. Kumar, K.S. Subramanian, M. Kannan, K. Raja, Green synthesis of copper oxide nanoparticles and mosquito larvicidal activity against dengue, zika and chikungunya causing vector A*edes aegypti*, IET Nanobiotechnol. 12 (8) (2018) 1042–1046.

[52] E.A. Sharon, K. Velayutham, R. Ramanibai, Biosynthesis of copper nanoparticles using *Artocarpus heterophyllus* against dengue vector *Aedes aegypti*, Int. J. Life Sci. Res. 4 (4) (2018) 1872–1879.

[53] N.K. Mondal, A. Hajra, Synthesis of copper nanoparticles (CuNPs) from petal extracts of marigold (*Tagetes* sp.) and sunflower (*Helianthus* sp.) and their effective use as a control tool against mosquito vectors. J. Mosq. Res. 6 (19) (2016) https://doi.org/10.5376/jmr.2016.06.0019.

[54] P. Deepak, R. Sowmiya, G. Balasubramani, D. Aiswarya, D. Arul, M.P.D. Josebin, P. Perumal, Mosquito-larvicidal efficacy of gold nanoparticles synthesized from the seaweed, *Turbinaria ornata* (Turner) J. Agardh 1848, Particul. Sci. Technol. 36 (8) (2018) 974–980.

[55] G. Angajala, P. Pavan, R. Subashini, One-step biofabrication of copper nanoparticles from Aegle *Marmelos correa* aqueous leaf extract and evaluation of its anti-inflammatory and mosquito larvicidal efficacy, RSC Adv. 93 (2014) 2–20.

[56] D. Kalaimurugan, P. Sivasankar, K. Lavanya, M.S. Shivakumar, Antibacterial and larvicidal activity of *Fusarium proliferatum* (YNS2) whole cell biomass mediated copper nanoparticles, J. Clust. Sci. 30 (4) (2019) 1071–1080.

[57] S. Vijayakumar, G. Vinoj, B. Malaikozhundan, S. Shanthi, B. Vaseeharan, *Plectranthus amboinicus* leaf extract mediated synthesis of zinc oxide nanoparticles and its control of methicillin resistant *Staphylococcus aureus* biofilm and blood sucking mosquito larvae. Spectrochim. Acta A Mol. Biomol. Spectrosc. 137 (2014) 886–891, https://doi.org/10.1016/j.saa.2014.08.064.

[58] B. Banumathi, B. Vaseeharan, R. Ishwarya, M. Govindarajan, N.S. Alharbi, S. Kadaikunnan, J.M. Khaled, G. Benelli, Toxicity of herbal extracts used in ethno-veterinary medicine and green-encapsulated ZnO nanoparticles against *Aedes aegypti* and microbial pathogens, Parasitol. Res. 116 (2017) 1637–1651.

[59] A.P. Ashokan, M. Paulpandi, D. Dinesh, K. Murugan, C. Vadivalagan, G. Benelli, Toxicity on dengue mosquito vectors through *Myristica fragrans* – synthesized zinc oxide nanorods, and their cytotoxic effects on liver cancer cells (HepG2). J. Clust. Sci. 28 (2017) 205–226, https://doi.org/10.1007/s10876-016-1075-y.

[60] N.A. Al-Dhabi, M.V. Arasu, Environmentally-friendly green approach for the production of zinc oxide nanoparticles and their anti-fungal, ovicidal, and larvicidal properties, Nanomaterials (Basel) 8 (7) (2018) 500.

[61] K. Murugan, M. Roni, C. Panneerselvam, A.T. Aziz, U. Suresh, R. Rajaganesh, R. Aruliah, *Sargassum wightii* -synthesized ZnO nanoparticles reduce the fitness and reproduction of the malaria vector *Anopheles stephensi* and cotton bollworm *Helicoverpa armigera*, Physiol. Mol. Plant Pathol. 101 (2018) 202–213.

[62] J. Subramaniam, K. Murugan, C. Panneerselvam, K. Kovendan, P. Madhiyazhagan, D. Dinesh, P.M. Kumar, Multipurpose effectiveness of *Couroupita guianensis*-synthesized gold nanoparticles: high antiplasmodial potential, field efficacy against malaria vectors and synergy with *Aplocheilus lineatus* predators, Environ. Sci. Pollut. Res. 23 (8) (2016) 7543–7558.

[63] N. Soni, S. Prakash, Efficacy of fungus mediated silver and gold nanoparticles against *Aedes aegypti* larvae. Parasitol. Res. 110 (2012) 175–184, https://doi.org/10.1007/s00436-011-2467-4.

[64] R.T. Yoganandham, R. Rajasree, G. Sathyamoorthy, R.R. Remya, Toxicity of biogenic gold nanoparticles fabricated by *Hybanthus enneaspermus* aqueous extract against *Anopheles stephensi* and *Culex tritaeniorhynchus*, Res. J. Biotechnol. 13 (9) (2018) 26–35.

[65] N. Soni, S. Prakash, Synthesis of gold nanoparticles by the fungus *Aspergillus niger* and its efficacy against mosquito larvae, Rep. Parasitol. 2 (2012) 1–7.

[66] M. Govindarajan, G. Benelli, Facile biosynthesis of silver nanoparticles using *Barleria cristata*: mosquitocidal potential and biotoxicity on three non-target aquatic organisms. Parasitol. Res. 115 (3) (2015) 925–935, https://doi.org/10.1007/s00436-015-4817-0.

[67] K. Murugan, P. Aruna, C. Panneerselvam, P. Madhiyazhagan, M. Paulpandi, J. Subramaniam, G. Benelli, Fighting arboviral diseases: low toxicity on mammalian cells, dengue growth inhibition (in vitro), and mosquitocidal activity of *Centroceras clavulatum*-synthesized silver nanoparticles. Parasitol. Res. 115 (2) (2015) 651–662, https://doi.org/10.1007/s00436-015-4783-6.

[68] W. Bai, Z. Zhang, W. Tian, X. He, Y. Ma, Y. Zhao, Z. Chai, Toxicity of zinc oxide nanoparticles to zebrafish embryo: a physicochemical study of toxicity mechanism. J. Nanopart. Res. 12 (5) (2009) 1645–1654, https://doi.org/10.1007/s11051-009-9740-9.

[69] J. Farkas, P. Christian, J.A.G. Urrea, N. Roos, M. Hasellöv, K.E. Tollefsen, K.V. Thomas, Effects of silver and gold nanoparticles on rainbow trout (*Oncorhynchus mykiss*) hepatocytes. Aquat. Toxicol. 96 (1) (2010) 44–52, https://doi.org/10.1016/j.aquatox.2009.09.016.

[70] M. Hejazy, M. Kazem Koohi, A. Bassiri, M. Pour, D. Najaf, Toxicity of manufactured copper nanoparticles – a review, Nanomed. Res. J. 3 (1) (2018) 1–9.

4

Biosorption-driven green technology for the treatment of heavy metal (loids)-contaminated effluents

Anirudha Paul and Jatindra N. Bhakta

DEPARTMENT OF ECOLOGICAL STUDIES & INTERNATIONAL CENTER FOR ECOLOGICAL ENGINEERING, UNIVERSITY OF KALYANI, KALYANI, WEST BENGAL, INDIA

1 Introduction

Effluents from industrial wastewater generally contain high amounts of heavy metal(loid), which pose serious threat to environmental health. Such soluble heavy metals are toxic in nature and major contaminant to aquatic system and are of major global concern. Different natural processes (volcanic eruption, leaching, soil erosion, metal corrosion) and anthropogenic activities (mining, electroplating, fossil fuel combustion, paper-textile-painting industries, refinery and polymer industries, nuclear power generator, agricultural processes, and refinery industry) contribute to heavy metal(loid) pollution mostly. Furthermore, migration of such contaminants to noncontaminated areas by dust or water causes spreading and magnification of heavy metal contamination [1]. According to Ross [2], heavy metal(loid) enters into the environment by five different ways through anthropogenic activities: (i) natural deposition, (ii) waste disposal, (iii) industrial process, (iv) hydrometallurgy, and (v) agricultural disposal. Industrial waste such as those from metallurgic engineering, production of dye, battery production house, electroplating, film and photography making, and mining of metals contributes mostly to water contamination.

Heavy metal(loid)s are generally a group of elements having atomic density more than 5 g/cm^3, possessing metallic properties with atomic number greater than twenty. Such heavy metals are toxic to humans along with other life forms and nonbiodegradable causing biomagnification [3]. Such toxic contaminants include arsenic (As), zinc (Zn), copper (Cu), lead (Pb), chromium (Cr), cadmium (Cd), nickel (Ni), iron (Fe), mercury (Hg), cobalt (Co) uranium (U), and manganese (Mn). In human, heavy metal(loid)s toxicity is associated with different host maladies, including birth defect, damages of central nervous system, kidney, lungs, gastrointestinal and cardiovascular system, along with mutation of cellular organelles leading to cancer [4].

Entry of heavy metal(loid)s into food chain occurs through mainly metallotherapy, drinking of contaminated water, and crop field irrigation with untreated wastewater [5]. Improper waste disposal of metal ores contributes mostly in metal contamination.

Such waste materials are further being carried out by water and air to different location. Therefore, there is an urgent need in removing heavy metal from industrial effluents. The common method used in the heavy metal removal process from wastewater involves reverse osmosis, electro-coagulation, ion exchange, chemical precipitation [6]. However such processes are cost effective and some may produce secondary pollution by producing excess sludge.

To combat the mentioned scenario, new advanced, cheap, and sustainable technologies are needed for future perspective. This has drawn attention of scientist toward biosorption, where cheap, easily available, hazard-free, biodegradable biomasses are being utilized for toxic metal removal from aqueous streams. In general biosorption is a metabolically independent process, where different biological biomasses have the ability to remove/ accumulate heavy metals from solutions [7]. Many different mechanisms are involved in biosorption such as precipitation, ion exchange, and adsorption. Such mechanisms are greatly influenced by certain parameters, such as (a) pH, (b) temperature, (c) types of biosorbent materials, (d) presence of competitive metals, (e) metal concentration [7]. Various types of biosorbent materials can be used due to its metal binding property, easy availability, high efficiency, and cost effectiveness [8,9]. Different bacteria, algae, fungi, and other waste material have proved to be potential metal biosorbent [10–13]. Apart from metal removal, such biosorption technology is also effective for metal recovery [14]. Wide variety of microorganisms such as bacteria, algae, fungi, different low cost materials, plant parts, industrial wastes, and agricultural wastes are widely used for biosorption of heavy metal(loid)s in laboratory or industrial set ups. Different biomass such as *Azadirachta indica* bark [15], citrus pectin [16], gram husk [17] has shown potency as good biosorbent. Biomass of bacteria such as *Bacillus jeotgali*, *Arthrobacter* sp., *Pseudomonas* sp., and *E. coli* has shown promise as biosorbent with the biosoprtion capacity (mg/g^{-1}) 222.2, 17.87, 95, and 6.9, respectively [18–21]. Algae like *Spirulina platensis*, *Ulva lactuca* sp., *Spirogyra* sp. can remove different heavy metals with the biosorption capacity (mg/g^{-1}) 67.93, 43.02, 140, respectively [22–24].

However, a vivid literature regarding the biosorption methods in removing heavy metal (loid)s is still lacking to get in depth knowledge. Therefore, keeping in mind the mentioned scenario, the objectives of this review are to critically interpret and summarize the up-to-date works carried out in this regard by different scientists, such as different parameters of biosorbent materials along with optimum treatment conditions, modeling for industrial application of biosorption, different isotherm models. This can be helpful for future studies in exploring the novel biosorbent to treat the industrial wastewater using biosoprtion technology.

2 Heavy metal(loid)s

A group of metals and metalloids are collectively known as heavy metals, having atomic density more than 5 g/cm^3. Heavy metals are distributed throughout the ecosystem at varied degree. Most of the heavy metals are essential micronutrients, although presence in high concentration is toxic. Coordination mechanism of heavy metal(loid)s is the key factor that contributes to the poisoning and toxicity. In general, heavy metal(loid)s bind with the cellular molecules such as enzymes or proteins and by doing so, alter their normal activities resulting in cellular dysfunctions [25] along with mitochondrial damage and cell membrane disintegration [26]. Heavy metals generally bind to the active site of the enzymes such as sulfydryl groups [27]. Most of the heavy metals upon accumulation within the cell cause production of Reactive Oxygen Species (ROS). Such compounds are the main reason for the destruction of several essential elements such as proteins, fats, nucleic acids along with other organelles [28]. The ROS is also responsible for alteration of DNA structure leading to host mutation and death. As most of the metals do not undergo biodegradation, metal(loid)s pollution is a global problem and has devastating effect on biological system. Since heavy metals are not biodegradable, further accumulation promotes biomagnifications where the higher tropic level organisms exist. Being toxic such heavy metals show harmful effects on plant system as well as on lower organisms, including microbes [29].

Toxicological impacts of heavy metal(loid)s are associated with malfunction of cellular organs such as kidney, liver mostly along with birth defect, skin damage, and cancer. Natural processes such as such as volcanic erosion or forest fires, microbial leaching from metal(loid)s ores, and other anthropogenic activities contribute to arsenic pollution [30–35]. Arsenic poisoning is associated with cancer, multiorgan dysfunction, and hyperkeratosis [36–38].

Nickel pollution occurs mainly due to anthropogenic activities. Other natural processes such as volcanic eruption, meteoric dust, forest fire, and windblown dust also contribute to nickel pollution [39,40]. The main anthropogenic activities include electroplating, mining, battery manufacturing, and smelting. In contrast to such, other processes (municipal sludge and sewage treatment plants, cement manufacturing plants, fossil fuel combustion, etc.) are also involved in nickel emission [41]. Nickel contamination contributes to skin rashes, tumors, pulmonary fibrosis, encephalopathy [42–45].

Anthropogenic activities such as mining, manufacturing, and burning of fuel causing leaded gasoline [46,47] are the primary sources of lead pollution. Lead contamination is mainly responsible for encephalopathy, nausea, anemia, anorexia, abdominal pain, nephropathy, liver and kidney damage [48–50].

Copper is mostly used in fiber production, electroplating industries, pipe making, printed circuits, and electrical industries. Untreated effluents associated with such industries contribute to copper pollution. Although copper acts as an essentially important cofactor in the enzymatic reaction of cell, high level is toxic. Copper poisoning leads to vomiting, gastrointestinal hemorrhage, multiorgan dysfunction, neurotoxicity, dizziness, Wilson disease [51].

Mercury is another toxic metal. Mercury is mainly used in oil refining, paints, paper, rubber processing industries along with, batteries, thermometers, pesticides, light tubes and lamps, cosmetics, and pharmaceuticals [52,53]. Mercury pollution contributes to nervous system disturbance, protoplasm poisoning, ataxia, eye dysfunction, dermatitis [50,54].

Zinc is another essential trace metal [55]; however, elevated level is toxic and detrimental [56,57]. Zinc toxicity is associated with gastrointestinal disorder, osteoporosis, and anemia.

The main source of cadmium pollutions is paints and plastics, welding, nuclear fission plants, electroplating, and fertilizers [58]. Health hazards associated with excess cadmium exposure include kidney damage, pneumonitis, lung cancer, high blood pressure, proteinuria, osteoporosis, destruction of testicular tissue, lung insufficiency, hypertension, Itai-Itai disease [58–65].

In general, such heavy metal(loid)s pose negative impact on all forms of life forms and, therefore, removal or decontamination of those contaminants is indispensable for the better green future regarding environmental safety.

3 Conventional treatment process for metal(loid)s removal from wastewater

Considering the global heavy metal(loid)s pollution, a variety of conventional (physico-chemical and electrochemical) treatment procedures were developed earlier to decontaminate or remediate heavy metal(loid)s form industrial wastewater. Such processes are:

(1) *Chemical precipitation*: In this technique selective chemicals are added to the waste water in removing the heavy metals [66]. Additions of chemicals react with heavy metals and decrease their solubility resulting in precipitation of heavy metals [67]. In subsequent steps the insoluble heavy metals are separated by filtration or sedimentation [68]. Such chemical precipitation process is highly dependent on the pH of solution. Decrease in pH of solution inhibits the precipitation process. The important disadvantage in this process includes production of greater amount of sludge, which is the source of secondary pollution and is of high cost.

(2) *Reverse osmosis*: It is a membrane filtration process where selective membranes are used for the separation of heavy metal ions. High pressure is applied for forceful movement of water through membrane leaving the contaminant on the other side. This process can selectively remove high amount of metals from solution [69] and had been applied for verification. Major disadvantages include membrane cost and operation cost along with maintenance.

(3) *Ion exchange*: In ion exchange process certain ions present on the biosorbent matrix are subjected to replace by heavy metal ions present in the solution. This is a reversible process and the bound heavy metals can be recovered by change in ionic strength or pH of the solution. Such process is greatly influenced by pH of solution, strength of ion, temperature, concentration of metal ion, and concentration of biosorbent materials [70]. Different kinds of ion exchanger materials are used, which include zeolite or synthetic hydrocarbon-derived polymeric substances [71]. Major disadvantages are that the resins get entangled easily with different organic substances present in the wastewater and such process is unable to deal with high metal concentration along with continuous maintenance of solution pH.

(4) *Electrocoagulation*: In this process electrical current is applied to remove metal ions from wastewater and widely used in wastewater treatment industries. Generally, the metal ions remain in solution by forming hydrogen bonding or through electrostatic interaction. Application of electrical current or electrical field causes neutralization of metal ions. Such neutralized metal ions get precipitated easily forming sludge [72].

(5) *Cementation*: In this process metal ions are displaced by other metals positioned higher in the electrochemical series. Electrical power is applied for long time to perform such process and mainly limited to small-scale waste treatment. For example, cementation process is applied to replace lead metal by iron, which is less toxic in nature [73].

Such conventional methods have a variety of disadvantages, including certain limitation and high cost. Therefore, researches and applications are nowadays oriented toward a better, eco-friendly, and cheap alternatives such as biosorption.

4 Biosorption

Biosorption is a multidimensional, physicochemical, metabolically independent and/or independent (passive) process [7,74]. This biosorption technique is a combination of solid phase (adsorbent) and a liquid phase (solvent) containing an adsorbate (metals or other substrates). High affinity of the adsorbent toward adsorbate confers selective binding of adsorbate involving different mechanisms. Depending on the binding affinity, the distribution pattern or adsorbate concentration varies in solid-to-liquid phases and reaction continues until it reaches equilibrium [75]. Both live-cell biomass and dead-cell preparation along with biological waste materials can be used in this technique. Although the live cell participates in bioaccumulation process, it works differently than in biosorption [76]. In biosorption mechanism metal(loid)s are generally attached rapidly through a different mechanism such as ion exchange, precipitation, physical adsorption, chemical sorption, and chemical complexation. Different microbial biomass and biological waste materials contain a variety of chemical components such as proteins, lipids, polysaccharides with different functional groups that specifically involves in metal binding [77–79]. It was well documented that the different metabolic process involved in live cell machinery may also interfere with biosorption process along with adsorbate availability and therefore dead cell biomass is preferentially being utilized in biosorption process. In most of the cases heavy metal(loid)s were assumed as divalent cations but anionic forms of heavy metal(loid)s such as arsenic, molybdenum, selenium, chromium, and vanadium are also available, which also exerts toxicity. Biosorption was considered as an effective cleaning process in case of such anionic form also [80].

5 Mechanisms of biosorption

As discussed earlier, biosorption is a physicochemical, rapid, and passive process, where different mechanisms are involved to carry out the process effectively. Several researches

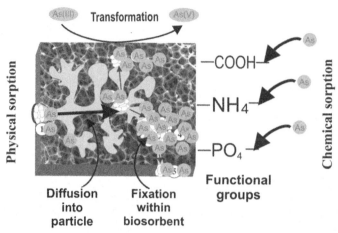

FIG. 1 Arsenic removal mechanisms of biosorbents [82].

have been conducted and which provide some basic idea about the mechanisms involved in biosorption [7,81]. Bhakta and Ali [82] proposed a simple arsenic removal mechanism of biosorbents (Fig. 1). Ion exchange, complexation, and adsorption are the basic mechanisms associated with biosorption. The biomass used in biosorption process contains several functional groups such as thiol, hydroxyl, carboxyl, amino, phosphate, which prove binding site for metal ions [79]. The binding of metal ions to the ligands present in biosorbent follows hard-soft-acid-base (HSAB) principle. Oxygen-containing ligands are binding site for hard acids, whereas sulfur- and nitrogen-containing ligands allow soft acids to bind. According to such rule ionic bonds are formed between hard acid and hard ligand, whereas covalent bonds are formed between soft acid and soft ligand. However, the practical results differ slightly with a few exceptions, such as binding of Sr^{2+} with live cells of yeast is of covalent type but the binding with dead denatured cell exhibits ionic interaction [7,83].

In complex formation/coordination mechanism different cations can bind to the biosorbent material either by sharing electron and forming covalent bonds or by ionic interaction with negatively charged molecules. Certain microbes produce different kinds or organic acids. Such acids can form complexes with metals and chelates them down. Another type of sorption, i.e., hydrophilic sorption also occurs in presence of hydrophobic biomass [84]. Physical adsorption mainly occurs through van der Waals' interaction. According to Kuyucak and Volesky [85], biosorption of different heavy metals by dead microbial biomass occurs through electrostatic interaction, for example copper biosorption by dead biomass of *Chiarella vulgaris* was due to electrostatic interaction [86]. In case of ion exchange mechanism, metal ions replace the similarly charged ions present in the microbial biomass. For example, marine algae contains some kinds of alginate that consist of Na, Ca, K, and Mg. Such ions can participate in the exchange process of different heavy metal ions from solution [85].

6 Advantages of biosorption process

For last few decades heavy metal(loid)s decontamination or removal process-oriented researches are shifted toward biosorption, which is effective in many ways and provides certain advantages over other processes

1. Rapid reaction kinetics provides removal of heavy metals from large amount of wastewater with ease.
2. Based on choice of the biosorbent different heavy metal(loid)s and waste can be removed by single treatment.
3. By using high-affinity biosorbent material specific choice of metal(loid)s can be removed with little or no residue.
4. Ecofriendly technique, as biodegradable materials is used in this technique.
5. No such adverse environmental impacts as that of the chemical process
6. Low operation cost as the energy need is less and sometimes biosorbent materials have reusable property.
7. Almost no or little hazardous sludge generation.
8. Low cost, renewable, and naturally occurring abundant cheap materials are used.

In this regard, it can be mentioned that biowaste materials can be used as biosorbent to treat the polluted water (Fig. 2) [79].

7 Factors affecting biosorption

The physicochemical process of biosorption of metal(loid)s depends on a variety of factors:

A. *pH*: pH is considered one of the most important factors in biosorption mechanism. Similar to ion exchange process, the biosorbent materials contains different basic or acidic groups, to which different metallic groups adhere. Therefore, the different functional groups present in the biosorbent material as well as the metals present in the solution and its solubility are strongly affected by the pH of the medium. Different

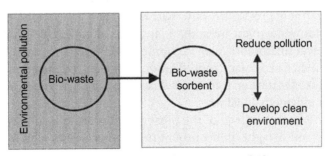

FIG. 2 Application of biowaste as biosorbent for cleaning the environment [79].

experimental studies indicate that the biosorption process is well operated in the pH range between 3 and 5. Low pH seems to affect biosorption antagonistically [87]. It was observed that at very low pH (pH 2) there was little or no metal removal occurred. The biosorption ability reduces with the decrease of medium pH [88]. However, in case of anionic metallic groups such as CrO_4^{2-} and SeO_4^{2-}, reduction in pH values increases the removal efficiency of the biosorbent. It was also documented that some metallic biosorption is pH independent such as in case of Hg. This is might be due to formation of strong covalent complex between metal ions and biosorbent materials [7].

B. *Temperature*: Biosorption is also a temperature-dependent process to some extent. In most of the cases, at temperature range between 20°C and 35°C, the efficiency of biosorption remains constant [86]. In some exceptional cases, high temperature (50°C) promotes biosorption but causes permanent damage to the biosorbent materials such as living biomass. As a result biosorption frequency drops below normal [75].

C. *Biomass concentration*: Scientific studies indicates that low amount of biomass promotes the biosorption process because high biomass concentration interferes with metal-binding sites [89]. It seems electrostatic interaction plays a crucial role in this case.

D. *Initial metal(loid)s concentration*: Initial metal(loid)s concentration plays crucial role in biosorption. Increase in initial metal concentration causes enhancement of total metal quantity but subsequently the removal efficiency decreases. In other words, initial metal concentration helps to overcome the mass transfer resistance [90].

E. *Presence of multiple heavy metal(loid)s*: Presence of other heavy metals decreases the biosorption of target heavy metal by competing with target binding site present on the biosorbent, although the exact influential pattern may vary depending on the presence of other metal ions. Some experimental data indicate that the presence of certain cations supports the binding of certain anions [7], whereas other experimental research indicates that certain cations increase the biosorption of other cations [91]. The increase or decrease in the biosorption of target heavy metal is dependent on the formation of complex or precipitation between the metal ions present in the solution and thus varies accordingly [7].

F. *Biomass characteristics*: Biosorption property is widely influenced by the nature of the biomass and treatment procedure. Physical or chemical treatments cause modification in the metal-binding site of the biosorbent and thus alter the biosorption capacity [75]. Generally chemical treatments (alkali treatment) improve the metal binding ability of the biosorbent [92]. Cell surface modification, cell wall composition, extracellular polymeric product formation, treatment doses, and change in cell size also influence the biosorption efficiency [7,93,94].

G. *Agitation speed*: Agitation speed of the biosorption process is directly proportional to the removal per unit weight of biosorbent or in other words high agitation may increase the metal removal capacity of the biosorbent by decreasing the resistance of mass transfer, although there may be structural disruption of biosorbent [93].

8 Biosorption isotherm and model

The industrial application of biosorption has some disadvantages due to its complexity of mechanisms and the applications are limited till now. On the contrary, a proper mathematical model may provide proper information about the complexity of the process to some extent. Different isotherm models are developed and exploited for industrial trial purposes. Among the different isotherms Langmuir and Freundlich isotherms are widely used to perform the batch studies [95].

To describe biosorption kinetics isotherms are developed over the year and have been widely used [95,96]. Isotherms are useful for the comparative study of different biosorbent for a single pollutant and also for the comparison between different pollutants for a single biosorbent. Generally, the equilibrium relationship of adsorbate concentration in different phases such as in liquid phase and in adsorbent particles is represented by such isotherms. In the graphical representation, the removal amount of adsorbate (q_e) at equilibrium stage and the equilibrium adsorbate concentration (C_e) remained in the solution is plotted [7]. The two most widely used simple single-component models are Freundlich and Langmuir models.

Freundlich Model: Freundlich model (Eq. 1).

$$q_e = KC_e^{1/n},\tag{1}$$

where K = Freundlich adsorption capacity constant and n = Freundlich adsorption intensity constant.

Freundlich model is mostly applied for biosorption at particular pH. Generally to determine the biosorption kinetics using single type of metal ions, such isotherm is very fruitful. According to this model, an increase in the concentration of adsorbate increases the concentration of adsorbate on adsorbent surface along with decrease in sorption energy.

Langmuir model: To describe the adsorption kinetics of liquid or gas on solid matrix, Irving Langmuir proposed a second model of isotherm on 1918. Such model is applicable for single-layer adsorption and where no interaction occurs between target sites and adjacent sites [97] (Eq. 2).

$$q_e = q_m bC_e/1 + bC_e,\tag{2}$$

where q_m = maximum specific uptake corresponding to saturation site and b = adsorption and desorption ratio.

However, uses of such models are limited. Exact biosorption mechanisms, adsorption behavior in changing ionic strength, and pH cannot be predicted from such isotherms [98].

9 Biosorbent

Biosorbent is a collective term to symbolize any biological substrate that has special affinity for any organic or inorganic pollutants; in other words, it is a biological substrate that has high biosorption potential [7]. In thousands of research articles different types of economically feasible bio-based materials were tested for pollution control and hazard management purpose or to some extent, metal recovery, which includes microbial biomass, industrial waste, agricultural waste, macroalgae, natural material [93,99].

Although a variety of materials can be used as biosorbent choosing an effective biosorbent is a major challenge for large-scale application. Considering the large-scale application such cheap and easily accessible biomass can be used from (a) industrial waste materials, as it is low charge or free of cost; (b) easily cultivated organism with high growth rate; (c) easily available natural organisms that can be found in high amounts [93].

9.1 Types of biosorbent

Apart from different mechanisms and factors involved in biosorption, the chemical nature and types of biosorbent are also crucial for biosorption. As we discussed earlier biosorbent is of different types. A few major types of biosorbent are:

9.1.1 Bacteria

Bacterial biomass can be used as an excellent biosorbent material as being abundant in nature with high growth rate and available everywhere in the ecosystem. Different functional groups present in the cell wall of both Gram-positive and Gram-negative bacteria are the major source of metal-binding site. Generally, carboxylic groups, phosphate groups, and other negatively charged groups present in the cell surface are mainly responsible for metal cations binding [7]. Some other polymeric materials such as S layer and bacterial sheath also contribute to metal ion binding along with peptidoglycan and polysaccharides. Metal ions with positive charges might have the ability to bind with the negatively charged groups such as OH groups, NH groups, SH groups, and PO^{2-} and $P(OH)_2$ groups associated with bacterial cell wall [100]. Blue green algae or cyanobacteria contain peptidoglycan as cell wall component. Some species of cyanobacteria also produce extracellular polymeric substances (EPS). This peptidoglycan along with EPS provides major metal-binding sites. The primitive archae bacteria contain modified peptidoglycan known as pseudopeptidoglycan or pseudomurein. Other cell wall components include glycoprotein and sulfonated polysaccharide, which provide excellent metal-binding anionic groups such as sulfate and carboxyl group (Table 1).

9.1.2 Fungi

Fungi are other diverse groups of microorganisms. Fungal cell wall is quite different from those of bacteria, consisting mainly of chitins and different polysaccharides. Other than those, mannans, glucans, different lipids, proteins, pigments are also found in the fungal cell wall [107]. Due to presence of different metal-binding functional groups, such substances are able to bind with the metal ions and can be exploited for biosorption of heavy metals [8]. Chitin and chitosan found in fungal cell wall are impressive biosorbent materials [108]. Different carboxyl, carbonyl, phenolic, and alcoholic groups present in fungal phenolic polymers provide excellent metal-binding site [7,107] (Table 2).

Table 1   Bacteria as biosorbent.

Biomass type	Metal involved	pH	Temperature (°C)	Contact time (h)	q (mg/g) or % of removal	References
Enterobacter	Cu	5	25	2	78.9	[101]
cloacae	Pb	5			67.9	
	Cd	5			58.9	
	Cr	4			55.8	
Geobacillus	Cu	5	25	12	65	[102]
themocatenulatus	Pb	4			54	
	Zn	5			12.3	
Bacillus licheniformis	Cr	3.5	28	48	95	[103]
	Fe	3.5			52	
	Cu	2.5			32	
Aspergillus niger	As	5		1.67	75	[104]
Turbinaria vulgaris	As	4	30	1	88.31	[105]
Staphylococcus	Pb	4.5	27	4	100	[106]
saprophyticus	Cr	2		3	24.1	
	Cu	3.5		2	14.5	

Table 2   Fungi as biosorbent.

Biomass type	Metal involved	pH	Temperature (°C)	Contact time (h)	q (mg/g) or % of removal	References
Aspergillus niger	Cd	4.75	25	6	13	[109]
Aspergillus awamori	Cu	5	20	3	28.75	[110]
Schizosaccharomyces pombe	Cu	4	25		96	[111]
Saccharomyces cerevisiae	U	5	27	1.25	7.89	[112]
Saccharomyces cerevisiae	Cd	6	28	24	55	[113]
Mucor rouxii	Pb	5	25	15	17.13	[114]
	Cd				10.07	
	Ni				6.28	
	Zn				6.07	
Aspergillus niger	As	5		100 min	75	[104]

9.1.3 Algae

Algae are autotropic microorganisms, capable of producing large biomass in low nutrient conditions and are available all over the ecosystem. Apart from such potential most of the algae are not involved in toxin production. Such interesting features of algae drew attention of scientists to use it as effective biosorbent materials in recent years [115–117]. Three major groups of algae include (i) microalgae, (ii) macroalgae, and (iii) red algae. Algal cell wall is mainly composed of xylan, alginic acid, mannan, and chitin. Such substances provide binding sites for heavy metals. Algal proteins and polysaccharides consist of phosphate, sulfate, amine, carboxyl, imidazole groups, to which heavy metal cations can

Table 3   Algae as biosorbent.

Biomass type	Metal involved	pH	Temperature (°C)	Contact time (h)	q (mg/g)	References
Spirogyra hyalina	Hg		25	2	39.212	[119]
	Pb				15.471	
	Cd				9.832	
	As				8.719	
	Co				7.856	
Sargassum sp.	Cd	5	25	2	84.7	[120]
Sargassum sp.	Cr	4	30	6	68.9	[121]
Green algae	Zn	5	25	1	7.62	[122]
Calotropis procera	Pb	4	25	6	22.8	[123]
	Cu	5			14.5	

Table 4   Different biological waste as biosorbent.

Biomass type	Metals involved	pH	Temperature (°C)	Contact time (h)	q_{max} (mg/g)	References
Azadirachta indica Bark	Zn	6		0.75	33.49	[15]
Coconut shell	Cu	5–7	50	0.5	92.3	[126]
Cocoa shell	Pb	2	22	<2	6.2	[127]
Corncob	Cr	4	80	2	90	[128]
Tea waste	Ni	4	25		15.26	[129]
Mangos teen	Cr	4	20	2	24.5	[130]
Rice husk	Ni	6	25	3	51.8	[131]
Palm fuel ash	Pb	5	25	10	75.48	[132]

bind easily [118]. Some species of microalgae are able to produce phytochelatins, a stress compound produced in the presence of high heavy metal concentration. Such compounds are also involved in heavy metal biosorption (Table 3).

9.1.4 Agricultural waste

Agricultural waste materials are mainly composed of cellulose, hemicelluloses, lignin, different proteins, lipids, simple sugars. Such components having different functional groups provide binding site for heavy metal ions. The different functional groups include carbonyl, carboxyl, amido, amino, immino, esters, phenolic, sulfydryl, and acetamido [124]. Such functional groups facilitate metal biosorption [8,125] (Table 4).

10 Modification of biosorbent

Having low density, small molecular size, and vulnerability to mechanical stress, most microbial biosorbent materials fail to improve scale-up process. To improve biosorption efficiency for industrial application purposes different modification or treatment of biosorbent has been done. One of such modification includes immobilization of biosorbent materials into packed column or fluidized bed bioreactors [7,93]. In immobilization process, entrapment within polymeric materials such as alginate, agarose is beneficial to some extent. Introduction of different magnetic materials such as alginate, chitosan, and alginate is also helpful for better biosorption [133]. Different physical modification can be introduced such as cutting, grinding, physical drying, and lyophilization. Surface modification also been done to promote biosorption efficiency, as biosorption mainly occur at surface region of the biomass materials [134]. Chemical modification is another alternative approach. In such chemical modification, functional groups of biosorbent molecules are more exposed to metal ions of solution employed [135]. By creating positive influence on microbial growth or by genetic engineering, biological manipulation can also be introduced [94,136]. Such genetical modification includes overproduction of cell surface membrane proteins, including metallothioneins, alpha agglutinin in yeast, SpA (Staphylococcal protein A) in Gram-positive bacteria, and maltose-binding protein and different cell wall-binding proteins (FimH, CS3 pilli) in Gram-negative bacteria. However, care should be taken during such modification regarding the use of (a) hazardous chemicals, (b) harmful microorganisms (wild type or genetically modified), and (c) ionizing and non-ionizing radiation for effective biosorption.

11 Instrumentation involved in analysis of biosorption

To study the biosorption process different researches have been carried out through time. The binding pattern and the identification of active binding site can be deduced though different analytical tools [80,92,93,137,138]. Such analytical tools include Atomic absorption spectroscopy (AAS), Scanning electron microscopy (SEM), Transmission electron microscopy (TEM), Infrared spectroscopy (IR), UV-visible spectroscopy, Fourier transformed Spectroscopy (FTIR), Energy dispersive X-ray spectroscopy (EDX), Nuclear magnetic resonance (NMR) spectroscopy, and X-ray diffraction (XRD). Application of each technique is listed in Table 5.

Table 5 Application of different analytical tools in biosorption.

Tools	Application
SEM	Visual confirmation of surface morphology of the biosorbent
TEM	Visual confirmation of surface morphology of the biosorbent
EDX	Element analysis and chemical characterization of metal bound on the biosorbent
XRD	For crystallographic structure and chemical composition of metal
UV-VIS	Determination of metal concentration
IR/FTIR	Determination of active sites of the biosorbent
NMR	Determination of active sites of the biosorbent
AAS	Determination of metal concentration

12 Conclusion

Toxic metal(loid) contamination of industrial wastewater and its discharge and distribution in the environment has broad deleterious impacts on environment and is of major global concern. Remediation and decontamination of polluted water is therefore a serious challenge for better future. Different laws and guidelines have emerged over time to restrict the unusual discharges containing excess heavy metal(loid)s for public health safety. To minimize the toxic metal(loid) load in industrial effluent, different processes are applied. Conventional physico-chemical methods are cost effective and to some extent associated with secondary pollution by producing excess sludge. This has demanded a more economically feasible and greener approach to restrict discharge of excess metal load in aquatic system. Biosorption has, therefore, drawn attention and emerged as a biologically safe alternative to conventional techniques. Biosorption offers various advantages over conventional process, as being cheaper, effective, high potency of metal(loid) removal, rapid and safe process along with having no such adverse effects on environment. Recycling of biomass is also feasible to some extent. Potential biosorbent includes biomass of bacteria, algae, fungi, and waste materials. Although most of the biosorption application is limited to batch processes, extensive research is needed for upgradation, designing, and continuous flow for industrial and commercial purposes. Therefore, the green technological process of biosorption has a great potential in future for cleaning the polluted environment. Finally, it could be concluded that the present review would be a comprehensive information to find new and novel avenue and thinking for forthcoming researchers in order to develop a cleaner and greener environment for future earth.

References

[1] B.V. Tangahu, S. Abdullah, H. Basri, M. Idris, N. Anuar, M. Mukhlisin, A review on heavy metals (As, Pb, Hg) uptake by plants through phytoremediation, Int. J. Chem. Eng. (2011) 1–31.

[2] S. Ross, Toxic Metals in Soil-Plant Systems, John Wiley and Sons, Chichester, UK, 1994.

[3] A.K. Chopra, C. Pathak, Biosorption technology of metallic pollutants – a review, J. Appl. Nat. Sci. 2 (2010) 318–329.

[4] H.K. Allure, S.R. Ronda, V.S. Settaluri, V.S. Bandili, V. Suryanarayan, P. Venkateshwar, Biosorption: an ecofriendly alternative for heavy metal removal, Afr. J. Biotechnol. 6 (2007) 2924–2931.

[5] M. Fazeli, F. Khosravan, M. Hossini, S. Sathyanarayan, P.N. Satish, Enrichment of heavy metal in paddy crops irrigated by papermill effluents near Najangud, Mysore District, Karnataka, India, Environ. Geol. 34 (1998) 297–302.

[6] G. Rich, K. Cherry, Hazardous Waste Treatment Technologies, Pudvan Publishers, New York, 1987.

[7] G.M. Gadd, Biosorption: critical review of scientific rationale, environmental importance and significance for pollution treatment, J. Chem. Technol. Biotechnol. 84 (2009) 13–28.

[8] S.E. Bailey, T.J. Olin, M. Bricka, D.D. Adrian, A review of potentially low-cost adsorbents for heavy metals, Water Res. 33 (1999) 2469–2479.

[9] G.M. Gadd, Fungi and yeasts for metal binding, in: H. Ehrlich, C. Brierley (Eds.), Microbial Mineral Recovery, McGraw-Hill, New York, 1990, pp. 249–275.

[10] G. Annadurai, R.S. Juang, D.L. Lee, Adsorption of heavy metals from water using banana and orange peels, Water Sci. Technol. 47 (2002) 185–190.

[11] K. Mohanty, M. Jha, M.N. Biswas, B.C. Meikap, Removal of chromium (VI) from dilute aqueous solutions by activated carbon developed from *Terminalia arjuna* nuts activated with zinc chloride, Chem. Eng. Sci. 60 (2005) 3049–3059.

[12] Z. Reddad, C. Gerente, Y. Andres, M.C. Ralet, J.F. Thibault, P.L. Cloirec, Ni (II) and Cu (II) binding properties of native and modified sugar beet pulp, Carbohydr. Polym. 49 (2002) 23–31.

[13] B. Volesky, Biosorbent materials, Biotechnol. Bioeng. Symp. 16 (1986) 121–126.

[14] Y.N. Mata, E. Torres, M.L. Blazquez, A. Ballester, F. Ganzalez, J.A. Munoz, Gold(III) biosorption and bioreduction with the brown alga *Fucus vesiculosus*, J. Hazard. Mater. 166 (2009) 612–618.

[15] P. King, K. Anuradha, S. Beena Lahari, Y. Prasanna Kumar, V.S.R.K. Prasad, Biosorption of zinc from aqueous solution using *Azadirachta indica* bark.: equilibrium and kinetics studies, J. Hazard. Mater. 152 (2007) 324–329.

[16] B. Ankit, S. Silke, Assessment of biosorption mechanism for Pb binding by citrus pectin, Sep. Purif. Technol. 63 (1998) 577–581.

[17] N. Ahalya, R.D. Kanamadi, T.V. Ramachandra, Biosorption of Chromium (VI) from aqueous solution by the husk Bengal gram (*Cicer Arientinum*), Environ. J. Biotechnol. 8 (2005) 258–264.

[18] C. Green-Ruiz, V. Rodriguez-Tirado, B. Gomez-Gil, Cadmium and zinc removal from aqueous solutions by *Bacillus jeotgali*: pH, salinity and temperature effects, Bioresour. Technol. 99 (2008) 3864–3870.

[19] S.H. Hasan, P. Srivastava, Batch and continuous biosorption of Cu(2+) by immobilized biomass of *Arthrobacter sp*, J. Environ. Manag. 90 (2009) 3313–3321.

[20] M. Ziagova, G. Dimitriadis, D. Aslanidou, X. Papaioannou, E.L. Tzannetaki, et al., Comparative study of Cd (II) and Cr (VI) biosorption on *Staphylococcus xylosus* and *Pseudomonas sp.* in single and binary mixtures, Bioresour. Technol. 98 (2007) 2859–2865.

[21] C. Quintelas, Z. Rocha, B. Silva, B. Fonseca, H. Figueiredo, et al., Removal of Cd (II), Cr (VI), Fe (III) and Ni (II) from aqueous solutions by an *E. coli* biofilm supported on kaolin, Chem. Eng. J. 149 (2009) 319–324.

[22] A. Celekli, M. Yavuzatmaca, H. Bozkurt, An eco-friendly process: predictive modelling of copper adsorption from aqueous solution on *Spirulina platensis*, J. Hazard. Mater. 173 (2010) 123–129.

[23] L. Bulgariu, M. Lupea, D. Bulgariu, C. Rusu, M. Macoveanu, Equilibrium study of Pb (II) and Cd (II) biosorption from aqueous solution on marine green algae biomass, Environ. Eng. Manag. J. 12 (2013) 183–190.

[24] V.K. Gupta, A. Rastogi, Biosorption of lead from aqueous solutions by green algae Spirogyra species: kinetics and equilibrium studies, J. Hazard. Mater. 152 (2008) 407–414.

[25] S.J.S. Flora, M. Mittal, A. Mehta, Heavy metal induced oxidative stress and its possible reversal by chelation therapy, Indian J. Med. Res. 128 (2008) 501–523.

[26] S. Wang, X. Shi, Molecular mechanisms of metal toxicity and carcinogenesis, Mol. Cell. Biochem. 222 (2001) 3–9.

[27] M.O. Ogwuegbu, M.A. Ijioma, Effects of certain heavy metals on the population due to mineral exploitation, in: International Conference on Scientific and Environmental Issues in the Population, Environment and Sustainable Development in Nigeria, University of Ado Ekiti, Ekiti State, Nigeria, 2003, pp. 8–10.

[28] J.E. Klaunig, Y. Xu, J.S. Isenberg, S. Bachowski, K.L. Kolaja, J. Jiang, The role of oxidative stress in chemical carcinogenesis, Environ. Health Perspect. 106 (1998) 289–295.

[29] S. Roy, S. Labelle, P. Mehta, A. Mihoc, N. Fortin, C. Masson, et al., Phytoremediation of heavy metal and PAH-contaminated brownfield sites, Plant Soil 272 (1) (2005) 277–290.

[30] J.M. Azcue, J.O. Nriagu, Arsenic: historical perspectives, in: J.O. Nriagu (Ed.), Arsenic in Environment. Part I: Cycling and Characterization, John Wiley and Sons, New York, 1994, pp. 1–15.

[31] D.K. Bhumbla, R.F. Keefer, Arsenic mobilization and bioavailability in soils, in: J.O. Nriagu (Ed.), Arsenic in Environment. Part I: Cycling and Characterization, John Wiley, New York, 1994, pp. 51–82.

[32] C.N. Cheng, D.D. Focht, Production of arsine and methylarsines in soil and in culture, Appl. Environ. Microbiol. 38 (1979) 494–498.

[33] R.S. Oremland, J.F. Stolz, The ecology of arsenic, Science 300 (2003) 939–944.

[34] H. Yan-Chu, Arsenic distribution in soils, in: J.O. Nriagu (Ed.), Arsenic in Environment. Part I: Cycling and Characterization, John Wiley and Sons, New York, 1994, pp. 17–49.

[35] United States Environmental Protection Agency (USEPA), Special report on ingested inorganic arsenic, Skin Cancer; Nutritional Essentiality, Washington, DC, United States, 1998.

[36] H.Y. Chiou, W.I. Huang, C.L. Su, S.F. Chang, Y.H. Hsu, C.J. Chen, Dose-response relationship between prevalence of cerebrovascular disease and ingested inorganic arsenic, Stroke 9 (1997) 1717–1723.

[37] M. Hendryx, Mortality from heart, respiratory, and kidney disease in coal mining areas of Appalachia, Int. Arch. Occup. Environ. Health 82 (2) (2009) 243–249.

[38] A.H. Smith, C. Hopenhayn-Rich, M.N. Bates, H.M. Goeden, H.M. Hertz-Picciotto, H.M. Duggan, et al., Cancer risks from arsenic in drinking water, Environ. Health Perspect. 97 (1992) 259–267.

[39] J.O. Nriagu, Global inventory of natural and anthropogenic emissions of trace metals to the atmosphere, Nature 279 (1979) 409–411.

[40] J.O. Nriagu, J. Pacyna, Quantitative assessment of worldwide contamination of air, water and soil by trace metals, Nature 333 (1988) 134–139.

[41] P. Grandjean, Human exposure to nickel, IARC Sci. Publ. 53 (1984) 469.

[42] C.Y. Chen, Y.F. Wang, Y.H. Lin, S.F. Yen, Nickel-induced oxidative stress and effect of antioxidants in human lymphocytes, Arch. Toxicol. 77 (2003) 123–130.

[43] K.K. Das, V. Buchner, Effect of nickel exposure on peripheral tissues: Role of oxidative stress in toxicity and possible protection by ascorbic acid, Rev. Environ. Health 2 (2007) 133–149.

[44] K. Salnikow, T. Davidson, M. Costa, The role of hypoxia-inducible signaling pathway in nickel carcinogenesis, Environ. Health Perspect. 110 (5) (2002) 831–834.

[45] F.W. Sunderman Jr., B. Dingle, S.M. Hopfer, T. Swift, Acute nickel toxicity in electroplating workers who accidentally ingested a solution of nickel sulphate and nickel chloride, Am. J. Ind. Med. 14 (1988) 257–266.

[46] S. Bhowmick, S. Chakraborty, P. Mondal, W. Van Renterghem, S. Van den Berghe, G. Roman-Ross, et al., Montmorillonite supported nanoscale zero-valent iron for removal of arsenic from aqueous solution: kinetics and mechanism, Chem. Eng. J. 243 (2014) 14–23.

[47] E.M. Jouad, F. Jourjon, G. Le Guillanton, D. Elothmani, Removal of metal ions in aqueous solutions by organic polymers: use of a polydiphenylamine resin, Desalination 180 (1–3) (2005) 271–276.

[48] S. Hernberg, J. Nikkanen, Enzyme inhibition by lead under normal urban conditions, Lancet 1 (1970) 63–64.

[49] H.L. Needleman, Lead and neuropsychological deficit: finding a threshold, in: H.L. Needleman (Ed.), Low Level Lead Exposure: The Clinical Implications of Current Research, Raven Press, New York, 1980, pp. 43–51.

[50] K. Schumann, The toxicological estimation of the heavy metal content (Cd, Hg, Pb) in food for infants and small children, Zeitschrift Fur Ethnologie 29 (1) (1990) 54–73.

[51] K. Nolan, Copper toxicity syndrome, J. Orthomol. Psychiatry 12 (4) (1983) 270–282.

[52] A.K. Krishnan, T.S. Anirudhan, Removal of mercury(II) from aqueous solutions and chlor-alkali industry effluent by steam activated and sulphurised activated carbons prepared from bagasse pith: kinetics and equilibrium studies, J. Hazard. Mater. 92 (2002) 161.

[53] C. Namasivayam, N. Kanchana, Removal of Congo red from aqueous solution by waste banana pith, Pertanika 1 (1993) 33–42.

[54] J.K. Piotrowski, M.J. Inskip, Health Effects of Methylmercury, MARC Report No. 24, Monitoring and Assessment Research Centre, Chelsea College, University of London, 1981.

[55] Y. Zhang, H. Chi, W. Zhang, Y. Sun, Q. Liang, Y. Gu, et al., Highly efficient adsorption of copper ions by a PVP-reduced graphene oxide based on a new adsorptions mechanism, Nano-Micro Lett. 6 (1) (2014) 80–87.

[56] H. Kozlowski, A. Janicka-Klos, J. Brasun, E. Gaggelli, D. Valensin, G. Valensin, Copper, iron, and zinc ions homeostasis and their role in neurodegenerative disorders (metal uptake, transport, distribution and regulation), Coord. Chem. Rev. 253 (21) (2009) 2665–2685.

[57] M. Omraei, H. Esfandian, R. Katal, M. Ghorbani, Study of the removal of Zn(II) from aqueous solution using polypyrrole nanocomposite, Desalination 271 (1) (2011) 248–256.

[58] C. Moreno-Castilla, M.A. Alvarez-Merino, M.V. Lopez-Ramon, J. Rivera-Utrilla, Cadmium ion adsorption on different carbon adsorbents from aqueous solutions. Effect of surface chemistry, pore texture, ionic strength, and dissolved natural organic matter, Langmuir 20 (19) (2004) 8142–8148.

[59] A. Bernard, J.P. Buchet, H. Roels, P. Masson, R. Lauwerys, Renal excretion of proteins and enzymes in workers exposed to cadmium, Eur. J. Clin. Investig. 9 (1979) 11–22.

[60] H. Horiguchi, H. Teranishi, K. Niiya, K. Aoshima, T. Katoh, N. Sakuragawa, et al., Hypoproduction of erythropoietin contributes to anemia in chronic cadmium intoxication: clinical study on Itai-itai disease in Japan, Arch. Toxicol. 68 (10) (1994) 632–636.

[61] L. Jarup, L. Hellström, T. Alfvén, M.D. Carlsson, A. Grubb, B. Persson, et al., Low level exposure to cadmium and early kidney damage: the OSCAR study, Occup. Environ. Med. 57 (10) (2000) 668–672.

[62] R. Karthik, S. Meenakshi, Chemical modification of chitin with polypyrrole for the uptake of Pb(II) and Cd(II) ions, Int. J. Biol. Macromol. 78 (2015) 157–164.

[63] M. Kasuya, H. Teranishi, K. Aohima, T. Katoh, N. Horignchi, Y. Morikawa, et al., Water pollution by cadmium and the onset of "itai-itai" disease, Water Sci. Technol. 25 (1992) 149–156.

[64] M. Kobya, E. Demirbas, E. Senturk, M. Ince, Adsorption of heavy metal ions from aqueous solutions by activated carbon prepared from apricot stone, Bioresour. Technol. 96 (13) (2005) 1518–1521.

[65] M. Yasuda, A. Miwa, M. Kitagawa, Morphometric studies of renal lesions in "Itai-itai" disease: chronic cadmium nephropathy, Nephron 69 (1995) 14–19.

[66] K. Juttner, U. Galla, H. Schmieder, Electrochemical approches to environmental problems in the process industry, Electrochim. Acta 45 (15) (2000) 2575–2594.

[67] F. Fu, Q. Wang, Removal of heavy metal ions from wastewaters: a review, J. Environ. Manag. 92 (2011) 407–418.

[68] S.M. Nomanbhay, K. Palanisamy, Removal of heavy metal from industrial wastewater using chitosan coated oil palm shell charcoal, Electron. J. Biotechnol. 8 (1) (2005) 43–53.

[69] Z. Pawlak, S. Zak, L. Zablocki, Removal of hazardous metals from groundwater by reverse osmosis, Pol. J. Environ. Stud. 15 (4) (2005) 579–583.

[70] S.Y. Kang, J.U. Lee, S.H. Moon, K.W. Kim, Competitive adsorption characteristics of Co^{2+}, Ni^{2+} and Cr^{3+} by IRN-77 cation exchange resin in synthesized wastewater, Chemosphere 56 (2004) 141–147.

[71] M.V. Vaca, R.L.P. Callejas, R. Gehr, B.J.N. Cisneros, P.J.J. Alvarez, Heavy metal removal with mexican clinoptilolite: multi-component ionic exchange, Water Res. 35 (2) (2001) 373–378.

[72] J.A. Gomes, P. Daida, M. Kesmez, M. Weir, H. Moreno, J.R. Parga, et al., Arsenic removal by electro-coagulation using combined Al-Fe electrode system and characterization of products, J. Hazard. Mater. 139 (2007) 220–231.

[73] T. Angelidis, K. Fytianos, G. Vasilikiotis, Lead recovery from aqueous solution and wastewater by cementation utilizing an iron rotating disc, Resour. Conserv. Recycl. 2 (2) (1989) 131–138.

[74] B. Volesky, Biosorption of Heavy Metals, CRC Press, USA, Boca Raton, 1990.

[75] N. Ahalya, T.V. Ramachandra, R.D. Kanamadi, Biosorption of heavy metals, Res. J. Chem. Environ. 7 (2003) 71–79.

[76] A. Malik, Metal bioremediation through growing cells, Environ. Int. 30 (2004) 261–278.

[77] P. Desale, D. Kashyap, N. Nawani, N. Nahar, A. Rahman, B. Kapadnis, et al., Biosorption of nickel by *Lysinibacillus* sp. BA2 native to bauxite mine, Ecotoxicol. Environ. Saf. 107 (2014) 260–268.

[78] W. Jiang, A. Saxena, B. Song, B.B. Ward, T.J. Beveridge, S.C.B. Myneni, Elucidation of functional groups on gram positive and gram negative bacterial surfaces using infrared spectroscopy, Langmuir 20 (2004) 11433–11442.

[79] S. Rana, N. Bhakta, Heavy metal(loid) remediation using bio-waste: a potential low-cost green technology for cleaning environment. in: J.N. Bhakta (Ed.), Handbook of Research on Inventive Bioremediation Techniques, IGI Global, Hershey, 2017, pp. 394–415, https://doi.org/10.4018/978-1-5225-2325-3.ch017.

[80] I. Michalak, K. Chojnacka, A. Witek-Krowiak, State of the art for the biosorption process – a review, Appl. Biochem. Biotechnol. 170 (2013) 1389–1416.

[81] P. Kotrba, Microbial biosorption of metals—general introduction, in: P. Kotrba, M. Mackova, T. Macek (Eds.), Microbial Biosorption of Metals, Springer, Dordrecht, Netherlands, 2011, pp. 1–6.

[82] J.N. Bhakta, M.M. Ali, A. Fares, S.K. Singh, Biosorption of arsenic: an emerging eco-technology of arsenic detoxification in drinking water, in: Arsenic Water Resources Contamination, Advances in Water Security, 2019, pp. 207–230.

[83] S.V. Avery, G.A. Codd, G.M. Gadd, Biosorption of tributyltin and other organotin compounds by cyanobacteria and microalgae, Appl. Microbiol. Biotechnol. 39 (1993) 812–817.

[84] W. Stumm, J.J. Morgan, Aquatic Chemistry: Chemical Equilibria and Rates in Natural Waters, 3rd ed, Wiley, New York, 1996.

[85] N. Kuyucak, B. Volesky, Biosorbents for recovery of metals from industrial solutions, Biotechnol. Lett. 10 (2) (1988) 137–142.

[86] Z. Aksu, Y. Sag, T. Kutsal, The biosorption of copper by *C. vulgaris* and *Z. ramigera*, Environ. Technol. 13 (1992) 579–586.

[87] G.M. Gadd, C. White, Copper uptake by *Penicillium ochrochloron*: influence of pH on toxicity and demonstration of energy-dependent copper influx using protoplasts, J. Gen. Microbiol. 131 (1985) 1875–1879.

[88] B. Greene, D.W. Darnall, Microbial oxygenic photoautotrophs (cyanobacteria and algae) for metal-ion binding, in: H. Ehrlich, C. Brierley (Eds.), Microbial Mineral Recovery, McGraw-Hill, New York, 1990, pp. 227–302.

[89] Y. Nuhoglu, E. Malkoc, Investigations of nickel (II) removal from aqueous solutions using tea factory waste, J. Hazard. Mater. 127 (1–3) (2005) 120–128.

[90] A.L. Zouboulis, K.A. Matis, I.C. Hancock, Biosorption of metals from dilute aqueous solutions, Sep. Purif. Method 26 (2) (1997) 255–295.

[91] E. Fourest, C. Canal, J.C. Roux, Improvement of heavy metal biosorption by mycelial dead biomasses (*Rhizopus arrhizus, Mucor miehei* and *Penicillium chrysogenum*): pH control and cationic activation, FEMS Microbiol. Rev. 14 (1994) 325–332.

[92] J.L. Wang, C. Chen, Biosorption of heavy metals by *Saccharomyces cerevisiae*: a review, Biotechnol. Adv. 24 (2006) 427–451.

[93] D. Park, Y.S. Yun, J.M. Park, The past, present, and future trends of biosorption, Biotechnol. Bioprocess Eng. 15 (2010) 86–102.

[94] P.S. Li, H.C. Tao, Cell surface engineering of microorganisms towards adsorption of heavy metals, Crit. Rev. Microbiol. 41 (2) (2013) 140–149.

[95] Y. Liu, Y.J. Liu, Biosorption isotherms, kinetics and thermodynamics, Sep. Purif. Technol. 61 (2008) 229–242.

[96] B. Volesky, Biosorption process simulation tools, Hydrometallurgy 71 (2003) 179–190.

[97] J. Febrianto, A.N. Kosasih, J. Sunarso, Y. Ju, N. Indraswati, S. Ismadji, Equilibrium and kinetic studies in adsorption of heavy metals using biosorbent: a summary of recent studies, J. Hazard. Mater. 162 (2009) 616–645.

[98] S. Goldberg, L.J. Criscenti, Modeling adsorption of metals and metalloids by soil components, in: A. Violante, P.M. Huang, G.M. Gadd (Eds.), Biophysico-Chemical Processes of Heavy Metals and Metalloids in Soil Environments, Wiley, New Jersey, 2008, pp. 215–264.

[99] R. Dhankhar, A. Hooda, Fungal biosorption – an alternative to meet the challenges of heavy metal pollution in aqueous solutions, Environ. Technol. 32 (2011) 467–491.

[100] P. Anand, J. Isar, S. Saran, R.K. Saxena, Bioaccumulation of copper by *Trichoderma viride*, Bioresour. Technol. 91 (2006) 1018–1025.

[101] J. Suriya, S. Bharathiraja, R. Rajasekaran, Biosorption of heavy metals by biomass of *Enterobacter cloacae* isolated from metal-polluted soils, Int. J. ChemTech Res. 5 (3) (2013) 1329–1338.

[102] L. Babak, P. Šupinova, M. Zichova, R. Burdychova, E. Vitova, Biosorption of Cu, Zn and Pb by thermophillic bacteria – effect of biomass concentration on biosorption capacity, Acta Univ. Agric. Silvic. Mendel. Brun. 60 (5) (2012) 9–18.

[103] D.P. Samarth, C.J. Chandekar, R.K. Bhadekar, Biosorption of heavy metals from aqueous solution using *Bacillus Licheniformis*, Int. J. Pure Appl. Sci. Technol 10 (2) (2012) 12–19.

[104] W.N. Xue, Y.B. Peng, Biosorption of arsenic (III) from aqueous solutions by industrial fermentation waste *Aspergillus niger*, Appl. Mech. Mater. 448-453 (2014) 791–794.

[105] B. Sumalatha, K.Y. Prasanna, P. King, Removal of Arsenic from aqueous solutions using *Turbinaria vulgaris* sp. as biosorbent, Int. J. Adv. Res. Sci. Eng. 6 (7) (2017) 217–223.

[106] S. Iihan, M. Nourbakhsh, S. Kilicarslan, H. Ozdag, Removal of chromium, lead and copper ions from industrial waste waters by *Staphylococcus saprophyticus*, Turk. Elec. J. Biotechnol. 2 (2004) 50–57.

[107] G.M. Gadd, Interactions of fungi with toxic metals, New Phytol. 124 (1993) 25–60.

[108] M. Tsezos, B. Volesky, The mechanism of uranium biosorption by *R. arrhizus*, Biotechnol. Bioeng. 24 (1982) 965–969.

[109] L.M. Jr Barros, G.R. Macedo, M.M.L. Duarte, E.P. Silva, A.K.C.L. Lobato, Biosorption of cadmium using the fungus *Aspergillus niger*, Braz. J. Chem. Eng. 20 (3) (2003) 229–239.

[110] V. Zdravka, S. Margrita, G. Velizar, Biosorption of Cu (II) onto chemically modified waste mycelium of *Aspergillus awamori*: equilibrium, kinetics and modeling studies, J. BioSci. Biotechnol. 1 (2) (2012) 163–169.

[111] S. Subhashini, S. Kaliappan, M. Velan, Removal of heavy metal from aqueous solution using *Schizosaccharomyces pombe* in free and alginate immobilized cells, in: 2nd International Conference on Environmental Science and Technology IPCBEE, vol. 6, IACSIT Press, Singapore, 2011.

[112] R. Dhankhar, A. Hooda, R. Solanki, P.A. Sainger, *Saccharomyces cerevisiae*: a potential biosorbent for biosorption of uranium, Int. J. Eng. Sci. Technol. 3 (6) (2006) 5397–5407.

[113] F. Ghorbani, H. Younesi, S.M. Ghasempouri, A.A. Zinatizadeh, M. Amini, A. Daneshi, Application of response surface methodology for optimization of cadmium biosorption in an aqueous solution by *Saccharomyces cerevisiae*, Chem. Eng. J. 145 (2008) 267–275.

[114] G. Yan, T. Viraraghavan, Effect of pretreatment on the bioadsorption of heavy metals on *Mucor rouxii*, Water SA 26 (1) (2000) 119–124.

[115] R. Flouty, G. Estephane, Bioaccumulation and biosorption of copper and lead by a unicellular algae *Chlamydomonas reinhardtii* in single and binary metal systems: a comparative study, J. Environ. Manag. 111 (2012) 106–114.

[116] M.H. Khani, Uranium biosorption by *Padina sp.* algae biomass: kinetics and thermodynamics, Environ. Sci. Pollut. Res. 18 (2011) 1593–1605.

[117] A.H. Sulaymon, A.A. Mohammed, T.J. Al-Musawi, Competitive biosorption of lead, cadmium, copper, and arsenic ions using algae, Environ. Sci. Pollut. Res. 20 (2012) 3011–3023.

[118] R.H. Crist, K. Oberholser, K. Shank, M. Nguyen, Nature of bonding between metallic ions and algal cell walls, Environ. Sci. Technol. 15 (1981) 1212–1217.

[119] I.N. Kumar, C. Oommen, Removal of heavy metals by biosorption using freshwater alga *Spirogyra hyaline*, J. Environ. Biol. 33 (2012) 27–31.

[120] M. Hajar, Biosorption of cadmium from aqueous solution using dead biomass of brown alga *Sargassum Sp*, Chem. Eng. Trans. 17 (2009) 1173–1178.

[121] E.S. Cossich, C.R.G. Tavares, T.M.K. Ravagnani, Biosorption of chromium(III) by *Sargassum* sp. biomass, Electron. J. Biotechnol. 5 (2) (2002) 133–140.

[122] D. Sheikha, I. Ashour, F.A. Abu Al-Rub, Biosorption of zinc on immobilized Green algae: equilibrium and dynamics studies, J. Eng. Res. 5 (1) (2008) 20–29.

[123] T.A. Oyedepo, Biosorption of lead (II) and copper (II) metal ions on *Calotropis procera* (Ait.), Sci. J. Pur. Appl. Chem. 1 (2011) 1–7.

[124] V.K. Gupta, I. Ali, Utilization of bagasse fly ash (a sugar industry waste) for the removal of copper and zinc from wastewater, Sep. Purif. Technol. 18 (2000) 131–140.

[125] A. Hashem, A. Abou-Okeil, A. El-Shafie, M. El-Sakhawy, Grafting of high-cellulose pulp extracted from sunflower stalks for removal of Hg (II) from aqueous solution, Polym.-Plast. Technol. Eng. 45 (2006) 135–141.

[126] O.O. Abdulrasaq, O.G. Basiru, Removal of copper (II), iron (III) and lead (II) ions from mono-component simulated waste effluent by adsorption on coconut husk, Afr. J. Environ. Sci. Technol. 4 (2010) 382–387.

[127] N. Meunier, J. Laroulandie, J.F. Blais, R.D. Tyagi, Cocoa shells for heavy metal removal from acidic solutions, Bioresour. Technol. 90 (3) (2003) 255–263.

[128] A.B. Sallau, S. Aliyu, S. Ukuwa, Biosorption of chromium (VI) from aqueous solution by corn cob powder, Int. J. Environ. Bioeng. 4 (2012) 131–140.

[129] E. Malkoc, Y. Nuhoglu, Investigations of Ni(II) removal from aqueous solutions using tea factory waste, J. Hazard. Mater. 127 (2005) 120–128.

[130] K. Huang, Y. Xiu, H. Zhu, Selective removal of Cr(VI) from aqueous solution by adsorption on mangosteen peel, Environ. Sci. Pollut. Res. 20 (2013) 5930–5938.

[131] M. Bansal, D. Singh, V.K. Garg, P. Rose, Use of agricultural waste for the removal of nickel ions from aqueous solutions: equilibrium and kinetics studies, Int. J. Civil Environ. Eng. 1 (2009) 108–114.

[132] O.S. Bello, M.A. Oladipo, A.M. Olatunde, Sorption studies of lead ions onto activated carbon produced from oil-palm fruit fibre, Stem Cells 1 (2010) 14–29.

[133] I. Safarik, K. Horska, M. Safarikova, Magnetically responsive biocomposites for inorganic and organic xenobiotics removal, in: P. Kotrba, M. Mackova, T. Macek (Eds.), Microbial Biosorption of Metals, Springer, Dordrecht, Netherlands, 2011, pp. 301–320.

[134] W.S. Wan Ngah, M.A.K.M. Hanafiah, Removal of heavy metal ions from wastewater by chemically modified plant wastes as adsorbents: a review, Bioresour. Technol. 99 (2008) 3935–3948.

[135] K. Vijayaraghavan, Y.S. Yun, Bacterial biosorbents and biosorption, Biotechnol. Adv. 26 (2008) 266–291.

[136] K. Kuroda, M. Ueda, Yeast biosorption and recycling of metal ions by cell surface engineering, in: P. Kotrba, M. Mackova, T. Macek (Eds.), Microbial Biosorption of Metals, Springer, Dordrecht, Netherlands, 2011, pp. 235–247.

[137] B.T. Ngwenya, Enhanced adsorption of zinc is associated with aging and lysis of bacterial cells in batch incubations, Chemosphere 67 (2007) 1982–1992.

[138] M. Fomina, G.M. Gadd, Metal sorption by biomass of melanin producing fungi grown in clay-containing medium, J. Chem. Technol. Biotechnol. 78 (2002) 23–34.

5

A comprehensive review of glyphosate adsorption with factors influencing mechanism: Kinetics, isotherms, thermodynamics study

Kamalesh Sen[a] and Soumya Chattoraj[b]

[a]ENVIRONMENTAL CHEMISTRY LABORATORY, DEPARTMENT OF ENVIRONMENTAL SCIENCE, THE UNIVERSITY OF BURDWAN, BARDHAMAN, WEST BENGAL, INDIA [b]UNIVERSITY INSTITUTE OF TECHNOLOGY, GENERAL SCIENCE AND HUMANITIES, THE UNIVERSITY OF BURDWAN, BARDHAMAN, WEST BENGAL, INDIA

1 Introduction

Indiscriminate disposal of pesticides such as herbicides, pesticides, pesticides etc., brings attention to public health, policy debates, and damage to the ecosystem. It's necessary to mitigate this pollution in sustainable ways [1,2]. The fact is that this contamination possibility goes into the food chain, so monitoring it is essential to avoid effects on human health [3,4]. Decontamination strategies make sustainable removal through chemical precipitation, microbial techniques [5], adsorption, degradation [5], and electrocoagulation [6]. Pesticides are distinguished by their chemical nature, which indicates their activities such as structural, toxicity, degradation, and other processes. Agrochemical persistence depends on its half-life in soil and water, and the contaminant's mobility into water and soil. As is known, organic and ionic elements in agricultural chemicals mainly contain metallic, halogenic organophosphate and carbonate. Herein, we focus only on organophosphate as a glyphosate, as there is a possibility of the formation of secondary pollutants. Xenobiotics is a major environmental concern because of an elicit ecological impact at low concentrations and food chain exchange through biomagnification [7,8]. Every year, up to 5 billion kilograms of pesticides are reportedly applied worldwide, and it's expected to reach approximately 10 billion kilograms in 2050 [9,10]. Decontamination occurs through arbitrary uses, accidental releases, and other anthropogenic courses, namely leakage, sprinklers, waste disposal, etc. [9] There are mainly two types of pollution released into the hydrosphere that possess serious roles for biodiversity and food chains, posing a high risk to the environment, living organisms, as well as human health. In rural areas, the health concerns of approximately 3–4 million farmers in different countries

Intelligent Environmental Data Monitoring for Pollution Management. https://doi.org/10.1016/B978-0-12-819671-7.00005-1

have suffered annually from unhygienic pesticide toxicity, due to misconceptions, illiteracy, and lack of control and education among farmers [9]. People have been exposed to pesticides poisoning by inhalation while spraying pesticides. It is therefore important for the environmental scientist to determine how to properly manage or sufficiently remove these pesticides from the Earth and nature.

Glyphosate IUPAC, known as N-(phosphonomethyl) glycine, is applied as a total nonselective type of herbicides (Roundup), with a formula $C_3H_8NO_5P$, molar mass of 169.07 g/mol, and solubility of 10.1 g/L (25°C). The crystal structure of glyphosate is a representation of zwitterion in amino groups, and its phosphonate acts as a nitrogenous trap via proton dislocation [11] (Fig. 1).

Acid dissociation constants have pKa1, pKa2, or pKa3; the determined value 2.32 (carboxylate), 5.6 (phosphonate), or 10.6 (amino) suggests possible coordination mode [12]. Glyphosate probable complexion, as well as divalent alkali earth metals and transition metals, easily chelate through tridentate/bidentate ligand's exchange of glyphosate [13]. Toxicity actually controls 5-enolpyruvylshikimic acid 3-phosphate, which produces and restrains metabolic activity by biochemical activity via aromatic amino acids such as tryptophan, tyrosine, phenylalanine, etc. [11,14]. Adsorption phenomena of water surface technology interacts with their charge; glyphosate negative ions are created with acidic pH, hence only the phosphonate acts as a combing bridge by ligand's exchange mechanism [15,16] (Fig. 2).

Glyphosate is toxic to plants and animals, and with respect to national and international regulatory agencies [17], its reported toxicity is basically carcinogenic to humans [18,19]. Glyphosate kills plants by inhibiting the biosynthesis of aromatic amino acids needed for plant growth. One obstacle in the way is the enzyme 5-enolpyruvylshikimate-3-phosphate synthase of the shikimate pathway [20,21]. Glyphosate acts as an antagonistic analog of phosphoenolpyruvate, which activates with a substrate for 5-enolpyruvylshikimate-3-phosphate synthases. Aromatic amino acids lead to metabolic functions of the plants, increases the enzyme's scarcity, and effectively inhibits the plant's growth as well as destroys it [19]. Overuse of glyphosate has produced serious environmental toxicity, e.g., reproduction decreases, biomass losses, reduced surface casting activity, etc. Aquatic plants and animals are also affected; basically invertebrates are lethality affected, and gastropoda and animalia classes

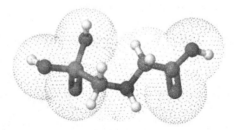

FIG. 1 N-(phosphonomethyl) glycine molecular structure.

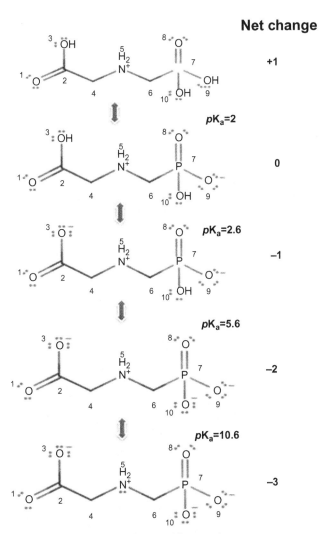

Net change

FIG. 2 Different forms of glyphosate and the net charge with *p*Ka values.

are largely damaged. This pesticide not only indirectly affects humans by causing epilepsy, but there are other harmful effects, such as acting as endocrine disrupters in humans, damaging placental cells, and reducing the enzyme aromatase [3,4,19].

Because a large amount of the environment is affected by hazardous materials, it's essential to remove it in suitable ways; remediation by chemical reagents are engaged to perform as chemical precipitates, such as alum, lime, resin, EDTA salt, and other biopolymers [22,23]. After precipitation, this method can separate the material and easily clean the adsorbents. After this process, the cleaner substance has many management problems because, although the glyphosate is removed, other byproducts increase, and these other products create new pollutants according the respective reagent. Electrodialysis (ED) is a very easy way to membrane filter as ions are transported through a semipermeable membrane [24–26]. If the membranes are cation-selective and polyelectrolytes have a negative charge, it rejects the negative ions and allows the positive ions to pass; the membrane has selective potential as cationic and anionic, to accept or reject ions during liquid transport. The particles coagulate by small aggregation, basically a colloidal material, which is mixed and easily settled out. Ultrafiltration is a separation technique that basically uses membranes with pores as small as 0.1 to 0.001 μm. It may depend on an ionic charge with membrane potential. The reverse osmosis process uses a cellophane-type membrane to force ionic diffusion of different contaminants [27]. There was a very interesting investigation of glyphosate depletion using microbes. The microbes (fungus) have the potential for pesticide degradation as well as organophosphate herbicides (glyphosate); according to the report, aminomethylphosphonic acid molecules are divided into two groups, namely sarcosine and glycine. *Aspergillus oryzae* A-F02 pathway is shown in Fig. 3. According to the report, there was a 1500-mg/L concentration in glyphosate decay, but it took about 96 days, and its kinetic rate was much slower than other processes [28].

In a preferred investigation of the photocatalytic degradation of glyphosate by semiconductor under UV/visible conditions, the degrading mechanism of the glyphosate molecules are divided into two parts by reducing toxin levels with two products, such as sarcosine and glycine, under photo-oxidation. Most semiconductor materials have the potential to reduce complex organic pollutants [5,29] (Fig. 4).

The schematic assumes TiO_2 variance under a conducting band or valance band to produce a photo-generated electron trap and holes, which easily depletes C—N bonds by free hydroxyl radicals, and ultimately produces glycine molecules [12]. As investigations for the byproducts of glycine ions may use the same catalytic agent, this study regarding water purification is lacking.

The chapter aims to review the following: (1) an adsorption study with factors influencing the mechanism with various adsorbent materials; (2) glyphosate adsorption with a comparative study; and (3) the lack of research about glyphosate along with other pesticides, as

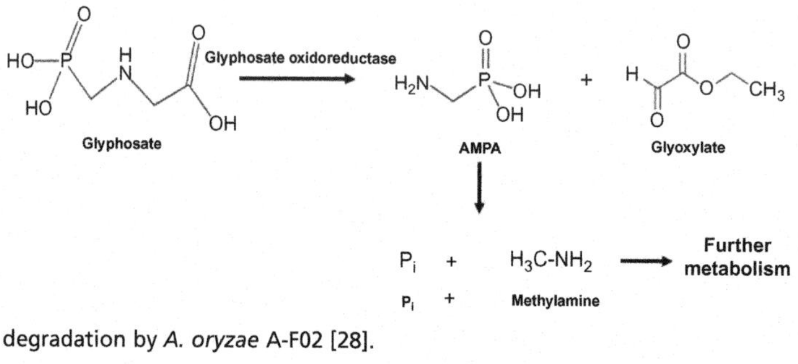

FIG. 3 Glyphosate degradation by *A. oryzae* A-F02 [28].

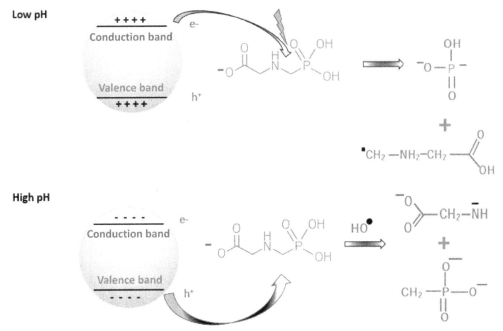

FIG. 4 The interaction of glyphosate with the surface of TiO$_2$, illumination with ultraband gap irradiation as photocatalytic degradation [12].

well as easily available adsorbents, sustainable adsorbents, easily manageable adsorbents, and low cost production of adsorbents.

2 Adsorption study

Adsorption is a process that easily interacts with and attributes gas or liquid molecules to produce the surface complexity of a solid (adsorbent), forming molecular films. It is found in most natural, biological, physical, and chemical systems and is widely used in industrial purposes and column chromatography [30–33]. Basically, the adsorption mechanism is described by isotherms, kinetics, and thermodynamics; the characterization of adsorbents are examined by zero point charge (ZPC), proton adsorption, scanning electron microscopy and energy-dispersive X-ray spectroscopy (SEM-EDS), Fourier-transform infrared spectroscopy (FTIR), X-ray powder diffraction (XRD), etc. [1,16,34–36]. Use of this adsorbent is economically viable during wastewater treatments, when there is a possibility of interaction with surface phenomena; if adsorbents are mesoporous or microporus, there is great advantage for maximum interaction. The interaction is quantified via equilibrium capacity, which is calculated [37], and different adsorbents are relatively observant, which shows more efficiency when focusing on working efficacy during adsorption with suitable regeneration after adsorption. A new type of adsorbent are greener adsorbents, which originate at a lower cost, and such separation is discussed in the following sections for sorption and their sustainable management.

2.1 Adsorbent preparation

Adsorbent preparation is a tremendous work, which are segregated into types, such as agricultural waste materials with their products, natural adsorbents, biogenic preparation of the adsorbents, soils, geomineral, nanomaterial, and composite. However, preparation must meet current demands, to create minimal waste product and utilizing maximum materials.

2.1.1 Agricultural waste materials and their product

Agrowaste is generally found in the agroindustry where collected materials is used to produce the absorbent, basically natural activated char, made from impregnated cellulose contents, composite materials, and food peels. Preparation and activation techniques have many differences to motivate the adsorption; when char has the same constituent production, adsorption capacity has no difference, because of the binding and uptake of different activation chemicals [1,38,39]. Normally agrowaste are constituents with lignin, hemicellulose extractives, lipids, proteins, simple sugars, starches, water, hydrocarbons, ash contents, and many compounds, have functional groups to accelerate the binding process [40,41]. Cellulose has a homopolymer of crystalline as a monomer of glucose, with ß1–4 glycosidic linkage by acting hydrogen bonds. When any metal ions are impregnated, they attach to substation by hydrogen bonds [42,43] (Fig. 5).

2.1.2 Soil and geominerals

Soil has good adsorbent materials for pesticide adsorption, because it presently contains chelated metal complexes, humic substances, and other materials. Therefore, pesticide adsorption emphasizes previous research. However, it depends on the properties of the containing adsorbents, such as cation exchange capacity, anion exchange capacity, organic matter, ZPC, proton adsorption capacity, FTIR, and texture as particle-belonging categories [37]. Upon review, soil has the potential to act as mobility of leaching performed on the pesticides [45]. Soil components include organic matter, clay minerals, and hydrous metal oxides; some polar functional groups contain molecules that interact through the cation exchange mechanism [46,47]. This interaction via ions dipole is held through cations and intermediated water molecules formed by hydrogen interactive bonds; if cation interaction is enhanced, their anionic adsorbate is strongly influenced. So the pesticide adsorption on montmorillonite, in decreasing order, is $Al^{3+} > Mg^{2+} > Ca^{2+} > Li^+ > Na^+$ [32].

2.1.3 Activated char materials from mass

Activated char produced from biomass of plant materials is mostly used, achieving its activation site or absorbing surface from activating materials; the activated materials may be acidic, alcoholic, alkali, etc., and the performance check supports their adsorbate uptake

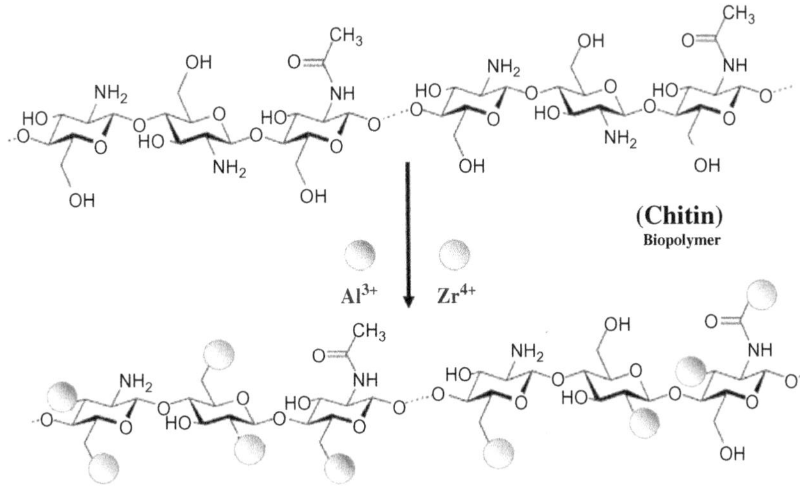

FIG. 5 Schematic diagram showing the interaction of metal ions with cellulose biopolymer surface [44].

efficacy. Activated carbon has more durable adsorption properties, essentially organic pollutants interact through carbon-carbon bonding, so adsorbent activation is necessary to increase adsorption. The mechanism that carbon uses is not cationic or anionic, which is why carbon makes polar bonds between positive and negative sites of the adsorbents. Basically, there are two parts of activation, which are physical activation and chemical activation. Physical activation generally increases the porosity as a carbonaceous precursor or intermediated minerals are oxidized by burnt transforms into carbon linked under pressure [48,49]. During chemical activation, activated carbon increases the activated site during adsorption by activating agents. The activating agent may increase its surface area. Table 1 provides a previous review of the chemical activation of rice husk.

2.1.4 Nanomaterial adsorbents

We know that increases in the surface area increases the potential to uptake pollutants by adsorption. There is now growing attention toward nanomaterial due to its characteristic features such as large surface area, small size, high reactivity, and binding interferences. The large surface area of metal nanoparticles (NPs) have several properties such as optical, chemical, mechanical, and catalytic, which properties depend on their size [43,60,61]. It basically forms at the nanoscale with a diameter range of 1–100 nm, which introduces several properties such as charge transferring, surface area, mechanical phenomena, etc. Nanomaterials play a vital role in the adsorption process due to availability of adsorbate's interaction and binding potential. However, the mechanism is such that corresponding adsorption increases or decreases, which depend on their synthesis. Formation/synthesis of nanomaterials have vital research, which has biological, chemical, and mechanical approaches. Previous reports support adsorption that depends on nanosynthesis

Table 1 Review of chemical and physical activation of different chars.

Materials	Preparation process	Components	Surface area	References
Rice husk	Physical activation	Ferric nitrate, 750°C, 1.5 h	527 m^2 g^{-1}	[50]
Rice husk	Physical activation	Stem heated, 800°C, 15 min	1365 m^2 g^{-1}	[51]
Rice husk	Physical activation	CO_2 activation, 800°C, 45 min	1514 m^2 g^{-1}	[51]
Doreylas fir pellets	Chemical activation	Activation orthophosphoric acid (81.21%), N_2 microwave assist pyrolysis 450°C, 48 min	1725.7 m^2 g^{-1}	[52]
Rice husk	Chemical activation	Orthophosphoric acid, N_2 assist, 500°C, 1 h	1016 m^2 g^{-1}	[53]
Rice husk	Chemical activation	N_2 assist, 450°C, 180 min	677 m^2 g^{-1}	[54]
Rice husk	Chemical activation	NaOH activated, 800°C, 120 min,	2841 m^2 g^{-1}	[55]
Sugar cane bagasse	Chemical activation	KOH activated, 687°C, 120 min, N_2	99.94 m^2 g^{-1}	[56]
Coconut shells	Chemical activation	Hydrothermal, $ZnCl_2$, 850°C, 120 min	1744 m^2 g^{-1}	[57]
Mango husk	Chemical activation	Nitrogen gas, 105°C for 24 h	1718 m^2 g^{-1}	[58]
Kernel shell	Chemical activation, Carbonized apricot	N_2, 850°C, 1 h	328.570 m^2 g^{-1}	[59]

protocols. For pollutant removal, nanoparticles used such as ZnO, TiO$_2$, Ag NPs, Au NPs, carbon nanotube (CNT), iron NPs etc., present demands of nano-nano composite, doped bases, and/or bio-nano composite, etc.

2.1.5 Biofabricated nanomaterials

Regarding current demands for synthesis of NPs, synthesis mechanism have been deferred. During the synthesis of NPs, metal can be reduced by biochemical compounds, which are found in respective plants. Table 2 shows which compounds are specifically involved with the respective metals. Catechins, theaflavins, thearubigins, curcacycline A, curcacycline B, azadirachtin, catechins, theaflavins, thearubigins, phenolic compounds, terpenoids or proteins, flavonoids, phenolic acids, polyphenols, etc. are compounds generally attributed toward NPs' formation [65]. These compounds are found plant extracts, and these types of biochemicals are used as capping agents. Actually, metal ions are reduced by these types of biochemical agents that fabricate the NPs. As for zinc oxide NPs, plant phenolic compounds and terpenoids help in the formation of zinc oxide NPs [66,67] (Fig. 6).

2.1.6 Composite materials

Composite preparation, either nano-nano, bio-nano, bio-bio, or nano-bio-nano preparation and utilization has a major demand because they increase the efficacy of adsorption. Previous literature introduced glyphosate adsorption as basically a composite and shows

Table 2   Different organic substances adsorption with different adsorbents.

Adsorbent materials	Chemicals	pH	Temperature	Q_e (mg/g)	Mechanism	References
Agricultural waste						
Bagasse	2,4-D	3.5	55	7.14	Van der Walls	[31]
Typha orientatis presl	Phenol	5.0	40	7.23	Electrostatic interaction, electron exchange	[62]
Rice straw	Tetracyclic	5.5	20	14.16	π-π interaction	[63]
Waste materials						
Carbon slurry	2,4-dichlorophenoxyacetic acid	7.5	35	212	Van der Walls	[64]
Carbon slurry	Carbofuran	7.5	45	208	Van der Walls	[64]
Activated carbon prepared from waste rubber tire	Methoxychlor	2	25	112.0	Physical in nature involving weak Van der Waals forces	[39]
Activated carbon prepared from waste rubber tire	Atrazine	2	25	104.9	Physical in nature involving weak Van der Waals forces	[39]
Activated carbon prepared from waste rubber tire	Methyl parathion	2	25	88.9	Physical in nature involving weak Van der Waals forces	[39]
Coal-based bottom ash	Vertigo Blue 49	6.8–7.0	25	13.51	Hindering free diffusion to the internal porosity, electrostatic interaction	[38]
Soil and geominerals						
Clay loam soil	Chlorantraniliprole	9	25	1.492	Molecular configuration and numbers of double and triple bonds, strong of H-bonding interactions of the molecules with the sorbent	[32]
Clay loam soil	Dinotefuran	9	50	1.210	Molecular configuration and numbers of double and triple bonds, strong of H-bonding interactions of the molecules with the sorbent	[32]

Continued

Table 2   Different organic substances adsorption with different adsorbents—cont'd

Adsorbent materials	Chemicals	pH	Temperature	Q_e (mg/g)	Mechanism	References
Forest soil	Glyphosate	12	30	161.29	Ions exchange through chemisorptive behaviors	[37]
Soil	Glyphosate	5.9	20	21.4	Electron sharing through electrostatic interaction	[45]
Carbon and biochar						
Woody (dendro) biochar	Glyphosate	5	20	44.0	Electron donor-acceptor interaction	[16]
Nanomaterial and composite						
Nano-zero-valent iron (NZVI)	Glyphosate	2	20	35.39	Negative charge interaction, electron donor-acceptor interaction	[34]
Magnetic MnFe$_2$O$_4$-graphene	Glyphosate	4.7	5	39	Negatively charged glyphosate through columbic interaction, electron donor acceptor	[36]

an increase in the adsorption efficacy [34]. Polymer chains have the potential of high adsorption capacity, and that capacity increases their electrostatic force by precursor interaction. Excessive coverage of the surface due to molecular binding site presents as NP incorporate (Fig. 7).

2.2 Characterization

2.2.1 Zero point charge (ZPC)

With regard to biochar's surface during adsorption, the ionic molecules are given surface attraction through columbic forces; however, it may be dissociated with deficits of electrical charges of —OH, —H, and —COOH. When the adsorption process runs into the cationic portion, it is easily sheared by anionic molecules [60,70]. ZPC is one of the characterizations that basically depends on an adsorbent surface. Surface density depends on their surface releases, either protonated or deprotonated, while binding adsorbate anions/cations [71] (Fig. 8).

ZPC is below the adsorptive pH, which means the more protons it gives, the less it donates to the hydroxyl groups. The surface becomes positively charged with attracting

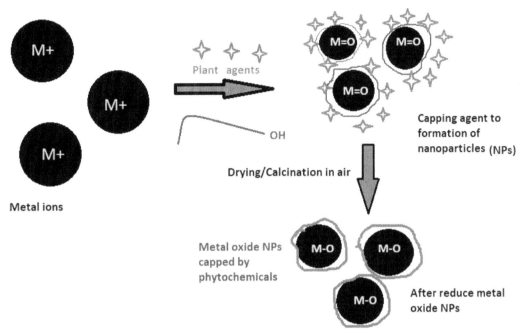

FIG. 6 Nanofabricated [68].

anions. Above the ZPC value, the surface is negatively charged, attracting cations and repelling anions. It is a vital parameter to characterize the adsorbent materials and how they bind adsorbate molecules, either cationic or anionic portions. Glyphosate generally forms the anionic by phosphonate dissociation with protons and negative charge from oxygen atoms, which tend toward adsorption on a positive surface.

2.2.2 Fourier-transform infrared spectroscopy (FTIR)

FTIR study has more important parameters than adsorption study, because of what is known about the molecular tendency to attach to an adsorbent surface. Adsorbent

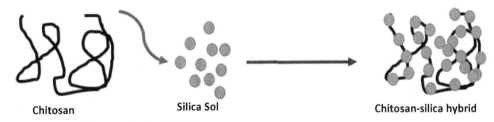

FIG. 7 Schematic of chitosan-silica hybrid [69].

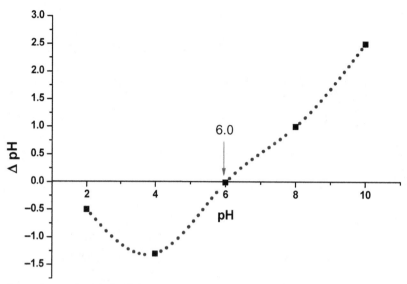

FIG. 8 Point of zero charge of GO (20)/cellulose (100) hydrogel [72].

functional groups are described by FTIR study. Peaks stretch the differences of different adsorbents. Because biosorbents are essential as binding/interactive molecules, peaks were obtained at 3200 to 3400 cm^{-1} due to the carboxylic and hydroxyl functional groups that are present. CH, CH$_2$, and CH$_3$ bonds stretching vibration by hydroxyl groups, their peaks represented 2800 to 3200 cm^{-1}, because the SP$_2$ hybridized. If peaks do not change clearly during adsorption due to Fermi resonance displacement, the peak bands slightly as same is obtained. Stretching on glyphosate molecules with phenolic groups of biochars, the peaks show as 3422 cm^{-1}, suggested that chemical interaction is held there. Such peaks as 1799 cm^{-1} and 1089 cm^{-1} are attributed to biochar peaks that vibrate as C—O—C [73], 2926 cm^{-1} for —OH interaction, 875 cm^{-1} for out-of-plane deformation of aromatic plane, and 1315 cm^{-1} and 1109 cm^{-1} for P—O bonds [16]. Another performed as synthesized to NPs or composite preparation, in which molecules properly attribute the formation of NPs. However, this parameter is more important to describe the adsorbent preparation when the glyphosate is loaded or unloaded by molecular vibrations.

2.2.3 SEM and EDS
SEM is a useful technique for adsorbent characterization; a high-energy electron beam scans across samples, and normally the samples are coated with gold films for adjustment of contrast and signal noise. Surface topography is analyzed and taken for these parameters. Simply described by Nasrollahzadeh et al., SEM provides the pore structure of activated carbon [34]; another characterization also described is the nanocomposite,

and because of this, the specimen surface further provides the nature of the composite surface, and the scale denotes the diameter of this composite [60,74]. The surface composition of the sample has to be detected by low-energy excitation of secondary electrons. Also, backscattered electrons with high energy and X-rays are transmitted from below the sample to collect the sample's morphology. EDS (or EDX) is a microanalysis of a chemical composition of elements; it is another characterization of surface morphology and is associated with SEM imagery of a specified area. EDS analysis is the vital characterization of adsorbents because it reveals the element constitutes, or which elements accelerated the adsorption. It's basically determined by lateral distribution for surface maps from sample SEM imagery. Previous literature supported surface morphology and EDS analysis, as well as describing the NPs and adsorption phenomena [60,75–77].

2.2.4 X-ray diffraction pattern

XRD is a powerful characterization that describes crystalline information as structures, phases, texture, strains, crystal defects, and lattices. XRD diffraction peaks are produced through monochromatic beam of X-rays scattering through the lattice plane by diffraction angles (θ). X-ray intensity is obtained from the atomic position present in each lattice plane. An online database give the information of the plane types of each sample (JCPDS). Peaks are noted by Miller indices of crystallography in Bravais lattices. Basically, particular lattice planes are determined from three integer numbers: h, k, and ℓ; these integer numbers are called the Miller indices. The plane (hkl) has been collected from orthogonal plane as reciprocal lattice vectors. The integer number is represented by 3 to -3, with negative vectors expressing as Miller indices bar (303), and greatest divisors as 1. Suppose the ZnO NPs as a crystalline after producing NPs, the nature of the crystalline by D spacing is shown by the particular area:

$$D = K\lambda/(W\cos\theta)$$

Here, W is the breadth of the diffraction peaks FWHM, K is the shape factor (approximately 0.89), and λ is the X-ray wavelength. However, the XRD clarified that specific surface belongs to their crystalline nature, if amorphous nature XRD pattern is not specified at their peaks [61,78].

2.2.5 Miscellaneous characterization for adsorption relates

Other characterizations are mentioned, such as XPS, Ramon spectroscopy, NMR, UV-visible, Boehm's titration, and XRF. XPS is a spectroscopy technique that quantifies metal composition present in specimens at parts per trillion level and calculates the binding energy from a core region of elements. However, the adsorption process is essential as an interaction of adsorbate-adsorbent [78]. UV-visible analysis needs to know about surface plasmon resonance (SPR) for NPs. SPR is actually a resonant oscillation during the interface between negative and positive electrical conductivity of the material through the incident light. Raman spectroscopy detects vibrations involving a change in polarizability [79].

Dynamic light scattering (DLS) and zeta potential are actually one of characterizations accordingly based on colloidal, nanoparticulate, and macromolecular. It is necessarily applied to what is known about the size of the NP; the particle zeta potential relates to the magnetic and electrical charge on surface, and molecular weight of polymeric dispersed of complex fluids. It is based on the phenomenon of inelastic light scattering known about structural elucidation, which polymorphs quantification. Boehm's titration analyzed surface functional groups (CSF) as well as those present in carboxylic group, phenolic groups, and hydroxyl groups, which are calculated by the following equation. Herein, the description of V_{HCl}, V_{NaOH}, and V_B are depicted as hydrochloric acid, sodium hydroxide, and the reaction base added to each sample, respectively. V_a is the volume used from V_B, $\frac{n_{HCl}}{n_B}$ is the molar ratio of n_{HCl} and n_B, n_{CSF} is represented as moles for CSF, m_C is the amount of adsorbents and respective CSF as denoted η_{CSF} (mol/g):

$$n_{CSF} = \frac{n_{HCl}}{n_B}[B]V_B - ([HCl]V_{HCl} - [NaOH]V_{NaOH})\frac{V_B}{V_a} \tag{1}$$

$$\eta_{CSF} = \frac{n_{CSF}}{m_C} \tag{2}$$

The specific neutralization of different functional groups against titration are as base $[n_{CSF}(NaHCO_3) - n_{CSF}(Na_2CO_3) = \text{lactonic groups}]$, $[n_{CSF}(Na_2CO_3) - n_{CSF}(NaOH) = \text{phenolic groups}]$, and $[n_{CSF}(NaHCO_3) - n_{CSF}(NaOH) = \text{carboxylic groups}]$, respectively.

X-ray fluorescence analyzed different metal elements as quantities and qualities. Electrical conductivity (EC) estimated adsorbent conductance as an aqueous form, where solid and liquid of polar interacted. Thermogravimetric analysis is a technique of precursor mass that changes temperature with time. It can give information about the likely physical attributes as adsorption-desorption, different phase transition, and chemisorption with thermal decomposition are also provided, if adsorbent is oxidized to formation NP to obtained and how much oxidized there [80]. Zeta potential quantified the electrokinetic colloidal dispersion of stationary and mobile phase. To estimate the surface charge of a nanomaterial using electrophoretic mobility of nanocarriers, NPs are very important as charge distributors of aggregation of surface or repulsion of surface. Proton adsorption capacity ($Q_{protons}$) are calculated by Eq. (3), where ($V_{EL(o)} - V_{EL(ti)}$) is the difference between initial and final electrolyte (KCL or NaCl) with $m_{adsorbent}$ (in g); i and e are denoted as initial and equilibrium concentration, respectively. In this parameter, protons conform uptake in mg/g of the adsorbent.

$$Q_{protons} = \frac{(V_{EL(o)} - V_{EL(ti)})}{m_{adsorbent}}([H]_i - [OH]_i - [H]_e + [OH]_e) \tag{3}$$

Brunauer-Emmett-Teller (BET) surface area is specified on the surface by N_2 adsorption-desorption, which is directly proportional to the mesoporous or microporous materials [81].

2.3 Influence the adsorption factors

Adsorption factors such as initial concentration, pH, contact time, dose, temperature, and agitation have been investigated, which deal with the equilibrium and mechanism of the process. Initial concentration describes how much uptake occurs with a single dose of adsorbent. If the above-the-surface binding of available sites is exceeded, the process gradually turns and interacts with fewer adsorbates. The adsorption is enhanced with increasing concentration because a higher driving force for mass transfer become increased with concentration. Contact time is performed to describe the kinetics rate. Whether the kinetics are faster or slower are mentioned, and equilibrium overcomes then releases adsorbates, because if the multilayer adsorption of their molecular binding ability is poor, then it switches to monolayer. Temperature is an important parameter for performing the adsorption mechanism, as adsorption energy changes while uptake performance runs. This parameter can derive the adsorption nature, either it is exothermic or endothermic. If the adsorption percentage increases with asserting temperature, the mechanism of the adsorbate-adsorbent interaction is endothermic, and inversely if it shows decreasing temperature in terms of adsorption, it is called exothermic sorption [82]. When temperature drives the adsorption attributes, more activation sites are formed; on the other hand, adsorption decreases by temperature, a fact that makes the repulsive forces act as barriers and easily eliminates the adsorbate [83].

Different precursor's importance was overlooked as minimum doses with maximum amount of output efficiency. Adsorbate and adsorbent interactions run until the adsorbents are saturated by adsorbate molecules. The adsorption does not respond while adsorbents increase due to the adsorbate molecules reaching maximum uptake. Uptake by the adsorbent also depends on equilibrium time; after equilibrium, adsorbent releases adsorbate molecules, and interactions become weak forces [80,84]. pH can deal with surface phenomena, and glyphosate could be attributed of interactive functional groups such as phosphate, and amino and carboxylic acids, and as these groups are protonated or deprotonated, it influences their *p*Ka values [15,74]. Similarly, in wood char, which is protonated as the ZPC (here, ZPC > pH), the glyphosate negative ions influences wood char to act as surface complexion [16].

2.4 Adsorption kinetics

During adsorption, the kinetic phenomena in different adsorbents occur, and the controlling interaction between adsorbent and adsorbate have possible mechanisms, such as chemical reaction phenomena, internal or external diffusion mechanisms, and surfactant nature determined through kinetic models. In general, kinetics models are employed as (1) pseudo-first order kinetics, (2) pseudo-second order kinetics, (3) intraparticle diffusion kinetics, and (4) Boyd model for external mass transfer or film diffusion mechanism.

2.4.1 Pseudo-first order kinetics model

This model is in a linear form, and it is generally expressed through Eqs. (4)–(6). The corresponding differential equation [85,86] is written as:

$$\frac{dQ_t}{dt} = K_{1st}(Q_e - Q_t) \tag{4}$$

Integrating term on applying condition as $t = (0$ to $t)$ and $Q_t = (0$ to $Q_t)$:

$$\int_{t=t}^{t=0} \frac{dQ_t}{dt} = \int_{Q_{t=Q_t}}^{Q_{t=0}} k_s(Q_s - Q_t) \tag{5}$$

Eq. (5) is reformed as:

$$\ln(Q_e - Q_t) = \ln Q_e - K_{1st}t \tag{6}$$

After respective generally linear forms, plotting against $\ln(Q_e - Q_t)$ vs. t, and K_{1st} and Q_e were predicted according to slope and intercept. Applying the mechanism determined glyphosate adsorption rate in minutes. It was observed that interaction of adsorption rate depends on possible equilibrium, at nonlinear fitted to deactivate the linear form. However, the adsorption mechanism derived the chemisorption mechanism. If the possible interaction is physisorptive, it can alias the linear fitted [87].

2.4.2 Pseudo-second order kinetics model

In methods developed for the pseudo-second order kinetics model, the adsorption process is described as rate limiting or chemisorption mechanism. Ho and McKay [88] introduced the following equations:

$$\frac{dQ_t}{dt} = K_{2nd}(Q_s - Q_t)^2 \tag{7}$$

And integrating the following condition at time variable starting at $t = 0$, $Q_t = 0$ extending limit t and Q_t, respectively:

$$\int_{t=t_e}^{t=0} \frac{dQ_t}{dt} = \int_{Qe=t_e}^{Q_{t=0}} K_{2nd}(Q_s - Q_t)^2 \tag{8}$$

Occupying the nonlinear pseudo-second order kinetics $Q_t = K_{2nd}Q_e^2 t/(1 + K_{2nd}Q_e t)$.

$$\text{At linear form as } \frac{t}{Q_t} = \frac{1}{K_{2nd}Q_e^2} + \frac{1}{Q_e}t \tag{9}$$

In Eq. (8), K_{2nd} (g/(mg min)) is the pseudo-second order rate constant, and Q_e and Q_t (mg/g) are the values of the amount adsorbed per unit mass at equilibrium and at any time t, respectively.

2.4.3 Elovich kinetics model

Elovich kinetics model [89], often used in adsorption surface interacting of chemisorption, satisfies the equation derived as:

$$\frac{dQ_t}{dt} = a\exp(-bQ_t) \tag{10}$$

After integrating a form where t value at $t_o \rightarrow t$ and Q_t value point in saturate adsorbent capacity (mg/g):

$$Q_t = \frac{1}{b}\ln ab + \frac{1}{b}\ln(t + t_0) \tag{11}$$

To evaluate kinetic surface interference with adsorbent/soil plot linearly against Q_t vs. $\ln(t + t_0)$ and a and b, calculate slope and intercept; b is the surface coverage and energy to interact by the chemisorption process.

2.4.4 Intraparticle diffusion

To understand intraparticle diffusion mechanism of kinetics, this model basically shows transport of aqueous solution according to the Weber and Morris [86] equation:

$$Q_t = K_{id}t^{1/2} + C \tag{12}$$

The model calculates from linear plot against of Q_t vs. $t^{1/2}$ is the interparticle rate constant (mg L^{-1} min$^{-1/2}$), and C is the thickness of boundary layer. Intraparticle diffusion model are described as multilinearity of a two-step adsorption process; their diffusion effect on the boundary layer in external resistance interacts with the particle's surface. The intraparticle diffusion model can introduce the process, whether there are diffusion controls or not. Herein only calculating through a graph, if intercept alias is zero, it means $C = 0$, where adsorption is controlled by diffusion of intraparticles. When $C \neq 0$, adsorption interaction not only controls diffusion but also controls chemical or physical phenomena that their complex mechanism initiates [74,90].

2.5 Adsorption isotherm

Adsorption isotherm is an empirical model to describe adsorption equilibrium. To understand the adsorption mechanism above the quantified distribution of adsorbates, the consistency of the theoretical assumption must be verified by a different model. Concomitant as computation is elaborated as following the models as studied in the previous literature, such as Langmuir isotherm, Freundlich isotherm, Dubinin-Radushkevich isotherm, and Tempkin isotherm.

2.5.1 Langmuir isotherm

Langmuir isotherm [91] is described as surface coverage about adsorptive capacity by monolayer. Langmuir equation is essentially characterized as a dimensionless constant named as a separating factor, which is expressed as R_L (Fig. 9). Dimensionless separating

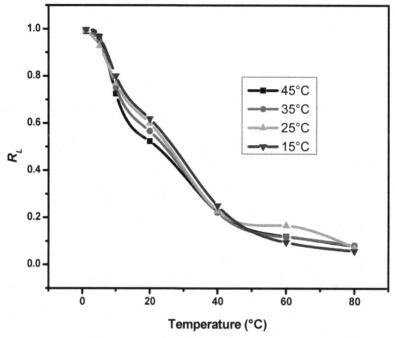

FIG. 9 Glyphosate-GO-α-γ-Fe$_2$O$_3$ composite R$_L$ plot at different temperature [74].

constant indicates whether the adsorption process is favorable or not. $R_L \geq 1$ indicates unfavorable adsorption; if $0 > R_L < 1$, the nature of adsorption is favorable, and irreversible when $R_L = 0$. Uptake capacity of different adsorbents was determined by equilibrium phenomena as monolayer with homogeneous attributes [34]. As shown in previous literature, regarding glyphosate adsorption, GO-α-γ-Fe$_2$O$_3$ adsorbent is well-described by the R_L, R^2 showing maximum adsorption fitted by Langmuir isotherm, and mentioned the adsorption capacity on 46.844 mg/g at 15°C, R_L factors are below 1 and tends to decrease order at increasing temperatures [74]. Possible interaction of the adsorption, chemisorption behavior is shown at lower temperature; when temperature increase, glyphosate molecules are released, and the surface nature of adsorbent shows multilayer and physisorption phenomena.

2.5.2 Freundlich isotherm
Freundlich isotherm is defined as heterogeneity with exponential distribution of binding sites at different energies. Linear plot was calculated with the Freundlich affinity (K_F), where n is the Freundlich exponent. When $1/n$ value always higher than 1, it means adsorption is favorable. Its characteristics revealed the multilayer adsorption, with a linear plot fitted with high correlated value (R^2). The glyphosate adsorption is described using biochar of adsorbent interaction through heterogeneous and amorphous surfaces; and $n < 1$ (0.406) [15] indicates the favored adsorption and heterogeneous surface biochar.

2.5.3 Tempkin isotherm

Temkin isotherm model [92] elaborates the interference solid/liquid interaction by the adsorption process, where all molecular layers take the binding energy (B_T), which are derived as chemisorption and physisorption behaviors. However, the heat adsorption is estimated by A_T (L/mg), which probably occurs from a linear plot. This model indirectly interacted in the adsorption process between adsorbate/adsorbate and assumed the layer decreases of linear ways by increasing the surface coverage [15,35,93].

2.5.4 Dubinin-Radushkevich isotherm

Dubinin-Radushkevich isotherm model [94] is an empirical adsorption model; its mechanism is expressed by Gaussian energy distribution into free energy to compute the Polanyi potential (ε). E represents the free energy involved in the uptake during adsorption mechanism D-R capacity. $E = \frac{1}{\sqrt{2B}}$ and $\varepsilon = RT \ln\left(1 + \frac{1}{C_e}\right)$ are computed from the D-R equation. Indeed, glyphosate adsorption nature is shown as a specified uptake by monolayer with respective free energy as 87.1737 mg/g and 3.9×10^{-6} KJ/mol [15,95,96] (Table 3).

2.5.5 Empirical multivariate isotherm model

Other models are well established in mechanical paths and verified by the theoretical assumptions of the linear and nonlinear isotherm models, such as the "one-parameter" isotherm described by Henry's model ($Q_e = K_{he}C_e$, $K_{he} =$ Constant). Equilibrium can described while in low concentration. And in a lateral adsorption study of the Hill-Deboer model [99], lateral energy with interaction for Fowler-Guggenheim model, Radke-Prausniiz isotherm describes the nonlinear model with low concentration adsorbate and heterogeneity interaction [98]. Four-parameter and five-parameter isotherms described broad range of prediction, and large adequacy was predicted. Reviewers described spreading energy with binding molecules with surface sites occupied. A five-parameter isotherm model such as Fritz-Schlunder [98] precisely describes a wide range of equilibrium by the following nonlinear equation:

$$Q_e = \frac{L_m K_1 C_e^{\alpha FS}}{1 + K_2 C_e^{\beta FS}} \tag{17}$$

where Q_e is maximum capacity (mg/g), and K_1, K_2, α_{FS}, and β_{FS} followed the Fritz-Schlunder parameters. Validity tested of this isotherm as L_m, alias this model if $L_m \leq 1$, Langmuir model are exponent $\alpha_{FS} = \beta_{FS}$, Freundlich model is taken to reduce this parameter.

2.6 Adsorption thermodynamics

During the adsorption process, the thermodynamics elaborate the tendency of the molecular concentration to interact with the bulk of the adsorbent until it reaches equilibrium, is in the following:

$$\text{Adsorbent}_{(solid)} + \text{Adsorbate}_{(aq)} \underset{\text{desorption}}{\overset{\text{adsorption}}{\rightleftharpoons}} \text{Adsorbent} - \text{Adsorbate}_{(solid)}$$

Table 3   Isotherm modeling of glyphosate adsorption with possible assumption.

Isotherm model	Linear equation	Equation no.	Plot	Description	Assumption	References
Langmuir isotherm	$C_e/Q_e = 1/Q_m K_L +$ C_e/Q_m $R_L = 1/(1 + K_L C_o)$	(13)	C_e/Q_e vs. C_e	R_L = Separation factor, K_L = Langmuir constant, Q_e = adsorption capacity at equilibrium condition (mg/g)	Express the adsorption phenomena on monolayer, Estimated the unfavorable, favorable and irreversible, Surface homogeneity phenomena	[16]
Freundlich isotherm	$LogQ_e = LogK_F +$ $1/nLogC_e$	(14)	$LogC_e$ vs. $LogQ_e$	$K_F = ((\text{mg g}^{-1})/(\text{mg L}^{-1})^n)e$ the Freundlich affinity of capacity parameter, $1/n$ = the Freundlich exponent	Multilayer adsorption, Relative Energy distribution with heterogeneity, Equilibrium capacity	[97]
Tempkin isotherm	$Q_e = \frac{RT}{b} \ln A_T + \frac{RT}{b} \ln C_e$	(15)	Q_e vs. $\ln C_e$	B_T (binding energy) constant related to heat of sorption (J/mol)	Effects of indirect adsorbate/ adsorbate interactions, Linear layer decreasing during increasing surface coverage, while heat adsorption all molecules. Chemisorption mechanism.	[93]
Dubinin-Radushkevich isotherm	$\ln Q_e = \ln Q_m + \beta E^2$	(16)	$\ln Q_e$ vs. E^2	ε = Polanyi potential, β = D-R constant, E = free energy (KJ/mol)	Physisorption mechanism, attraction largely due to van der Waals forces	[98]

From the literature, the calculation changes in Gibbs free energy (ΔG), changes of enthalpy ($\Delta H°$), and changes in the entropy ($\Delta S°$) by Van de Hoff's equation, as follows:

$$\Delta G^0 = -RT \ln(K_c, K_L, K_e, ...) \tag{18}$$

where R and T are denoted as the universal gas constant, and temperature (K) and K_e are the thermodynamic equilibrium constants [72,100]. The review shown as K_c, is the ratio of concentration of adsorbate and equilibrium concentration of adsorbate; some papers follow the K_e described as the ratio of equilibrium capacity ($Q_{max} = mg/g$) and equilibrium concentration ($C_e = mg/L$). It has the correct description as to where the equilibrium constant is employed to describe the thermodynamic parameters [100]. From the equilibrium of adsorption obtained from Langmuir isotherm, dimensionless constant of K_L as the right way to describe thermodynamics study. As glyphosate adsorption mentions this way is well established on behalf of the thermodynamics constant K_L [72].

$$\Delta G^0 = \Delta H^0 + T\Delta S^0 \tag{19}$$

$$\ln(K_c, K_L, K_e, K_d, ...) = -\frac{\Delta H^o}{RT} + \frac{\Delta S^o}{R} \tag{20}$$

Assume the following: If Gibbs energy is zero, it describes the direction of adsorption, but also it derives the adsorption spontaneity. Then when Gibbs free energy is greater than zero ($\Delta G° > 0$), the process is nonfeasible and nonspontaneous. An explanation of this reaction is that the adsorption process is exothermic ($\Delta H° = -$negative). The interacting phenomena also show disorder; although $\Delta S°$ positive value occurs, the reaction ends spontaneously. If the adsorption takes place in an exothermic reaction, phenomena show a decreasing order of the solid-liquid interaction. The process is spontaneous ($\Delta G° < 0$) as $\Delta H° > T\Delta S°$. The enthalpy is positive as shown by increasing disorder at the solid-liquid interface. Whether $\Delta S°$ is negative or not, the process is spontaneous ($\Delta G° < 0$). On the basis of glyphosate, GO-α-γ-Fe$_2$O$_3$ nanocomposite as an exothermic nature decreases randomness with spontaneous nature; whereas glyphosate interaction with GO-α-γ-Fe$_2$O$_3$, possible binding is covalent as well as chemisorption behavior. According to the literature, glyphosate adsorption describes the adsorbent of resign D301, acts as a commercial adsorbent. The process has held physical and chemical processes together, and their supporting evidence is provided as the $\Delta H°$ and $\Delta G°$ values, followed by $\Delta H°$ as negative pursuing -92.64 KJ/Mol, with increasing temperature as well as increased free energy. It means $\Delta H°$ value mentions the physical adsorption because of the negative value, but the free energy is a positive value, which are chemisorption behaviors [90]. Conversely, the $\Delta S°$ nature of adsorbent as degree of freedom decreases, a possible phenomenon described by some structural change of the adsorbate as well. The exothermic nature clearly indicates, in this report, that the main driving forces integrate with the nonspontaneity of the chemical driving forces.

3 Comparative study of glyphosate in recent published paper

A comparative study of glyphosate adsorption in a recently published paper is shown in Table 4. Glyphosate adsorption by a bionanocomposite is shown, here specifically elaborating the palm biochar, zero valent iron, palm biochar @ zero valent iron, respectively, as Q_{max} value 40.29 mg/g, 35.29 mg/g, and 80.00 mg/g [34]. Afterward, adsorbents are basically integrated at pH 4.0, 2.0, and 4.0; however, respective adsorption showed low pH and basically phenomena anion attraction. Table 4 shows the maximum adsorption capacity on resin D301 (833.33 mg/g) at pH 4. From bioadsorbents such as woody (dendro) biochar, palm biochar, and rice husk-derived biochar are described as having maximum adsorption that are easily prepared. Herewith, all adsorptions are prepared as composites, maximum utilized toward nanobiocomposites, therefore focused on glyphosate adsorption efficacy. All composite precursors have a major demand as plant substances, highlighted as $MnFe_2O_4$ with cellulose-activated, carbon-magnetic hybrid, and palm biochar with nano zero valent iron acting as a waste material, which show significant uptake capacity that follows as 80.0 mg/g and 93.48 mg/g, respectively. Emphasis is placed on nanomaterials, waste materials, soils, resins, composites, and char/activated carbon compared with glyphosate adsorption; pH 4 to 6 greatly affects the surface charge and favorable absorption at normal temperature, which was divided as 278 K to 320 K. All reports therefore showed an electronic connection as the main mechanism. In addition to all the conclusions, in a majority of the NPs, there was a lack of a field for previous reviewer comments such as the bio-bio combination, which were not examined, nor emphasized on multivariant composite materials. There is a large field of study of natural materials such as biopolymer-based chitosan, which was not used in the previous context.

4 Possible mechanism of adsorption

A few possible mechanisms during adsorption are the mechanisms described as pore filling, hydrophobic interaction, and van der Waal electrostatic interaction during the process. Likewise, these possibilities are considered as interactions through adsorption; when intermolecular abilities are present, rate limiting occurred. Chemisorption behaviors make a stronger bond by electron sharing, specifically described and achieved as a function of calciferous, amides, metal complexion. The main driving forces of intermolecular interaction is how to interact with hydrogen bonds, π-interaction, and dipole-dipole [15,16,36]. The enthalpy change is higher to the possible mechanism as chemisorption, and they are involved in the attachment through covalent bonds. The covalent bonds are actual irreversible interactions, which are held by ion exchanges. Generally, the adsorption process postulates that, when the involving energy is higher (>50 kJ/mol), the process is called chemisorption. Chemisorption process corresponds with monolayer adsorption, as the system takes place during monolayer adsorption alternatively introduced by Langmuir isotherm (Fig. 10).

Table 4 Comparative study about glyphosate adsorption by different adsorbents.

Adsorbents	Qmax (mg/g)	Parametric condition	Reference
Palm biochar (BC)	40.49	pH 4, T 298 K, equilibrium time 1600 min, initial glyphosate 0.5–100 mg/L, solid/solution 0.015 g/25 mL	[34]
Forest soil	169.29	pH 12, T 303 K, initial concentration 5–40 mg/L, dose 1 g/50 mL	[37]
Nano-zero-valent iron (NZVI)	35.39	pH 2,T 298 K, equilibrium time 360 min, initial glyphosate 0.5–100 mg/L, solid/solution 0.015 g/25 mL	[34]
Palm biochar @ Nano zero valent iron	80.00	pH 4,T 298 K, equilibrium time 1600 min, initial glyphosate 0.5–100 mg/L, solid/solution 0.015 g/25 mL	[34]
Woody (dendro) biochar	44.00	pH 5,T 298 K, equilibrium time 240 min, initial glyphosate 5–100 mg/L, solid/solution 1 g/L	[16]
Soils	57–207	Solution 10 g/100 ml	
Forest soil	161.29	pH = 2 to 14, dose 0.01 to 2 g, initial concentration 5–40 mg/L, 5–120 min, 290 to 373 K	[37]
Magnetic $MnFe_2O_4$-graphene	39.00	pH –,T 278 K, equilibrium time 480 min, initial glyphosate 5–80 mg/L, solid/solution 0.08 g/80 mL	[36]
Mg@Al-layered double hydroxides	27.4–184.6	pH 5.6–13.1, T 298 K, equilibrium time 24 h	[5]
Biopolymer membranes	10.88	pH 6.5,T 298 K, equilibrium time 240 min, initial glyphosate 5–35 mg/L, solid/solution 0.0004 g/50 mL	[101]
Water treatment residual (alum sludge)	85.90	pH 5.2,T 295 K, equilibrium time 52 h, initial glyphosate 50–100 mg/L, solid/solution 0.5 g/100 mL	[102]
Dendrimer grafted adsorbent	14.0404	pH 3, $K_F = 3.1092$, Gibbs energy = negative, removal% = 96.87, $\Delta H = -13.88$ KJ/Mol, $\Delta S = -31.28$(J/mol K^{-1}), concentration 5 mg/L to 100 mg/L	[87]
Graphene oxide (GO) functionalized by magnetic nanoparticles of iron oxide (α-γ-Fe_2O_3)	46.844	pH 4 to 10, Concentration 0.5 to 3 mg/L, 0–24 h, $\Delta H = -37.65$ KJ/Mol, $\Delta S = -0.0100$, removal 92%.	[74]
Polysulfone membranes mixed by graphene oxide/ TiO_2 nanocomposite	26.59	4.5 pH, 7 days contact time, 25°C temperature, 61% removal.	[103]
Rice husk derived engineered biochar	123.03	Initial concentration 5 to 60 mg/L, pH = 2 to 12, contact time 5 to 270 min.	[104]
Nanosized copper hydroxide modified resin	113.7	3 to 12 (0 to 500 min, 0–80°C temperatures 288–318	[105]
Zr-MOF based smart adsorbent	256.54 mg/g	0–50 mg/L initial concentration, pH 2–12(4 pH), 0–25 min	[106]
$MnFe_2O_4$@cellulose activated carbon magnetic hybrid	93.48 mg/g	pH 2–12, initial concentration 5–200 mg/L 0–1200 min, 288, 298 and 308 K	[79]
Resin D301	833.33	303.15–318.15 K, 5–50 mg/L, pH 4	[90]

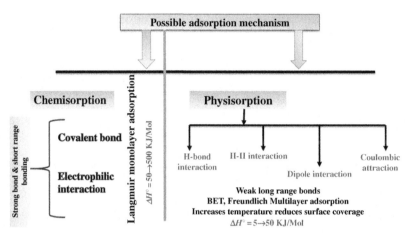

FIG. 10 Classification of adsorption mechanisms.

Hydrogen bonding interaction is a dipole interaction mechanism where H-bonds are bonded with acceptor sites of different atoms, such as nitrogen (N), oxygen (O), carbon (C), phosphate, etc. within a functional group's π-system electron enrichment [107]. In electrostatic interactions of adsorption process during adsorption tended by columbic forces, there are two charged moieties between the adsorbate and adsorbent. Glyphosate has a major role as introduced in columbic interaction and/or electrostatic interactions of phase transition as different pH. Maximum adsorption is shown in pH 4.0–6.0; if the process is isoelectric, the point below the pH influences the electron-sharing due to electrostatic interaction through proton dissociation. The π-interaction is weaker than H-bonds, as it helps to interact with polar molecules and aromatic compounds or C=C interaction. The π-π interactions are mainly obtained at low temperature when their physisorption mechanism has interacted. Basically, the pseudo-first order mechanism postulates as well, which means the adsorbed molecular interaction is obtained by π-π interactions [107]. Hydrophobic interaction shows nonspecific attachment of the adsorbates of different sites of adsorbent, primarily driven by entropy. Although the primary driven forces are through hydrophobic interaction, adsorbents interact with their nonpolar molecules. The π-π interactions are basically understood through FTIR studies (function groups interaction shown from different stretching peaks), where they involved functional groups of the organic substances. The adsorption mechanisms of total free energy are quantified by thermodynamics study. The rate limiting study for the adsorption occurring from pseudo-first order and pseudo-second order mechanisms conducting shared electron pairs through covalent interaction. The monolayer interaction and multilayer interaction are other phenomena that may primarily interact by monolayer, where contact time increases, as shown in multilayer phenomena [74]. This has become more contradictory from empirical research, but some hypotheses are likely. When adsorption capacity from the pseudo-first order results, obtained at faster interaction, phenomena such as the monolayer acquisition is the highest involvement in

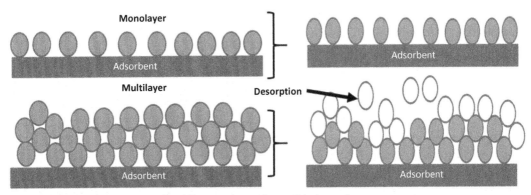

FIG. 11 Adsorption mechanisms of multilayer and monolayer with desorption.

uptake. Further desorption mechanisms situated at maximum in the second order show that physisorption easily bonded the adsorbate molecules. However, chemisorption actually relates to monolayer capacity during Langmuir with first order kinetic involvement [36,108]. The weak force molecules easily dissociate while recovery is achieved, because the molecular bonding capacity confirms that possible interaction "π-π," van der Waals forces, and hydrophobic [16,109] (Fig. 11).

5 Lack of area of glyphosate adsorption

Previous literature has mentioned exploitation by various adsorbents and chemicals used by many researchers, but not natural materials. The natural ingredients of green production are always preferred as a major benefit for research fields. Basically, the adsorption study for biological composites are almost completed. And modifying composites like clay or geominerals will present major demands. Herein, adsorption studies are described as a major contribution of isotherm, as previous literature has described the equilibrium by the Langmuir and Freundlich but have not yet used the multiparameter model. Multiparameter isotherm models also sufficiently describe the equilibrium requiring this type of isotherm [98]. The use of water after absorption has not been shown regarding glyphosate-related adsorption, only in one paper, and there were also water residues used that are less toxic to plant health [19,110]. On the other hand, weeds and crops are killed by glyphosate; therefore, adsorption toxicity tests of glyphosate are essential parameters for adsorption. Agricultural waste material only works as palm shell and rice husk, but current demands for waste material utilization, such as jack peels, bamboo, barks like guava, orange peel, coconut shell barks, wastage polythene, rubber, etc. are lacking in this research. So, investigation of glyphosate adsorption can take at a major demands as low cost, waste minimization, and sustainable. Composite as multimaterial substances is a new gap, which includes bio-nano (woody char-ZnO), nano-nano (Zinc-iron NPs), nano-bio-nano (e.g. ZnO-palm shell-zero valent iron), bio-nano-bio (e.g., rice husk char-zero valent iron-palm shell), bio-bio-bio (rice husk char-orange peel-palm shell). Desorption is informed about glyphosate recovery, which is an interesting part of this study, as it will give detail about the studies, such as how long adsorbents will be used by different desorption agents.

6 Conclusion

The fate of adsorption of glyphosate naturally has insights as to its interaction with adsorbent materials, to easily uptake through adsorption. However, molecular interaction while equilibrium is reached compared as resin D501, shown 833.33 mg/g. Here also we reported on adsorption preparation and its characterizations such as ZPC, proton adsorption, SEM-EDS, FTIR, XPS, XRD, Ramon spectroscopy, etc. The importance for each characterization was described. Adsorption isotherm are derived for mechanisms such as Langmuir for monolayer, Freundlich for multilayer, Temkin for chemisorption, and D-R isotherm for physisorption. Adsorption-influencing parameters have been described such as the importance of pH, dose, contact time, initial concentration, and temperature. Thus, while this chapter reviews glyphosate removal, there is a great demand for easy contaminant removal in a sustainable way even after adsorption. Here, the possible commands are summarized as (1) the adsorbent preparation and surface properties have a major role in adsorption; (2) comparison study between different adsorbents are illustrated; (3) categorizing the adsorbents as to removal for glyphosate; (4) describing the adsorption kinetics, isotherm, and thermodynamics; (5) describing the mechanism as possible interacted phenomena; and (6) discussing renewable and management ways derived for glyphosate adsorption. Furthermore, adsorbents are specifically studied regarding rewound methods as well for lack of areas, other adsorbates are needed to develop more efficacy for future aspects.

References

[1] W. Aktar, D. Sengupta, A. Chowdhury, Impact of pesticides use in agriculture: their benefits and hazards. Interdiscip. Toxicol. 2 (1) (2009) 1–12, https://doi.org/10.2478/v10102-009-0001-7.

[2] M.L. Ortiz-Hernández, E. Sánchez-Salinas, E. Dantán-González, M.L. Castrejón-Godínez, Pesticide biodegradation: mechanisms, genetics and strategies to enhance the process. Biodegrad. Life Sci. (2013), https://doi.org/10.5772/56098.

[3] M. Arias-Estevez, E. Lopez-Periago, E. Martinez-Carballo, J. Simal-Gandara, J.C. Mejuto, L. Garcia-Rio, The mobility and degradation of pesticides in soils and the pollution of groundwater resources, Agric. Ecosyst. Environ. 123 (4) (2008) 247–260.

[4] C.V. Waiman, M.J. Avena, M. Garrido, B. Fernández Band, G.P. Zanini, A simple and rapid spectrophotometric method to quantify the herbicide glyphosate in aqueous media. Application to adsorption isotherms on soils and goethite. Geoderma 170 (2012) 154–158, https://doi.org/10.1016/j.geoderma.2011.11.027.

[5] M. Wang, G. Zhang, G. Qiu, D. Cai, Z. Wu, Degradation of herbicide (glyphosate) using sunlight-sensitive MnO2/C catalyst immediately fabricated by high energy Electron beam. Chem. Eng. J. 306 (2016) 693–703, https://doi.org/10.1016/j.cej.2016.07.063.

[6] P. Rajasulochana, V. Preethy, Comparison on efficiency of various techniques in treatment of waste and sewage water – a comprehensive review. Resour.-Effic. Technol. 2 (4) (2016) 175–184, https://doi.org/10.1016/j.reffit.2016.09.004.

[7] E. Malaj, P.C. von der Ohe, M. Grote, R. Kühne, C.P. Mondy, P. Usseglio-Polatera, W. Brack, R. B. Schäfer, Organic chemicals jeopardize the health of freshwater ecosystems on the continental scale. Proc. Natl. Acad. Sci. 111 (26) (2014) 9549–9554, https://doi.org/10.1073/pnas.1321082111.

[8] W. Queyrel, F. Habets, H. Blanchoud, D. Ripoche, M. Launay, Pesticide fate modeling in soils with the crop model STICS: feasibility for assessment of agricultural practices. Sci. Total Environ. 542 (2016) 787–802, https://doi.org/10.1016/j.scitotenv.2015.10.066.

[9] A. Özkara, D. Akyıl, M. Konuk, Pesticides, environmental pollution, and health. in: Environmental Health Risk: Hazardous Factors to Living Species, IntechOpen, 2016, https://doi.org/10.5772/63094.

[10] J. Popp, K. Pető, J. Nagy, Pesticide productivity and food security. A review. Agron. Sustain. Dev. 33 (1) (2013) 243–255, https://doi.org/10.1007/s13593-012-0105-x.

[11] D.S. Sagatys, C. Dahlgren, G. Smith, R.C. Bott, J.M. White, The complex chemistry of N-(phosphonomethyl)glycine (glyphosate): preparation and characterization of the ammonium, lithium, sodium (4 polymorphs) and silver(I) complexes. J. Chem. Soc. Dalton Trans. 19 (2000) 3404–3410, https://doi.org/10.1039/B002748K.

[12] M. Muneer, C. Boxall, Photocatalyzed degradation of a pesticide derivative glyphosate in aqueous suspensions of titanium dioxide, Int. J. Photoenergy 2008 (2008), https://www.hindawi.com/journals/ijp/2008/197346/ref/. (Accessed 31 December 2019).

[13] D. Heineke, S.J. Franklin, K.N. Raymond, Coordination chemistry of glyphosate: structural and spectroscopic characterization of bis(glyphosate)metal(III) complexes. Inorg. Chem. 33 (11) (1994) 2413–2421, https://doi.org/10.1021/ic00089a017.

[14] V. Tzin, G. Galili, The biosynthetic pathways for shikimate and aromatic amino acids in arabidopsis Thaliana. Arab. Book Am. Soc. Plant Biol. 8 (2010), https://doi.org/10.1199/tab.0132.

[15] G.A. Dissanayake Herath, L.S. Poh, W.J. Ng, Statistical optimization of glyphosate adsorption by biochar and activated carbon with response surface methodology. Chemosphere 227 (2019) 533–540, https://doi.org/10.1016/j.chemosphere.2019.04.078.

[16] S.S. Mayakaduwa, P. Kumarathilaka, I. Herath, M. Ahmad, M. Al-Wabel, Y.S. Ok, A. Usman, A. Abduljabbar, M. Vithanage, Equilibrium and kinetic mechanisms of woody biochar on aqueous glyphosate removal. Chemosphere 144 (2016) 2516–2521, https://doi.org/10.1016/j.chemosphere.2015.07.080.

[17] G.M. Williams, R. Kroes, I.C. Munro, Safety evaluation and risk assessment of the herbicide roundup and its active ingredient, glyphosate, for humans. Regul. Toxicol. Pharmacol. 31 (2) (2000) 117–165, https://doi.org/10.1006/rtph.1999.1371.

[18] IARC, Some Organophosphate Insecticides and Herbicides, (2017).

[19] J.V. Tarazona, D. Court-Marques, M. Tiramani, H. Reich, R. Pfeil, F. Istace, F. Crivellente, Glyphosate toxicity and carcinogenicity: a review of the scientific basis of the European Union assessment and its differences with IARC. Arch. Toxicol. 91 (8) (2017) 2723–2743, https://doi.org/10.1007/s00204-017-1962-5.

[20] E. Schönbrunn, S. Eschenburg, W.A. Shuttleworth, J.V. Schloss, N. Amrhein, J.N.S. Evans, W. Kabsch, Interaction of the herbicide glyphosate with its target enzyme 5-enolpyruvylshikimate 3-phosphate synthase in atomic detail, Proc. Natl. Acad. Sci. U. S. A. 98 (4) (2001) 1376–1380.

[21] H.C. Steinrücken, N. Amrhein, 5-Enolpyruvylshikimate-3-phosphate synthase of *Klebsiella pneumoniae*. Eur. J. Biochem. 143 (2) (1984) 351–357, https://doi.org/10.1111/j.1432-1033.1984.tb08379.x.

[22] S.L.R.K. Kanamarlapudi, V. Kumar Chintalpudi, S. Muddada, Application of biosorption for removal of heavy metals from wastewater. in: Biosorption, IntechOpen, 2018, https://doi.org/10.5772/intechopen.77315.

[23] K.C. Khulbe, T. Matsuura, Removal of heavy metals and pollutants by membrane adsorption techniques. Appl Water Sci 8 (1) (2018) 19, https://doi.org/10.1007/s13201-018-0661-6.

[24] S. Moran, Chapter 7: Clean water unit operation design: physical processes. in: S. Moran (Ed.), An Applied Guide to Water and Effluent Treatment Plant Design, Butterworth-Heinemann, 2018, pp. 69–100, https://doi.org/10.1016/B978-0-12-811309-7.00007-2.

[25] S. Moran, Chapter 4: Engineering science of water treatment unit operations. in: S. Moran (Ed.), An Applied Guide to Water and Effluent Treatment Plant Design, Butterworth-Heinemann, 2018, pp. 39–51, https://doi.org/10.1016/B978-0-12-811309-7.00004-7.

[26] B. Van der Bruggen, Chapter 7: Ion-exchange membrane systems—electrodialysis and other electromembrane processes. in: P. Luis (Ed.), Fundamental Modelling of Membrane Systems, Elsevier, 2018, pp. 251–300, https://doi.org/10.1016/B978-0-12-813483-2.00007-1.

[27] A.E. González, Colloidal aggregation coupled with sedimentation: a comprehensive overview. in: Advances in Colloid Science, IntechOpen, 2016. https://doi.org/10.5772/65699.

[28] G. Fu, Y. Chen, R. Li, X. Yuan, C. Liu, B. Li, Y. Wan, Pathway and rate-limiting step of glyphosate degradation by *Aspergillus oryzae* A-F02. Prep. Biochem. Biotechnol. 47 (8) (2017) 782–788, https://doi.org/10.1080/10826068.2017.1342260.

[29] T. Long, Y. Xu, X. Lv, J. Ran, S. Yang, L. Xu, Fabrication of the annular photocatalytic reactor using large-sized freestanding titania-silica monolithic aerogel as the catalyst for degradation of glyphosate. Mater. Des. 159 (2018) 195–200, https://doi.org/10.1016/j.matdes.2018.08.047.

[30] S. Chattoraj, N.K. Mondal, B. Sadhukhan, P. Roy, T.K. Roy, Optimization of adsorption parameters for removal of carbaryl insecticide using neem bark dust by response surface methodology. Water Conserv. Sci. Eng. 1 (2) (2016) 127–141, https://doi.org/10.1007/s41101-016-0008-9.

[31] S.K. Deokar, D. Singh, S. Modak, S.A. Mandavgane, B.D. Kulkarni, Adsorptive removal of diuron on biomass ashes: a comparative study using rice husk ash and bagasse fly ash as adsorbents. Desalin. Water Treat. 57 (47) (2016) 22378–22391, https://doi.org/10.1080/19443994.2015.1132394.

[32] A.F. El-Aswad, M.I. Aly, M.R. Fouad, M.E.I. Badawy, Adsorption and thermodynamic parameters of chlorantraniliprole and dinotefuran on clay loam soil with difference in particle size and PH. J. Environ. Sci. Health Part B 54 (6) (2019) 475–488, https://doi.org/10.1080/03601234.2019.1595893.

[33] J.M. Salman, Batch study for insecticide carbofuran adsorption onto palm-oil-fronds-activated carbon, J. Chem. 2013 (2013), https://www.hindawi.com/journals/jchem/2013/630371/. (Accessed 31 December 2019).

[34] X. Jiang, Z. Ouyang, Z. Zhang, C. Yang, X. Li, Z. Dang, P. Wu, Mechanism of glyphosate removal by biochar supported nano-zero-valent iron in aqueous solutions. Colloids Surf. Physicochem. Eng. Asp. 547 (2018) 64–72, https://doi.org/10.1016/j.colsurfa.2018.03.041.

[35] K. Sen, J.K. Datta, N.K. Mondal, Glyphosate adsorption by eucalyptus camaldulensis bark-mediated char and optimization through response surface modeling. Appl Water Sci 9 (7) (2019) 162, https://doi.org/10.1007/s13201-019-1036-3.

[36] N. Ueda Yamaguchi, R. Bergamasco, S. Hamoudi, Magnetic $MnFe_2O_4$–graphene hybrid composite for efficient removal of glyphosate from water. Chem. Eng. J. 295 (2016) 391–402, https://doi.org/10.1016/j.cej.2016.03.051.

[37] K. Sen, N.K. Mondal, S. Chattoraj, J.K. Datta, Statistical optimization study of adsorption parameters for the removal of glyphosate on Forest Soil using the response surface methodology. Environ. Earth Sci. 76 (1) (2016) 22, https://doi.org/10.1007/s12665-016-6333-7.

[38] A.R. Dinçer, Y. Güneş, N. Karakaya, Coal-based bottom ash (CBBA) waste material as adsorbent for removal of textile dyestuffs from aqueous solution. J. Hazard. Mater. 141 (3) (2007) 529–535, https://doi.org/10.1016/j.jhazmat.2006.07.064.

[39] V.K. Gupta, B. Gupta, A. Rastogi, S. Agarwal, A. Nayak, Pesticides removal from waste water by activated carbon prepared from waste rubber tire. Water Res. 45 (13) (2011) 4047–4055, https://doi.org/10.1016/j.watres.2011.05.016.

[40] Y. Dai, Q. Sun, W. Wang, L. Lu, M. Liu, J. Li, S. Yang, Y. Sun, K. Zhang, J. Xu, et al., Utilizations of agricultural waste as adsorbent for the removal of contaminants: a review. Chemosphere 211 (2018) 235–253, https://doi.org/10.1016/j.chemosphere.2018.06.179.

[41] S. De Gisi, G. Lofrano, M. Grassi, M. Notarnicola, Characteristics and adsorption capacities of low-cost sorbents for wastewater treatment: a review. Sustain. Mater. Technol. 9 (2016) 10–40, https://doi.org/10.1016/j.susmat.2016.06.002.

[42] S. Jain, R.V. Jayaram, Adsorption of phenol and substituted chlorophenols from aqueous solution by activated carbon prepared from jackfruit (*Artocarpus heterophyllus*) peel-kinetics and equilibrium studies. Sep. Sci. Technol. 42 (9) (2007) 2019–2032, https://doi.org/10.1080/15275920701313608.

[43] S. Wang, B. Seiwert, M. Kästner, A. Miltner, A. Schäffer, T. Reemtsma, Q. Yang, K.M. Nowak, (Bio)degradation of glyphosate in water-sediment microcosms – a stable isotope co-labeling approach. Water Res. 99 (2016) 91–100, https://doi.org/10.1016/j.watres.2016.04.041.

[44] M. Barathi, A. Santhana Krishna Kumar, N. Rajesh, Efficacy of novel Al–Zr impregnated cellulose adsorbent prepared using microwave irradiation for the facile defluoridation of water. J. Environ. Chem. Eng. 1 (4) (2013) 1325–1335, https://doi.org/10.1016/j.jece.2013.09.026.

[45] A.L. Gimsing, C. Szilas, O.K. Borggaard, Sorption of glyphosate and phosphate by variable-charge tropical soils from Tanzania. Geoderma 138 (1) (2007) 127–132, https://doi.org/10.1016/j.geoderma.2006.11.001.

[46] L. Mamy, E. Barriuso, Glyphosate adsorption in soils compared to herbicides replaced with the introduction of glyphosate resistant crops. Chemosphere 61 (6) (2005) 844–855, https://doi.org/10.1016/j.chemosphere.2005.04.051.

[47] Chapter Four: Soil system. A.M.O. Mohamed, H.E. Antia (Eds.), Developments in Geotechnical Engineering, In: Geoenvironmental Engineering, vol. 82, Elsevier, 1998, pp. 59–83, https://doi.org/10.1016/S0165-1250(98)80027-8.

[48] G.A. Grant, P.R. Fisher, J.E. Barrett, P.C. Wilson, Removal of agrichemicals from water using granular activated carbon filtration. Water Air Soil Pollut. 230 (1) (2019) 1–12, https://doi.org/10.1007/s11270-018-4056-y.

[49] S. Rio, L. Le Coq, C. Faur, D. Lecomte, P. Le Cloirec, Preparation of adsorbents from sewage sludge by steam activation for industrial emission treatment. Process. Saf. Environ. Prot. 84 (4) (2006) 258–264, https://doi.org/10.1205/psep.05161.

[50] Z. Zhang, Z. Ouyang, J. Yang, Y. Liu, C. Yang, Z. Dang, High mineral adsorption of glyphosate versus diethyl phthalate and tetracycline, during visible light photodegradation with goethite and oxalate. Environ. Chem. Lett. 17 (3) (2019) 1421–1428, https://doi.org/10.1007/s10311-019-00877-x.

[51] J. Alvarez, G. Lopez, M. Amutio, J. Bilbao, M. Olazar, Physical activation of rice husk pyrolysis char for the production of high surface area activated carbons. Ind. Eng. Chem. Res. 54 (29) (2015) 7241–7250, https://doi.org/10.1021/acs.iecr.5b01589.

[52] E.M. Villota, H. Lei, M. Qian, Z. Yang, S.M.A. Villota, Y. Zhang, G. Yadavalli, Optimizing microwave-assisted pyrolysis of phosphoric acid-activated biomass: impact of concentration on heating rate and carbonization time. ACS Sustain. Chem. Eng. 6 (1) (2018) 1318–1326, https://doi.org/10.1021/acssuschemeng.7b03669.

[53] Y. Li, X. Zhang, R. Yang, G. Li, C. Hu, The role of H_3PO_4 in the preparation of activated carbon from NaOH-treated rice husk residue. RSC Adv. 5 (41) (2015) 32626–32636, https://doi.org/10.1039/C5RA04634C.

[54] A. Zubrik, M. Matik, S. Hredzák, M. Lovás, Z. Danková, M. Kováčová, J. Briančin, Preparation of chemically activated carbon from waste biomass by single-stage and two-stage pyrolysis. J. Clean. Prod. 143 (2017) 643–653, https://doi.org/10.1016/j.jclepro.2016.12.061.

[55] X. Song, Y. Zhang, C. Chang, Novel method for preparing activated carbons with high specific surface area from rice husk. Ind. Eng. Chem. Res. 51 (46) (2012) 15075–15081, https://doi.org/10.1021/ie3012853.

[56] N.B. Azmi, M.J.K. Bashir, S. Sethupathi, L.J. Wei, N.C. Aun, Stabilized landfill leachate treatment by sugarcane bagasse derived activated carbon for removal of color, COD and NH3-N – optimization of preparation conditions by RSM. J. Environ. Chem. Eng. 3 (2) (2015) 1287–1294, https://doi.org/10.1016/j.jece.2014.12.002.

[57] A. Jain, S. Jayaraman, R. Balasubramanian, M.P. Srinivasan, Hydrothermal pre-treatment for mesoporous carbon synthesis: enhancement of chemical activation. J. Mater. Chem. A 2 (2) (2013) 520–528, https://doi.org/10.1039/C3TA12648J.

[58] A. Dzigbor, A. Chimphango, Production and optimization of NaCl-activated carbon from mango seed using response surface methodology. Biomass Convers. Biorefinery 9 (2) (2019) 421–431, https://doi.org/10.1007/s13399-018-0361-3.

[59] B. Janković, N. Manić, V. Dodevski, I. Radović, M. Pijović, Đ. Katnić, G. Tasić, Physico-chemical characterization of carbonized apricot kernel shell as precursor for activated carbon preparation in clean technology utilization. J. Clean. Prod. 236 (2019) 117614, https://doi.org/10.1016/j.jclepro.2019.117614.

[60] C. Chen, J. Hu, D. Shao, J. Li, X. Wang, Adsorption behavior of multiwall carbon nanotube/iron oxide magnetic composites for Ni(II) and Sr(II). J. Hazard. Mater. 164 (2) (2009) 923–928, https://doi.org/10.1016/j.jhazmat.2008.08.089.

[61] Z. Lu, J. Yu, H. Zeng, Q. Liu, Polyamine-modified magnetic graphene oxide nanocomposite for enhanced selenium removal. Sep. Purif. Technol. 183 (2017) 249–257, https://doi.org/10.1016/j.seppur.2017.04.010.

[62] J. Feng, K. Qiao, L. Pei, J. Lv, S. Xie, Using activated carbon prepared from *Typha orientalis* presl to remove phenol from aqueous solutions. Ecol. Eng. 84 (2015) 209–217, https://doi.org/10.1016/j.ecoleng.2015.09.028.

[63] H. Wang, Y. Chu, C. Fang, F. Huang, Y. Song, X. Xue, Sorption of tetracycline on biochar derived from rice straw under different temperatures. PLoS ONE 12 (8) (2017) e0182776, https://doi.org/10.1371/journal.pone.0182776.

[64] V.K. Gupta, I. Ali, Suhas, V.K. Saini, Adsorption of 2,4-D and Carbofuran pesticides using fertilizer and steel industry wastes. J. Colloid Interface Sci. 299 (2) (2006) 556–563, https://doi.org/10.1016/j.jcis.2006.02.017.

[65] Z. Issaabadi, Chapter 6: Plant-mediated green synthesis of nanostructures: mechanisms, characterization, and applications. in: M. Nasrollahzadeh, M. Atarod, M. Sajjadi, S.M. Sajadi, M. Nasrollahzadeh, S.M. Sajadi, … M. Atarod (Eds.), Interface Science and Technology, An Introduction to Green Nanotechnology, vol. 28, Elsevier, 2019, pp. 199–322, https://doi.org/10.1016/B978-0-12-813586-0.00006-7.

[66] H. Agarwal, S. Venkat Kumar, S. Rajeshkumar, A review on green synthesis of zinc oxide nanoparticles – an eco-friendly approach. Resour.-Effic. Technol. 3 (4) (2017) 406–413, https://doi.org/10.1016/j.reffit.2017.03.002.

[67] M.M. Chikkanna, S.E. Neelagund, K.K. Rajashekarappa, Green synthesis of zinc oxide nanoparticles (ZnO NPs) and their biological activity. SN Appl. Sci. 1 (1) (2018) 117, https://doi.org/10.1007/s42452-018-0095-7.

[68] M. Imran Din, A. Rani, Recent advances in the synthesis and stabilization of nickel and nickel oxide nanoparticles: a green adeptness, Int. J. Analyt. Chem. 2016 (2016), https://www.hindawi.com/journals/ijac/2016/3512145/. (Accessed 1 January 2020).

[69] H. Hassan, A. Salama, A.K. El-ziaty, M. El-Sakhawy, New chitosan/silica/zinc oxide nanocomposite as adsorbent for dye removal. Int. J. Biol. Macromol. 131 (2019) 520–526, https://doi.org/10.1016/j.ijbiomac.2019.03.087.

[70] B.M. Babić, S.K. Milonjić, M.J. Polovina, B.V. Kaludierović, Point of zero charge and intrinsic equilibrium constants of activated carbon cloth. Carbon 37 (3) (1999) 477–481, https://doi.org/10.1016/S0008-6223(98)00216-4.

[71] Y. Yang, Y. Chun, G. Sheng, M. Huang, PH-dependence of pesticide adsorption by wheat-residue-derived black carbon. Langmuir 20 (16) (2004) 6736–6741, https://doi.org/10.1021/la049363t.

[72] X. Chen, S. Zhou, L. Zhang, T. You, F. Xu, Adsorption of heavy metals by graphene oxide/cellulose hydrogel prepared from NaOH/urea aqueous solution. Materials 9 (7) (2016) 582, https://doi.org/10.3390/ma9070582.

[73] P. He, Q. Yu, H. Zhang, L. Shao, F. Lü, Removal of copper (II) by biochar mediated by dissolved organic matter. Sci. Rep. 7 (1) (2017) 1–10, https://doi.org/10.1038/s41598-017-07507-y.

[74] T.R.T. Santos, M.B. Andrade, M.F. Silva, R. Bergamasco, S. Hamoudi, Development of α- and γ-Fe_2O_3 decorated graphene oxides for glyphosate removal from water. Environ. Technol. 40 (9) (2019) 1118–1137, https://doi.org/10.1080/09593330.2017.1411397.

[75] N. Mansouriieh, M.R. Sohrabi, M. Khosravi, Adsorption kinetics and thermodynamics of organophosphorus profenofos pesticide onto Fe/Ni bimetallic nanoparticles. Int. J. Environ. Sci. Technol. 13 (5) (2016) 1393–1404, https://doi.org/10.1007/s13762-016-0960-0.

[76] A. Ouali, L.S. Belaroui, A. Bengueddach, A.L. Galindo, A. Peña, Fe_2O_3–palygorskite nanoparticles, efficient adsorbates for pesticide removal. Appl. Clay Sci. 115 (2015) 67–75, https://doi.org/10.1016/j.clay.2015.07.026.

[77] V.W.O. Wanjeri, C.J. Sheppard, A.R.E. Prinsloo, J.C. Ngila, P.G. Ndungu, Isotherm and kinetic investigations on the adsorption of organophosphorus pesticides on graphene oxide based silica coated magnetic nanoparticles functionalized with 2-phenylethylamine. J. Environ. Chem. Eng. 6 (1) (2018) 1333–1346, https://doi.org/10.1016/j.jece.2018.01.064.

[78] Y. Yang, Q. Deng, Y. Zhang, Comparative study of low-index {101}-TiO_2, {001}-TiO_2, {100}-TiO_2 and high-index {201}-TiO_2 on glyphosate adsorption and photo-degradation. Chem. Eng. J. 360 (2019) 1247–1254, https://doi.org/10.1016/j.cej.2018.10.219.

[79] Q. Chen, J. Zheng, Q. Yang, Z. Dang, L. Zhang, Insights into the glyphosate adsorption behavior and mechanism by a MnFe2O4@cellulose-activated carbon magnetic hybrid. ACS Appl. Mater. Interfaces 11 (17) (2019) 15478–15488, https://doi.org/10.1021/acsami.8b22386.

[80] Z. Zaheer, W. AbuBaker Bawazir, S.M. Al-Bukhari, A.S. Basaleh, Adsorption, equilibrium isotherm, and thermodynamic studies to the removal of acid orange 7. Mater. Chem. Phys. 232 (2019) 109–120, https://doi.org/10.1016/j.matchemphys.2019.04.064.

[81] T.S. van Erp, J.A. Martens, A standardization for BET fitting of adsorption isotherms. Microporous Mesoporous Mater. 145 (1) (2011) 188–193, https://doi.org/10.1016/j.micromeso.2011.05.022.

[82] M.T. Yagub, T.K. Sen, S. Afroze, H.M. Ang, Dye and its removal from aqueous solution by adsorption: a review. Adv. Colloid Interf. Sci. 209 (2014) 172–184, https://doi.org/10.1016/j.cis.2014.04.002.

[83] B.I. Olu-Owolabi, A.H. Alabi, P.N. Diagboya, E.I. Unuabonah, R.-A. Düring, Adsorptive removal of 2,4,6-trichlorophenol in aqueous solution using calcined kaolinite-biomass composites. J. Environ. Manag. 192 (2017) 94–99, https://doi.org/10.1016/j.jenvman.2017.01.055.

[84] G. Sangeetha, S. Rajeshwari, R. Venckatesh, Green synthesis of zinc oxide nanoparticles by *Aloe barbadensis* Miller leaf extract: structure and optical properties. Mater. Res. Bull. 46 (12) (2011) 2560–2566, https://doi.org/10.1016/j.materresbull.2011.07.046.

[85] Y.S. Ho, G. Mckay, The kinetics of sorption of basic dyes from aqueous solution by Sphagnum moss peat. Can. J. Chem. Eng. 76 (4) (1998) 822–827, https://doi.org/10.1002/cjce.5450760419.

[86] W.J. Weber, J.C. Morris, Kinetics of adsorption on carbon from solution, J. Sanit. Eng. Div. 89 (2) (1963) 31–60.

[87] D. Guo, N. Muhammad, C. Lou, D. Shou, Y. Zhu, Synthesis of dendrimer functionalized adsorbents for rapid removal of glyphosate from aqueous solution. New J. Chem. 43 (1) (2019) 121–129, https://doi.org/10.1039/C8NJ04433C.

[88] Y.S. Ho, G. McKay, Sorption of dye from aqueous solution by peat. Chem. Eng. J. 70 (2) (1998) 115–124, https://doi.org/10.1016/S0923-0467(98)00076-1.

[89] I.S. Mclintock, The Elovich equation in chemisorption kinetics. Nature 216 (5121) (1967) 1204–1205, https://doi.org/10.1038/2161204a0.

[90] F. Chen, C. Zhou, G. Li, F. Peng, Thermodynamics and kinetics of glyphosate adsorption on resin D301. Arab. J. Chem. 9 (2016) S1665–S1669, https://doi.org/10.1016/j.arabjc.2012.04.014.

[91] N. Singh, 13: Significance of bioadsorption process on textile industry wastewater. in: O. Sahu, Shahid-ul-Islam, B.S. Butola (Eds.), The Impact and Prospects of Green Chemistry for Textile Technology. The Textile Institute Book Series, Woodhead Publishing, 2019, pp. 367–416, https://doi.org/10.1016/B978-0-08-102491-1.00013-7.

[92] H.I. Meléndez-Ortiz, B. Puente-Urbina, J.A. Mercado-Silva, L. García-Uriostegui, Adsorption performance of mesoporous silicas towards a cationic dye. Influence of mesostructure on adsorption capacity. Int. J. Appl. Ceram. Technol. 16 (4) (2019) 1533–1543, https://doi.org/10.1111/ijac.13179.

[93] K. Vijayaraghavan, T.V.N. Padmesh, K. Palanivelu, M. Velan, Biosorption of nickel(II) ions onto *Sargassum wightii*: application of two-parameter and three-parameter isotherm models. J. Hazard. Mater. 133 (1) (2006) 304–308, https://doi.org/10.1016/j.jhazmat.2005.10.016.

[94] N.D. Hutson, R.T. Yang, Theoretical basis for the Dubinin-Radushkevitch (D-R) adsorption isotherm equation. Adsorption 3 (3) (1997) 189–195, https://doi.org/10.1007/BF01650130.

[95] S. Chattoraj, N.K. Mondal, K. Sen, Removal of carbaryl insecticide from aqueous solution using eggshell powder: a modeling study. Appl Water Sci 8 (6) (2018) 163, https://doi.org/10.1007/s13201-018-0808-5.

[96] N.K. Mondal, S. Basu, K. Sen, P. Debnath, Potentiality of Mosambi (*Citrus limetta*) Peel dust toward removal of Cr(VI) from aqueous solution: an optimization study. Appl Water Sci 9 (4) (2019) 116, https://doi.org/10.1007/s13201-019-0997-6.

[97] H.K. Boparai, M. Joseph, D.M. O'Carroll, Kinetics and thermodynamics of cadmium ion removal by adsorption onto nano zerovalent iron particles. J. Hazard. Mater. 186 (1) (2011) 458–465, https://doi.org/10.1016/j.jhazmat.2010.11.029.

[98] N. Ayawei, A.N. Ebelegi, D. Wankasi, Modelling and interpretation of adsorption isotherms, J. Chem. 2017 (2017), https://www.hindawi.com/journals/jchem/2017/3039817/. (Accessed 26 December 2019).

[99] P. Senthil Kumar, S. Ramalingam, C. Senthamarai, M. Niranjanaa, P. Vijayalakshmi, S. Sivanesan, Adsorption of dye from aqueous solution by cashew nut shell: studies on equilibrium isotherm, kinetics and thermodynamics of interactions. Desalination 261 (1) (2010) 52–60, https://doi.org/10.1016/j.desal.2010.05.032.

[100] E.C. Lima, A. Hosseini-Bandegharaei, J.C. Moreno-Piraján, I. Anastopoulos, A critical review of the estimation of the thermodynamic parameters on adsorption equilibria. Wrong use of equilibrium constant in the Van't Hoof equation for calculation of thermodynamic parameters of adsorption. J. Mol. Liq. 273 (2019) 425–434, https://doi.org/10.1016/j.molliq.2018.10.048.

[101] R.T.A. Carneiro, T.B. Taketa, R.J. Gomes Neto, J.L. Oliveira, E.V.R. Campos, M.A. de Moraes, C.M.G. da Silva, M.M. Beppu, L.F. Fraceto, Removal of glyphosate herbicide from water using biopolymer membranes. J. Environ. Manag. 151 (2015) 353–360, https://doi.org/10.1016/j.jenvman.2015.01.005.

[102] Y.S. Hu, Y.Q. Zhao, B. Sorohan, Removal of glyphosate from aqueous environment by adsorption using water industrial residual. Desalination 271 (1) (2011) 150–156, https://doi.org/10.1016/j.desal.2010.12.014.

[103] N. Hosseini, M.R. Toosi, Removal of 2,4-D, glyphosate, trifluralin, and butachlor herbicides from water by polysulfone membranes mixed by graphene oxide/TiO$_2$ nanocomposite: study of filtration and batch adsorption. J. Environ. Health Sci. Eng. 17 (1) (2019) 247–258, https://doi.org/10.1007/s40201-019-00344-3.

[104] I. Herath, P. Kumarathilaka, M.I. Al-Wabel, A. Abduljabbar, M. Ahmad, A.R.A. Usman, M. Vithanage, Mechanistic modeling of glyphosate interaction with rice husk derived engineered biochar. Microporous Mesoporous Mater. 225 (2016) 280–288, https://doi.org/10.1016/j.micromeso.2016.01.017.

[105] C. Zhou, D. Jia, M. Liu, X. Liu, C. Li, Removal of glyphosate from aqueous solution using nanosized copper hydroxide modified resin: equilibrium isotherms and kinetics. J. Chem. Eng. Data 62 (10) (2017) 3585–3592, https://doi.org/10.1021/acs.jced.7b00569.

[106] Q. Yang, J. Wang, X. Chen, W. Yang, H. Pei, N. Hu, Z. Li, Y. Suo, T. Li, J. Wang, The simultaneous detection and removal of organophosphorus pesticides by a novel Zr-MOF based smart adsorbent. J. Mater. Chem. A 6 (5) (2018) 2184–2192, https://doi.org/10.1039/C7TA08399H.

[107] Y. Tong, P.J. McNamara, B.K. Mayer, Adsorption of organic micropollutants onto biochar: a review of relevant kinetics, mechanisms and equilibrium. Environ. Sci. Water Res. Technol. 5 (5) (2019) 821–838, https://doi.org/10.1039/C8EW00938D.

[108] H. Tutu, E. Bakatula, S. Dlamini, E. Rosenberg, V. Kailasam, E.M. Cukrowska, Kinetic, equilibrium and thermodynamic modelling of the sorption of metals from aqueous solution by a silica polyamine composite, Water SA 39 (4) (2013) 437–443.

[109] P.O. Bedolla, G. Feldbauer, M. Wolloch, S.J. Eder, N. Dörr, P. Mohn, J. Redinger, A. Vernes, Effects of van Der Waals interactions in the adsorption of isooctane and ethanol on Fe(100) surfaces. J. Phys. Chem. C 118 (31) (2014) 17608–17615, https://doi.org/10.1021/jp503829c.

[110] P. Garcia-Muñoz, W. Dachtler, B. Altmayer, R. Schulz, D. Robert, F. Seitz, R. Rosenfeldt, N. Keller, Reaction pathways, kinetics and toxicity assessment during the photocatalytic degradation of glyphosate and myclobutanil pesticides: influence of the aqueous matrix, Chem. Eng. J. 384 (2019) 123315.

Dyes and their removal technologies from wastewater: A critical review

Mouni Roy[a,b] and Rajnarayan Saha[a]

[a]DEPARTMENT OF CHEMISTRY, NATIONAL INSTITUTE OF TECHNOLOGY DURGAPUR (NITD), DURGAPUR, WEST BENGAL, INDIA [b]DEPARTMENT OF CHEMISTRY, BANASTHALI UNIVERSITY, BANASTHALI, RAJASTHAN, INDIA

1 Introduction

Dyes can be described as colored substances chemically bonded to an applied substrate and hence conveying a desired color to that material. The chemical bond (fastness) of the dye with the substrate can be improved using a mordant. Pigments are also dyeing agents. However, pigments do not chemically bind, rather they adhere by physical adsorption, by covalent bond formation, or mechanical retention [1]. The organic dye molecules are comprised of delocalized electronic systems with conjugate double bonds that possess exclusive chemistry. It usually consists of two components: a chromophore component, capable of absorbing a visible range of the light spectrum imparting toward color development for the dye molecules, and an auxochrome constituent that generally increases its affinity toward the substrate. The previous Witt theory has been substituted with an electronic theory. The new theory states that the color originates owing to visible light-mediated excitation of π electrons in the valance band [2]. The application of dyes are widespread. They find enormous application in various industries such as food, textile, pharmaceutical, cosmetics, color photography, and paper printing purposes. Nowadays, widespread application of different dyes in several purposes results in the discharge of excess unused dyes as industrial effluents. The aromatic, xenobiotic, carcinogenic synthetic dye effluents are usually stable to direct sunlight and hinder the reoxygenation capacity of water. Thus, it largely disturbs the aquatic ecology and is even notorious for causing allergic dermatitis, skin irritations, and dysfunction of major organs like kidneys, brain, liver, central nervous, and reproductive system of human beings [3]. Thus, to address this prime environmental concern of dealing with dye-contaminated wastewater, suitable treatment methods are needed to be employed as early as possible. This chapter is comprised of elaborate discussions related to the origin of dyes, classification of dyes by their structure, and usage and toxicity of synthetic dyes. The topic also includes detailed discussions on different removal technologies known so far to treat dye-contaminated wastewater. In this regard, special emphasis has been provided on the nanoparticle-based treatment processes.

Intelligent Environmental Data Monitoring for Pollution Management. https://doi.org/10.1016/B978-0-12-819671-7.00006-3

1.1 History about dyes and dye industries

Since the dawn of the human race, pigments and natural dyes have had important influence on people's lives and were used for day-to-day purposes such as dyeing their skin, clothes, and painting on cave walls. The colors then used were of natural origin, which generally originated from soot, or inorganic pigments like hematite, manganese oxide, and ocher [4]. There lies a remarkable history regarding natural dyes, their origin, and use. The application of dyes in a 30,000-year-old ancient Paleolithic rock paintings on the walls of the caves of Chauvet, France, offers remarkable prehistoric evidence. There is written evidence regarding the use of natural dyes in 2600 BCE in China. Even cloth wrappings used in Egyptian mummies were dyed using an extract from the madder plant. Red color madder juice, owning alizarin dye, was known to be used by Alexander the Grea, to sprinkle upon his soldiers to swindle Persians who thought them as wounded [5]. These dyes are all aromatic compounds, most of which are from vegetable extracts, plant sources such as indigo, wood, madder, and saffron, and a few animal products viz insects, mushrooms, and lichens. However, compared with the utility of dyes, the color varieties were limited. This provoked chemists to discover methods of dye synthesis with new colors. In this regard, a first attempt was made by Woulfe in 1771 who synthesized picric acid, by reacting indigo with HNO_3, resulting in the dyeing of silk cloth in a bright yellow color. The first synthetic dye was produced by an English chemist, named William Henry Perkin, in attempt to synthesize quinine from coal-tar chemicals in 1856. The obtained bluish material had outstanding dyeing characteristics, which later turned into violet aniline, which was also named as mauveine or purple tyrant [6]. Perkin patented his discovery and later established a production line near London to supply the synthetic dye. This idea of research and development on synthetic dyes soon gained high stimulation, and many dyes began to appear on the market. Another marked advancement in dye discovery occurred when Otto Witt produced and marketed an azo dye by the name of London Yellow in late 1870s. Although England had a prompt lead, in the early 1900s, Germany turned out to be a leading universal supplier of a vivid color range, meeting around 85% of world's dye necessities. In the first half of 20th century, a rainbow of synthetic dyes entirely superseded natural dyes [7]. In 1914, the outbreak of World War I forced the Germans to transfigure dye industries into explosive factories, leading to a sudden fall off in the supply of synthetic dyes in the global market and a corresponding hike in price. This synthetic dye dearth presented investment in the dye industries as the most luring one, which proliferated several large and small dye manufacturing companies across the globe [8]. Currently, around 1 million different types of dyes are known today. Presently, the annual production rate of dyes is about 700,000 tons of which textile industries consumes about 36,000 tons dye/year [9]. Certain European countries, such as France, Germany, Spain, Italy, and England, as well as Asian countries, like Japan, Korea, Taiwan, China, India, and Pakistan, are the major dye-synthesizing countries. However, the much lower capital cost for proper organization of a dye industry leads to a surge in the dye industry in many Asian countries, especially in India and China.

1.2 Types of dyes

According to an inspection conducted by a central pollution control board in India, it is concluded that there are roughly 1 million known dyes and dye intermediates, of which approximately 5000 are factory-made. Based on the classification of different research groups, dyes may be classified into different categories as shown in Fig. 1 [10,11]. However, the classification based on chemical properties and their applications finds much importance nowadays. The former classifications were generally used by chemists and accounts for the chromophoric group present in the molecule. The chromophoric groups are generally electron withdrawing ones, whereas auxochrome are electron donors. The most important chromophores are $-C=C-$, $-C=N-$, $-C=O$, $-N=N-$, $-NO_2$, and $-NO$, and auxochromes are $-NHR$, $-NH_2$, $-OH$, $-NR_2$, $-COOH$, $-SOH$, and $-OCH_3$. On the basis of the chromophore structure, dyes can be divided into 20 to 30 groups, of which some of the most important chromophores are azo, anthraquinone, phthalocyanine, and triarylmethane [12]. The classification based on usage or application has been adopted by the Colour Index (C.I.) and in principal is used widely by dye users and dye

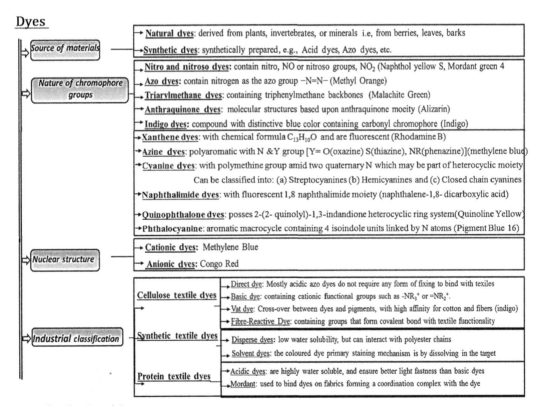

FIG. 1 Classification of dyes.

Table 1 Classification of dyes based on their usage [14–17].

Class of dye	Chromophoric group present	Applied on substrate	Method of application
Direct	Stilbene, phthalocyanine, azo, oxazine, and polyazo	Silk, cotton, paper, nylon, rayon and leather	In the midst of dye baths containing slightly alkaline or neutral conditions along with additional electrolyte
Basic	Diazahemicyanine, azine, oxazine diphenylmethane, anthraquinone, azo, thiazine, xanthine triarylmethane, cyanine, hemicyanine, and acridine	Modified nylon, inks, medicine, paper, tannin mordanted cotton, modified wool, polyacrylonitrile, polyester, silk, polyester, and medicine	Dye baths amidst acidic environments
Vat	Indigoids and anthraquinone (including polycyclic quinines)	Rayon, wool, cotton, polyester-cotton, and cellulosic fibers,	Dyes insoluble in aqueous media are solubilized on reduction with $NaHSO_3$, after which exhausted on fiber, finally reoxidized
Fiber reactive/reactive	Azo, phthalocyanine, basic, anthraquinone, oxazine, triphenylmethane, and formazan	Silk, cotton, cellulosic, nylon, and wool	Interaction between active site of dye with functional groups present in fiber, covalently bonding under alkaline pH in the influence of heat
Disperse	Nitro, azo, styryl, anthraquinone, and benzodifuranone	Polyester, acrylic fibers, nylon, cellulose acetate, plastic, Acetate, plastic, cellulose, polyester, polyamide, cotton	Padded on cloth followed by thermofixed or baked at high temperature and pressure or low temperature carrier process
Solvent	Triphenylmethane, azo, anthraquinone, and phthalocyanine	Plastic, waxes, gasoline, oils, lubricants, inks, fats, lacquers, stains and varnishes	Dissolution of substrate
Acid	Nitro, xanthene, azo, anthraquinone, nitroso, azine, and triphenylmethane	Cosmetics, leather, food, printing ink, paper modified acrylics, nylon, silk, wool	Under acidic to neutral conditions in dye bath
Mordant	Azo and anthraquinone	Natural fibers, wool, anodized aluminum and leather	Along with chromium salts
Azo	Triphenylmethane, xanthenes, and nitroso	Cotton, polyester, acetate, cellulose and rayon	Coupling materials used for infusing fiber also a solution of diazonium salt used in treatment
Sulfur	Indeterminate structures	Paper, silk, cotton, leather, polyamide fibers, wood and rayon	Aromatic vatted coupled with Na_2S and reoxidized to water insoluble sulfur-compounds on the fibers
Fluorescent	Naphthylamides, pyrazolos, and coumarin	All fibers, soaps as well as detergents, paints plastic	suspension or solution and mass dispersion
Cosmetics, food, and drug	Azo, triarylmethane, anthraquinone, carotenoid	Food, drug, cosmetics	Mixing

technologists [13]. Table 1 summarizes the classes of dyes, their chemical structures, substrates, and methods of application of such dye classes [18].

1.3 Toxicity of synthetic dyes

Synthetic dyes finds ample use in the textile industry, cosmetic industry, food technology, paper industry, leather tanning industry, hair coloring industry, light-harvesting solar cells, photoelectrochemical cells, agricultural research, etc. [19]. Although these industries impart a significant part of the economy for every country, there are too many obstacles from an environmental standpoint. The World Health Organization (WHO) presented a report that says dyeing treatment in several industries triggered 17%–20% of water pollution. The dyeing procedure leads to the release of 20% of the dye as industrial discharges, which eventually contaminates drinking water sources [20]. The upsurge in the concentration of dyes in water bodies disrupts the reoxygenating capacity of water and hinders sunlight, disconcerting biological activity in the aquatic ecology. This synthetic organic dye accumulation in the water sources increases biological oxygen demand (BOD) and chemical oxygen demand (COD) of water causing changes in pH. Varying chemical contents in a body of water adversely affect water resources, aquatic organisms, animals, and even mankind [3]. The toxic, xenobiotic, and carcinogenic dye effluents usually cause allergic dermatitis, skin irritations, and can cause dysfunction of the kidneys, brain, liver, central nervous system, and intense damage to the reproductive system of human beings [21]. Furthermore, salmonella/microsome assay report suggested a high percentage (67%) of mutagenic activity due to textile wastewater effluents [22]. Numerous studies on various dye-related industries, especially textile industry, also provided evidence. Table 2 shows the day-to-day applications and hazardous effects of a few common market available dyes.

2 Dye removal technologies

Recently, the council regarding environmental conservation had set rules stating limits to the nature and state of effluent discharge of the industries. The International Dye Industry Wastewater Discharge Quality Standards have implemented the Zero Discharge of Hazardous Chemicals Programme (ZDHC) [28]. Thus, textile industries are throttled with enormous stress to minimize the use of harmful carcinogenic, mutagenic, and allergenic chemicals for textile dyeing purposes. The international permissible level of dye effluent discharge in water is found in Table 3. The level of color, BOD, COD, hazardous chemical dissolved solid (TSS and TDS), pH, and temperature should be below the restricted limit.

The table signifies that the detoxification of textile sewage not only includes abolition of colors but also requires degradation and mineralization of the dye substrate. Furthermore, several features, such as composition of wastewater, nature of dye, environmental fate, handling, running costs, technical influence, and economic viability of dye removal process require attention. An extensive variety of technologies have been developed to

Table 2 List of a few market available dyes.

Commonly used dye	Structure	Chromophoric group present	Application	Hazardous effect	References
Reactive brilliant red		Azo	Finds extensive application in textiles	Probably bind to enzyme or protein causing alteration of function. Inhibits function of human serum albumin	[23]
Malachite Green		Triphenylmethane	Finds applications as dyeing element in leather, silk, paper industry, and also used as antimicrobial in aqua culture	Carcinogenic, respiratory toxic, chromosomal fractures, mutagenesis, teratogenicity. Substantial variations were observed in the blood biochemical factors of fishes exposed to dye. Histopathological implication of dye consist of multi-organ tissue damage	[24]
Direct Blue 15 (dimethoxybenzidine-based dye)		Azo	Replacement for trypan blue and used for certain biological staining purposes	Show mutagenic effect also having strong carcinogenic effect	[25]
Congo red		Azo	For dyeing cotton	Imparts mutagenic as well as carcinogenic effect	[26]
Acid Violet 7		Single azo	Used especially in textile industry. Also applied in cosmetic paper, food industries	Reveal acetylcholinesterase activity inhibition, chromosomal aberration, membrane lipid peroxidation	[27]

Table 3 International standard regarding discharge of dye effluent into the environment [29,30].

Factor	Standard allowed
Temperature	<42°C
COD	<50 mg/L
pH	Between 6 and 9
BOD	<30 mg/L
Noxious pollutants	Totally unacceptable
Color	<1 ppm
Suspended particulates	<20 mg/L

achieve satisfactory and acceptable quality of water levels by degradation or removal of synthetic dye products (Fig. 2).

These methods include biological methods (microbial and enzymatic degradation), physical methods (coagulation or flocculation, membrane filtration techniques, sorption processes), and chemical methods (conventional oxidation processes). Moreover, in previous decades, the role of nanotechnology in dye degradation or removal process has been marked as a promising technique compared with existing processes. Each of the previously mentioned dye removal methods has its own merits and demerits. Thus, a single technique might not be enough to accomplish total decolorization or dye removal from wastewater. However, to address this hurdle, dye degradation strategies consequently

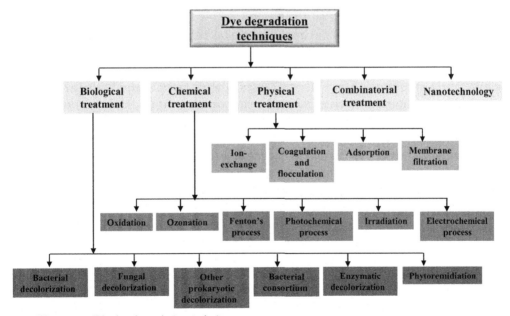

FIG. 2 Different possible dye degradation techniques.

involve various effective techniques. These methods have been illustrated in detail in the following sections.

2.1 Biological treatment method

Biological wastewater treatment method, also known as the conventional method, is a common and widely used method of treatment. It takes into account biodegradation bleaching by taking aid of several micro-organisms, fungi, bacteria, yeasts, and algae. This is a cheap and easy process that goes through a combination of aerobic and anaerobic processes. However, there are some major limitations regarding this process:

(a) Complete color removal is not possible
(b) Xenobiotic dyes with complex chemical structures are found to be recalcitrant toward degradation
(c) Biological method requires large land area, diurnal, and greater time for their functioning
(d) The process provides little flexibility in design and operation

Numerous studies depict the use of micro-organisms and leads to the removal of dye via a biosorption process. This property of micro-organisms exists because of the presence of cell wall components like lipids; heteropolysaccharide-containing functional groups like hydroxyl, carboxyl, amino, and phosphate; and many other charged substituents. The presence of such groups creates strong interactive forces among dye and cell wall of micro-organisms [31,32]. The following sections coupled with few literature reports tabulated in Table 4 provide insight on the topic.

Decolorization by bacteria: The capability of a bacterial entity to degrade azo dyes has widely been studied. This process ensures a higher degree of biodegradation and also produces less sludge [41]. The decolorization by bacterial process may occur by aerobic, anaerobic, or a combination of both processes. Under aerobic condition, azo dyes are not willingly metabolized. It creates more toxic aromatic amine compounds. Thus to achieve complete degradation, aerobic degradation is followed by an anaerobic degradation process. [42]. In the case of phthalocyanine chromophores, reversible decolorization and reduction occurs only via an anaerobic method. Horitsu et al. first reported degradation of azo dyes using *Bacillus subtilis* culture [43].

Decolorization by fungi: Numerous studies depict the application of fungi as a potential candidate for oxidization of soluble or nonsoluble, phenolic and nonphenolic dyes [44,45]. Fungi attribute few transformation reactions, such as hydroxylation of polyaromatic dyes. Researchers suggested that fungi-facilitated aromatic compound degradation happens when there is a scarcity of nutrients (C, N, and S) for the propagation of the cultures [46]. Even some non-white-rot mushrooms have the potential to successfully fade dyes [47].

Decolorization by other prokaryotes: Prokaryotes such as microalgae propose a solution to global environmental problems related to wastewater treatments. Algae such as

Table 4 Decolorization of different dyes by microbial culture.

Name of microbes	Type of microbe	Name of dye	Condition	% of Dye degraded	References
Aspergillus niger	Immobilized fungal biosorbent)	Malachite Green	At pH 5 and at 32°C	82.6	[33]
Chlorella vulgaris	Dried green algae	Remazol Black B	At pH 2 and at 35°C	85.2	[34]
Phanerochaete chrysosporium	Fungi source	Mixture of Nigrosin, Malachite green, and Basic fuchsin	At pH 1–2 and at 25°C	78.4	[35]
Aeromonas caviae, *Rhodococcus globerulus*, and *Protues mirabilis*	Consortia	Acid Orange 7 and many other azo dyes	At pH 7, in 16 h and at 37°C	90	[36]
Sphingomonas paucimoboilis	Bacterial strain	Methyl Red	At pH 9, in 10 h and at 30°C aerobic process	98	[37]
C. tropicalis	Yeast	Violet 3	At pH 4	–	[38]
Afilamentous fungi (*Umbelopsis isabellina*) and Yeast (*D. polymorphus*, *C. tropicalis*)	Consortia	Reactive Black 5	Within 16–48 h	Complete degradation	[39]
Staphylococcus hominis	Bacterial strain	*Acid Orange*	At pH 7, in 60 h under static condition and at 35°C	94	[40]

Chlorella, Oscillateria [48], and *Spirogyra* [49] offer remarkable efficiency toward the decolorization of dyes in wastewater. Besides, algae does not require constant contribution of carbon or other necessary supplements as required for bacteria and fungi cultures [50].

In current years, rigorous studies has been performed regarding dye depletion in wastewater using diverse yeast species, as it demonstrates several smart features and advantages over filamentous fungi and bacterial strains. It not only cultivates swiftly but also has enough potential to thrive in certain hostile environmental conditions, such as low pH [51]. Jinqi and Houtian in their report illustrate the application of *Candida zeylanoides* and *Ascomycete* yeast that are usually collected from contaminated soils [52]. It is to be borne in mind that both algal and yeast decolorization mechanisms involve adsorption, enzymatic degradation, or both pathways.

Decolorization by bacterial consortium: The Consortium-based biological treatment system attains advanced degradation and/or mineralization owing to synergistic metabolic characteristics of concerned microbial community. It possesses substantial advantages over the application of single microbial strains [53]. Consortia can be comprised of different bacteria, fungi strains, or combination of both. Detailed investigation demonstrate that in the case of microbial consortium, the individual microbial strains may target the dye molecule at different positions or may perform decomposition process on the produced metabolites formed by the co-existing strains [54].

Microbial enzymes toward dye degradation or decolorization: Attempts of identifying bacterial strains that may have efficient dye degradation properties was initiated back in 1970. The identification of three bacterial strains, viz. *Bacillus subtilis, Aeromonas hydrophila*, and *Bacillus cereus* was the result [55,56]. However, the dye reduction was found to be non-specific. Decolorization is rapid, using the bacterial enzymatic degradation process. [57]. Fungi indeed provide a great source of intra- and extracellular dye degrading enzymes. Enzymatic process using species such as *Schizophyllum, Pleurotus ostreatus, Neurospora crassa*, and *Sclerotium rolfsii* resulted in an increase up to 25% regarding degree of decolorization of certain textile dyes, i.e., triarylmethane, indigoid, and anthraquinone dyes [58]. There are numerous enzymes used in dye removal process evidenced as an operational molecular weapon for removal and/or decolorization purposes. Some of them have been discussed in the following:

Azoreductase is an enzyme that undergoes reductive degradation of azo dye into colorless amines. Azoreductase are found either membrane bound or in cytoplasm in microbes. This enzyme attacks the azo group ($-N=N-$) and cleaves them leading to transfer of four electrons. The resultant of cleavage is lethal aromatic amine that can be later decomposed by the aerobic process [59,60].

Laccase is a small molecular weight, multicopper-containing enzyme that degrades dye substrate with less specificity. It is incorporated in the oxidase family and is a potential candidate to degrade a wide range of xenobiotic compounds, aromatic and nonaromatic substrates. Thus, it is able to decolorize and degrade aromatic azo, phenolic compound, etc. The enzyme with the aid of Cu^{2+} ion as mediator oxidizes aromatic amine. Thus degrading azo dye via free radical mediated highly nonspecific mechanism, forming phenolic compounds and not toxic aromatic compound [61]. In the case of phenolic ring compounds, the enzyme initially oxidizes the substrate by one electron to produce a phenoxy radical followed by further enzymatic oxidation to yield carbonium ions. Further nucleophilic attack of water results in generation of unstable (under oxygen-containing environment) 4-Sulfophenyldiazene and benzoquinone compounds. *Peroxidases* are hemoproteins found in wide variety of micro-organisms, plants, and animals that aid to catalyze dye degradation process in the presence of hydrogen peroxide [62]. Two types of enzymes are known: lignin and manganese peroxidases (MnP). Each of them shows almost the same reaction mechanism. Peroxidase-assisted degradation targets lignin moieties by auxiliary enzymes and mediators. The low molecular weight substance improves lignin biotransformation by fast diffusing into the lignocellulosic matrix. This results in high redox potentials that enhance degradation ability of the enzymes.

Tyrosines are Cu-containing dioxygen-activating enzymes found in many bacterial species and are generally related to melanin production. They find applications in production of L-DOPA, protein cross-linking, phenolic biosensors, and phenol and dye removal purposes. These proteins show specificity toward phenolic and diphenolic substrates and produce activated quinone as product. This limits their applicability. Major studies have focused on *Streptomyces* sp. enzymes, whereas other proteins are also gaining importance regarding biotechnology.

Microbial enzymes ensure numerous benefits when compared with other processes because of inexpensive maintenance cost, easy culture, and downstream processing. A list of a few industrially important microbial enzyme has been incorporated to depict its significance in synthetic dye degradation purposes (Table 5).

Phytoremediation method: This is another biological method for wastewater treatment. The combination of two Latin words—plant and remedy—gave rise to the term *phytoremediation*. The plant, plant origin microbes, or associated microbiota are used to take up the contamination from soil or water. The remediation is achieved either by retaining, elimination, or degradation in a natural way as it happens in an ecosystem. Phytoremediation is a cheaper, eco-friendly, and feasibly sustainable method for removal of dye pollutants. Moreover, the process requires little nutrient cost and also has aesthetic demand. However, the phytoremediation method is also not a limitations-free method. The major disadvantage of the phytoremediation procedure is the sluggish rate of ecological cleanup process, which may even last for more than a decade. There may also be a

Table 5 List of some industrially important microbial enzymes on different dyes.

Name of enzyme	Producing microbes	Name of dye degraded	pH	Temperature (°C)	% of Dye degraded	References
Laccase	*Geobacillus stearothermophilus* (thermophilic origin) (bacterial)	Indigo carmine, Congo red, Brilliant Green, Remazole Brilliant Blue R	Stable in diverse pH	Temperature stable	99, 98, and 60, respectively	[63]
Azoreductase	*Shewanella* sp. Strain IFN4 (bacteria)	Acid Yellow 19, Acid Red 88, Reactive Black 5, Direct Red 81, Disperse Orange 3	8	45	Max. degradation with Reactive Black 5	[64]
Claccase (SmLac)	*Stenotrophomonas maltophilia* AAP56 (bacteria)	Reactive Black 5 dye	Optimal pH neutral or basic	–	Up to 99	[65]
Soybean peroxidase	*Immobilized Desmodesmus* sp. (algae source)	Methylene Blue	–	–	98.6	[66]

substantial decrease in phytoremediation, generally during winter (plant growth retards or hinders) and/or may be damage of vegetation as a result of weather, plant diseases, or pests [67]. Few reports regarding decolorization of pollutant in wastewater by phytoremediation have been briefly illustrated in this section. Researchers exploited the phytoremediation prospective of *Petunia grandiflora Juss.*, for remediation of dye mixture present in wastewater [68]. Individual effects of three plant species, *Sorghum vulgare*, *Phaseolus mungo*, and *Brassica juncea*, have been considered for decolorization efficacy of textile runoffs, which showed color removal up to 79%, 53%, and 57%, respectively [69].

2.2 Chemical treatment method

In chemical treatment processes, dye degradation is assisted by certain chemical substances, which may be added from an external source or generated in-situ. The dye removal by chemical treatment procedure exhibits premier removal percentage ranging from 88.8% to 99%. Table 6 depicts a few literature reports on chemical treatment process for wastewater management. However, chemical dye removal approaches have unbearable disadvantages.

(a) Often the process is expensive, requiring costly chemicals or high consumption of electrical energy.
(b) Except for electrochemical destruction technology, all the other process are sensitive to the pH of the solution.
(c) There is also a possibility of generation of a secondary pollutant.

The key chemical methods used for dye effluent management in wastewater are as follows:

Oxidation: In these processes, chemicals like sodium hypochloride, hydrogen peroxide, $(O_3 + H_2O_2)$, potassium ferrate (VI), or potassium permanganate, etc. are used to carry forward the degradation process. The synergistic action of O_3 and H_2O_2 results in faster generation of $^\bullet OH$ radicals, responsible for oxidation of pollutants in water [78]. The typical reaction is shown as:

$$2O_3 + H_2O_2 \rightarrow 2 \bullet OH + 3O_2 \tag{1}$$

The oxidation of textile effluent by these chemicals occurs at a shorter reaction time, but the method is costly and also depends largely on the pH of the medium. Chlorination of dye wastewater by active chlorine is rarely used due to its adverse effects upon its discharge into waterways [79].

Ozonation: Ozone is a powerful oxidative agent as ozone possesses high oxidation potential (2.07) in comparison to H_2O_2 (1.78) and chlorine (1.36). The reaction occurs quickly, and no sludge generation occurs. However, a major shortcoming of this process is the shorter half-life (20 min) of ozone and thus requires continuous flow resulting in an increase in operation cost. Besides, the process is sensitive toward the pH, presence of salts, and temperature. Ozone produces the oxidative radicals in the wastewater, which

Table 6   Decolorization of various dyes using chemical treatment process.

Chemical process	Name of dye	Condition	% Dye degraded	References
Electrochemical process within cylinder containing Ti/β-PbO$_2$ electrode	Textile wastewater	Crucial dye removal parameter: retention time of 4 h, and [salt] = 4000 mg/L, direct current of 5.6 V, pH 6	96	[70]
Ozonation	Acid Red 183	Ideal removal parameter: [initial dye] = 50 mg/dm^3, [ozone] = 300 mg/dm^3, acidic pH and reaction time is 15 min	97	[71]
Fenton process along with photooxidation method	Acid orange 24	Optimum condition: [initial dye] = 3 mg/L, [FeSO$_4$] = 0.75 g/L, [H$_2$O$_2$] = 0.75 g/L, pH 3, reaction time = 40 min	95	[72]
Electrochemical process (by modified TiO$_2$/PbO$_2$ electrode) along with active in-situ generated chlorine	Remazol Brilliant Blue	Best condition: pH between 5 and 10, reaction time 50–60 min, NaCl conc. 4000 mg/L	70.38	[73]
Oxidation	Orange II	Addition of hypochlorite to potassium ferrate (VI) results in better degradation in pH range from 3.0 to 11.0, reaction time 30 min.	95.2	[74]
Electrochemical oxidation using conductive boron-enclosed diamond anodes	Procion Red MX-5B	Ideal dye removal parameters: [initial dye] = 50 mg/dm^{-3}, [Cl −] = 100 mg/dm^3, pH 7, flow rate = 300 dm^3/h, current density = 10 mA/cm^2, and maximum reaction time required is 240 min	85	[75]
Acrylic grafted polysulfone ultrafiltration membrane and UV irradiation	Acid Red 301, Acid Orange 7, Acid Blue 92, and Acid Green 20	Dyes with lower molecular weights are targeted well. Ideal operating parameters: irradiation time > 30 min, pressure ~4 bars, sodium sulfate of 80 mM are added.	99.9	[76]
Fenton reaction	Direct Blue 71	Ideal parameters: [Fe^{2+}] = 3 mg/L, [H$_2$O$_2$] = 132 mg/L, [initial dye] = 100 mg/L, pH 3, reaction temperature is 20–60 °C and reaction time is 20 min.	94–50.7 COD removed	[77]

leads to the chemical breakdown of pollutants. Representative reaction mechanism occurs during ozonation has been reported as [80]:

$$3O_3 + OH^- + H^+ \rightarrow 2 \bullet OH + 4O_2 \tag{2}$$

Fenton's process: Fenton reagent (H$_2$O$_2$-Fe^{2+}) provides a suitable chemical means of handling both soluble and insoluble dyes in wastewaters, which are resilient to biological treatment. The Fenton reaction was initially studied by H.J. Fenton (1894). It was found that the oxidative potential of H$_2$O$_2$ is increased when iron (Fe) is used as a catalyst only under acidic conditions. One key drawback of this scheme is the unnecessary production of sludge, which creates problematic disposal. Besides, the process requires long working time and cannot target disperse and vat dye [81].

Photochemical process: In this treatment method, light (UV, visible, or solar radiation) is used with Fenton's process or without Fenton's reagent. In this process, light activates the generation of hydroxyl radicals to mineralize dye molecules to CO_2 and water [82]. The disadvantages include formation of colorless byproducts, which may be more noxious than the original dye. However, numerous research works undergoing in this arena might end up with a solution.

Irradiation: Among the numerous techniques for wastewater treatment, the irradiation approach with ultrasound or microwave is also known. Ultrasound frequencies of a certain power are emitted into the water by precise transducers. The transient cavitations, developed by continuous rarefaction and compression of bulk water, when collapsing creates high temperature and local pressure peaks. This splits water into •OH radicals and H atoms. The produced radicals, temperature, and pressure are capable of degrading dye pollutants. High-frequency ultrasound can significantly decay color of dyes. Adding TiO_2 or zero valent Cu, CCl_4, enhanced the performance of low-frequency ultrasound. In this scenario, a huge amount of dissolved oxygen is required, which limits its applicability in large scale [83].

Recently, microwave irradiation has acquired a great deal of attention toward wastewater treatment applications. Various microwave-absorbing materials, along with high surface area and wide range of pore size distribution, were developed for enhancing the degradation of organic pollutants under microwaves [84]. When such material is used, hot spots are formed on the surface resulting in selective heating, thus helping in degradation.

These methods require high energy input and may also produce toxic and carcinogenic aromatics by pyrolysis.

Electrochemical process: The electrochemical destruction process is a comparatively new technique where relatively little or no intake of chemicals is required. Apart from this, the process discards assembly of sludge. In the case of an electrochemical method, organic pollutants are adsorbed on the surface of the anode followed by oxidation. Cathodic reduction of oxygen to hydrogen peroxide can also oxidize numerous azo dyes efficiently. Another approach demonstrates the use of Fe catalyst to electrogenerate H_2O_2, which leads to improvement of the degradability of dyes by several folds. However, this process is accompanied by some problems such as requirement of electrolytic apparatus, high cost electricity, and reduced color removal for some dyes because of higher flow rate [85,86].

2.3 Physical treatment method

The physical methods are generally expensive and their performance is affected by the presence of other wastewater constituents, originating serious issues associated with concentrated sludge disposal. Furthermore, the toxicity of the dye molecules remains unchanged even after such treatment. The following section along with Table 7 illustrates the advancement in different physical treatment methods.

Table 7   Decolorization of various dyes using physical treatment process.

Physical process	Name of dye	Condition	% Dye removed	References
Activated carbon prepared from Brazilian pine-fruit (*Araucaria angustifolia*) shells	Remazol Black B textile dye	Favorable uptake occurs in pH range of 2.0–7.0, with contact time of 4 h at temperature 298 K	Max. uptake capacity of 446.2 mg g^{-1}	[87]
Ultrasound Irradiation	Acid brown 348	Optimum removal condition: [initial dye] = 40 mg/L, [Exfoliated graphite as sorbent] = 2 g/L, at pH 1 and temperature = 40°C	90	[88]
Ion-exchange method	Azo dyes as well as aromatic amines	Two types of calix [5] are Ne-based polymer, working in a wide pH range (2.0 to 8.0)	95–99 and 83–97	[89]
Reverse osmosis via Polyamide Membrane	Direct yellow	Feed concentration, contact time and temperature were varied from 75 to 450 ppm, 0.2 to 2.0 h, and 30 to 50°C, respectively. Best result is when both feed conc. And temperature is kept low	98.30	[90]
Coagulation/flocculation treatment	Acid red 119	FeCl$_3$ sludge generated by water treatment plant is used. Optimum parameter: [initial dye] = 65.91 mg/L, amount of coagulant = 236.68 mg and pH = 3.5	96.53	[91]
Nano filtration via Dalton membrane	Reactive, Acidic, Direct and Disperse dye	Ideal parameters: Contact time 2 h, [initial dye] = 50 mg/L, pressure 12 bars, [total dissolved solids] = 3000 mg/L	98	[92]
Adsorption unto biowaste substances (eichhornia charcoal and ground nut shells charcoal) separately	Congo Red	Optimum parameter: [initial dye] = 1 g/L, adsorbent dose = 1.2 g, reaction time 60 min, pH in acidic range, temperature = 318 K. Maximum adsorption capacity attained: ground-nut shell charcoal was 117.6 mg/g while Eichhornia charcoal was 56.8 mg/g	98	[93]
Adsorption by sawdust treated with formaldehyde and sulfuric acid, separately	Malachite green	Sulfuric acid treated sawdust provides best removal efficiency. Optimum parameter: adsorbent dose = 0.4 g/100 mL, temperature = 26 °C, pH 6–9	99.8	[94]

Ion-exchange: Ion exchange is a reversible water treatment method containing ion-exchange resin that swaps away one or more undesirable contaminants in exchange with another non-objectionable or less objectionable substance. Then the resin is regenerated by successive backwashing process with a concentrated solution of replacement ions to eradicate accumulated ions followed by dousing the flushing solution from the resin. Certain modified natural materials, for example, organofunctionalized layered silicate and cationic-polymer/bentonite complex, revealed interesting results [95]. The need of backwashing, flushing, and rinsing during restoration of ion-exchange media confines the practicality of ion exchange for wastewater treatment. Although charged dye substances can be removed by this process, the system is not extremely compatible for the removal of dye system [96].

Coagulation/flocculation: These cheap, robust processes are generally used for effective settling of suspended solids present in wastewater [97]. Suspended particulates, bearing the same type of surface charge when settling down, results in repulsion among each other. Proper coagulation and flocculation techniques arise out to be handy in this case. In the coagulation process, chemical coagulants (such as iron and aluminum salts, synthetic polyelectrolytes, fly ash, and clay) possessing opposite charges than those of suspended matters leads to better stacking [98]. Disperse, vat, and sulfur dye from textile effluents can be removed by these techniques. However, inorganic coagulants cannot effectively remove highly soluble dyes. Flocculation, on the other hand, uses larger hairs and inorganic or organic polymers, which can be removed easily and binds with the particles to form floc [99]. The utmost disadvantage of this process is the generation of secondary pollution [100].

Adsorption: Physical adsorption ensures feeble intermediate bonds between the adsorbent and the adsorbate. Weak physical forces such as hydrogen bonding, Van der Waals' interactions, and dipole-dipole interactions usually take part. Generally, physical adsorption is easily reversible, whereas irreversible chemisorption occurs when strong bindings exist between the adsorbate and the adsorbent by electron exchange. Most adsorbents are porous materials with high external surface area and optimum pore diameter. Among the many techniques of dye elimination, this technique can be appreciably used to treat dye wastewater [101]. This technique is influenced by several factors such as adsorbent surface, dye/adsorbent interaction, particle size, pH, temperature, and contact time. Such processes with properly designed system can decolorize dyes present in a wastewater system.

Membrane filtration: Membranes suggest exciting possibilities for separation of dyes and dyeing auxiliaries from water. Several substances, such as ceramic membranes prepared from clay and alumina, and nanofiltration polyamide-based composite membranes revealed decent color removal properties particularly when applied after coagulation-flocculation [102].The advantages of membrane filtration are rapid, low space-requiring process, and can be reused. The main disadvantage with membrane filtration method is its short lifetime, due to fouling, which excessively increases the cost effect of the process [103]. The membrane filtration can be different types such as reverse osmosis, ultrafiltration, and nanofiltration. In the former process, dye wastewater is allowed via extremely thin-pored membrane and, therefore, requires high pressure for separation of treated water from pollutants. Moreover, the membrane often gets clogged by dye molecules.

Typically, reverse osmosis is a pressure-driven system where liquid is allowed to pass from low solute concentration to high solute concentration by an extremely thin semipermeable membrane with retention rate of 90% or more. Reverse osmosis effectively separates contaminants in one side of the membrane and treated pollutant-free water on the other side. Bear in mind that the higher the solute concentration, the more significant the osmotic pressure, and so greater energy is mandatory for the separation procedure. This limits its applicability as there is a constant requirement of high pressure, making the process costly [104].

2.4 Combinatorial treatment method

None of the treatment methods discussed earlier is devoid of demerits. These methods also lack adequately wide substrate variety and vigor to remove polyaromatic dye in wastewaters. A possible solution to address this challenge is consecutive application of a blend of processes. These treatment methods must be designated crucially such that the limitation of one process can be overcome by the other [105]. Numerous combinatorial treatment processes are under trial and testing phases in the laboratories. The high sensitivity, lack of integrity, and sluggishness of the biological method mostly appreciate the use of physicochemical treatment methods in combinatorial purpose. Other literature recommend coalescing characteristic adsorbents (physical process) with a biocatalyst (biological enzymes-related process) as both are very cost-effective and require easy handling [106]. In Table 8, a few such combinatorial treatment methods has been depicted.

2.5 Nanotechnology-based treatment techniques

Nanoscience has drawn increasing attention of researchers worldwide to investigate the potential of nanomaterials in wastewater treatment. Several inorganic nanosized materials, such as metal, metal oxide, metal sulfide-based materials, etc., find application toward removal of dye pollutants. The following section provides brief insight on these materials and their applicability for wastewater treatment purposes. The controlling factors that lead to higher removal efficiency are high surface area, crystallinity, surface charge, band gap, and specific affinity. The key advantage of nanotechnology-based treatment techniques lie in several cycle reusability of the catalyst, bringing down the cost of the method. Besides, the use of nanocatalyst imparts minimal environmental impact, low solubility, and zero or little secondary pollutant generation. Table 9 summarizes a few examples of dye decolorization-applying nanotechnology.

Metallic nanoparticle: Zero-valent metal synthesis has drawn interest of researchers due to its highly active surface. Several literature reports are known for the synthesis of metallic nanoparticles via a green approach. Among the several promising natural

Table 8 Decolorization of various dyes using combinatorial treatment process.

Process	Name of dye	Condition	% Dye removed	References
Two separate processes coagulation-flocculation using $Al_2(SO_4)_3$ followed by either of membrane based routes such as ultrafiltration, microfiltration, or activated carbon mediated adsorption	Textile dye effluent	Application in an array; coagulation-flocculation trailed by ultrafiltration processes manifested best result	74	[107]
Membrane-free bio-catalyzed electrolysis combined with aerobic oxidation	Alizarin Yellow R	Optimum removal condition: [initial dye] = 100 mg/L, reaction time 2 h, External power = 0.5 V ~100% electron retrieval in cathode region	94.8	[108]
A sequence of nano-filtration by means of Hydracore 50 membrane trailed by Electrochemical technique using Ti and Pt electrodes	Cibacron Yellow S-3R	Crucial parameter: [initial dye] = 0.1 g/L, pH 3, [NaCl] = 60 g/L, current density 33 mA/cm^2	98	[109]
Activated carbon (VMK-1) adsorption coupled with ultrafiltration	Cationic violet	The process increases the lifetime of membrane	–	[110]
Advanced oxidation process (AOP) along with blend of UV light, H_2O_2, ozone	Dyeing effluents of acetate and polyester fiber	Optimum conditions: 500 mg/L of chemicals, [H_2O_2] 300 mg/L, pH 3, reaction time 90 min	99	[111]
AOP (UV light and H_2O_2)	Methylene Blue	[initial dye] = 3, 5, and 10 mg/L, contact time 3.5, 4.5, and 10.5 min [H_2O_2] = 1 mmol, 8 W UVC lamp with UV radiation intensity 2400 μW/cm^2	98	[112]
A combination of nanofiltration process and reverse osmosis	Acid red	Ideal parameters: [initial dye] = 65 mg/L, pH 3, [NaCl] = 60 g/L, temperature = 39°C, pH = 8.3, pressure 8 bar, contact time 2 h	93.77	[113]

products, biologically active plant-based products demonstrate superb podia for this purpose [125–127]. The numerous biologically active organic compounds (phytochemicals) present in these plants stands as a significant natural entity aiding metal nanoparticle synthesis. These phytochemicals show dual roles: stabilization as well as reduction of nanoparticles.

Bhakya et al. [128] reported silver nanoparticles (Ag NPs) synthesis following green method; different parts of Vishanika (Indian screw tree), an ayurvedic medicinal plant, was used for synthesis. They performed the catalytic activity of the biosynthesized

Table 9 Decolorization of various dyes using nanotechnology based treatment process.

Type of nanoparticle	Dye degraded	Optimum condition	% Removal of dye	References
TiO_2	Acid red 44	The impact of pH on the degradation was studied. The decolorization of dye onto TiO_2 nanocrystalline powder, $D_{av} = 30$ nm, under visible light	84	[114]
TiO_2/CdO-ZnO	Blue azo dye	Best condition: [dye] = 100 mg/L, pH = 3, temperature 85°C [H_2O_2] = 400 mg/L Catayst: TiO_2/CdO-ZnO films	73	[115]
ZnO	Methyl orange	ZnO powders, $D_{av} = 200$ nm, prepared via thermal evaporation process	95	[116]
TiON/Cu/Co (copper and cobalt co-modified nitrogen doped titania)	Erichrome black-T	Dye concentration = 37 mg/L Catalyst concentration = 0.05 g/L, pH = 4	95	[117]
L-Glutatione modified ultrathin SnS_2 nanosheets	Methyl orange	Dye concentration = 260 mg/g Time = 20 min	81	[118]
Copper ferrite ($CuFe_2O_4$) nanoparticle	Reactive Red 120, Reactive Red 198	Dye concentration = 50 mg/L Catalyst concentration = 0.03 g/L pH = 3	78	[119]
Pristine ZnO, ZnO/CNTs composites	Rhodamine B	Dye concentration = 30 mg/L Catalyst concentration = 0.06 g/L	84	[120]
SnO_2	Methylene blue	It was proved that the dye degradation was 0.0952/min, and the degradation reaction completed at 70 min exactly.	77	[121]
CeO_2/Bi_2MoO_6	Rhodamin B, Methyl orange	CeO_2/Bi_2MoO_6 heterojunction having CeO_2/Bi_2MoO_6 weight ratio 0.05, i.e., (0.05Ce-Bi) displayed highest photocatalytic efficiency degrading RhB in 75 mins.	100	[122]
α-Fe_2O_3	Methylene blue	α-Fe_2O_3 particles were prepared accompanied with various alkali metal (lithium, sodium, and potassium) acetates under hydrothermal method. Dye concentration = 2×10^{-5} (M) Catalyst dosage = 0.375 g/L Fe_2O_3-Na shows the best catalytic performance. The MB decolorization occurred at faster under natural pH condition. The intermediates of the degradation were identified.	90	[123]
Porous cobalt oxide (Co_3O_4)	Chicago Sky Blue 6B	Co_3O_4 prepared under hydrothermal system at 150°C maintained for 5 h shows maximum catalytic activity, rate constant = 56.8×10^{-3} min^{-1}.	98	[124]

nanoparticles toward the degradation of organic dyes, e.g., eosin, methyl violet, methylene blue, safranin, and methyl orange. It is reported that by increasing the Ag NPs catalyst dose from 20 to 100 μg/mL, the degradation rate of safranin and eosin methylene blue was increased up to 100%. Hike in concentration of Ag NPs presented increased degradation as with the increase in catalyst concentration number of reaction sites increases.

Nano zero-valent iron (nZVI) particles have a high efficiency in removing a variety of dye. Rahman et al. [129] reported the use of nZVI for the removal of three azo dyes, namely Sunset yellow, Acid blue A, and Methyl orange. It investigated that increasing the dosage of nZVI particles favor the decolorization of the dyes. The percentage of degradation decreased with increasing concentration of dyes. The removal efficiency of dyes is favorable in an acidic pH environment.

Bimetallic nanoparticles: Generally, bimetallic nanoparticles can be synthesized by a combination of two metal nanoparticles with several architectures [130]. Bimetallic nanoparticles possess characteristic patterns and geometrical architecture that enhances their functionality [131,132]. The synergistic effects of two metal nanoparticles show certain new properties that increases their function and application in many different fields [133,134].

Bokare et al. [135] examined the reductive degradation of Orange G, azo dye, with Fe-Ni bimetallic nanoparticles in aqueous solution. The dye degradation kinetics indicated that the rate of reaction majorly depends on initial dye concentration, Fe-Ni concentration, and pH of solution. The resulting products produced from degradation of this dye found to be nondegradable, noxious aromatic amines (naphthol amines, aniline) that require further attention and treatment.

Wang et al. [136] reported a postimpregnation followed by sodium borohydride reduction approach to synthesize a series of hollow mesoporous silica supported Fe-Cu bimetallic nanoparticles possessing different Fe/Cu ratios. They explored the difference of catalytic reactivity for orange II dye degradation with a synthesized catalyst containing varying composition. The optimal condition for degradation comprised of orange II dye with concentration 100 mg/L, H_2O_2 concentration of 27.4 mM, and catalyst of 1.0 g/L, pH 7.0, and 30°C. The results showed that the catalytic activity is largely dependent upon the Fe/Cu ratio. The most important fact of this study is the recoverability of the catalyst. It showed a good catalytic efficiency even after five consecutive runs. Thus, produced composite catalyst possesses a solid potential to act as a smart alternative toward dye wastewater treatment.

Zeolite/silicates nanoparticle: Zeolites are microporous, aluminosilicate minerals based on skeletal structure that possesses countless channels and pores with sizes ranging from nano- to micrometers. Their structure resembles a sponge whose holes are in the micro- or nanometer range. Researcher worldwide found particular interest in zeolite due to its excellent catalytic property that allows liquids to pass easily through the pores while accessing zeolite pore walls for furnishing degradation-based chemical reactions. The excellent well-organized porous structure in zeolites imparts better filtration process such that the nanopores can separate even very small substances. Nowadays, it has received growing attention for its effective ion exchange property [137].

Chong et al. [138] synthesized TiO_2-zeolite composite by sol-gel method and measured the photoactivity of synthesized TiO_2-zeolite nanocomposite. The results showed that the synthesized TiO_2-zeolite nanocomposite follows a more adsorption-oriented photocatalytic degradation of water pollutants, which is useful for removing trace and untreated dye compounds in the advanced industrial dye wastewater treatment stage.

Metal oxide nanoparticle: Metal and metal oxide-based nanoparticles are widely used nanoparticles for the degradation of organic pollutants present in wastewater. Porous transition metal oxides find special research interest as they have applications in catalysis, as electrode materials for sensors, batteries, etc. [139,140] because of their large internal surface area, nanosized walls, and d electrons in an open shell. Recently, a huge number of metal oxide nanoparticles like TiO_2, MnO_2 ZnO, CeO_2, Fe_2O_3, SnO_2, etc., are known to be useful for water remediation purposes.

Roy et al. [124] used a facile hydrothermal synthesis of Co_3O_4 nanostructures with tunable textural properties and morphology via hydrothermal route. Oxidative catalytic degradation of an azo dye, Chicago Sky Blue 6B, in presence of H_2O_2 as an oxidant, was performed with prepared Co_3O_4 particles. The effect of morphology and textural properties on the catalytic behaviors of Co_3O_4 was also examined. The experiments proved that mesoporous nanorod-like Co_3O_4 particles synthesized from 2 h and 5 h reaction time with higher pore volume and mesoporous surface area were catalytically more active compared with those of nanosheet-like particles attained from 12 h and 24 h reaction time with microporous surface area and smaller pore volume. Thus, catalyst prepared tuning the surface properties and morphology Co_3O_4 and other metal oxides could be applied to many other fields.

Suvanka Dutta et al. [141] synthesized ZnO nanorods by simple alkaline hydrolysis method at elevated temperature. Photocatalytic degradation of the model reactive diazo dye RB 31 by ZnO nanorod under UV irradiation in an aqueous medium was carried out in the context of response surface methodology (RSM). It was concluded that decolorization efficiency increases with increasing pH and doses of ZnO nanorods, whereas the removal efficiency is inversely related to initial dye concentration. Thus, optimizing the process by RSM based on CCD showed 96.9% removal efficiency under optimized operational conditions.

Magnetic nanoparticle: Iron can exist in variable oxidation states as it is one of the highly reactive metals, which allows it to coordinate with other elements. The three commonly used oxide forms in nature are magnetite (Fe_3O_4), hematite (γ-Fe_2O_3), and maghemite (α-Fe_2O_3) [142,143]. Based on properties like high surface area-to-volume ratio, high magnetic susceptibility, and excellent biocompatibility, these materials have more applicative importance [144]. Bare magnetite nanoparticles have the tendency of oxidation by air and aggregation in aqueous systems [145].

Shuang et al. [146] synthesized α-Fe_2O_3 nanoparticles by a hydrothermal route to study both magnetic as well as photocatalytic properties. The α-Fe_2O_3 powders with the smaller crystallite sizes show the highest photocatalytic degradation efficiency than that of the powders with larger crystallite sizes.

Cha et al. [147] prepared α-Fe_2O_3 nanorod thin films by metal organic chemical vapor deposition technique on Si substrates and studied the photocatalytic degradation of RhB. In the presence of α-Fe_2O_3 thin films, 99% of the RhB was found to degrade in 6 h.

Chunhua Liang et al. [148] synthesized maghemite (γ-Fe_2O_3) nanopowders under aeration (oxidizing) conditions. With an increase in the reaction pH, reaction temperature, and reaction time, γ-Fe_2O_3 was successfully synthesized with better crystallization and larger particle size. All the prepared γ-Fe_2O_3 powders had significant photocatalytic activities achieving up to 48.89% removal and 36.5% mineralization of 20 mg/L of Orange I solutions under UV light irradiation.

Transition metal oxide nanoparticle: Nowadays, transition metal oxides are used as photocatalysts for wastewater treatment purposes. They are actually semiconductor materials that possess high surface area-to-volume ratio, suitable band gap, crystallinity, photo and chemical stability, etc.

Maria Stylidi et al. [149] reported the photocatalytic degradation of Acid Orange 7 in TiO_2 suspensions with the use of a solar light simulating source. Decolorization and eventually complete mineralization of the solution was observed. A reaction pathway was also proposed stating that the degradation of the dye molecules is achieved via a series of successive oxidation steps.

Renu Sankar et al. [150] synthesized colloidal Cu-oxide nanoparticle via *Carica papaya* plant and showed the photocatalytic degradation of Coomassie brilliant blue dye. It was proved that more the incubation time of nanoparticles was increased, the rates of decolorization of dye also increased.

MgO nanoparticle: Nanosized non-toxic alkaline earth metal oxides, in particular magnesium oxide (nano-MgO), are an interesting multifunctional and exceptionally important material. It revels excellent electrical, optical, thermodynamic, mechanical, and special chemical properties. This material has also been known as bactericides. Furthermore, nano MgO has extensively been used as a destructive adsorbent for the removal of many toxic chemicals due to their high surface reactivity and adsorption capacity [151].

Dhal et al. [152] showed the ability of MgO nanomaterials with various morphologies (MgO hierarchical nanostructures, nanorods, and nanoflakes) to act as adsorbents for the removal of toxic dyes such as Congo red and Malachite green.

Spinel metal oxide nanoparticle: Transition metal oxides with spinel type structure, possessing a unique general chemical formula AB_2O_4, A (occupying tetrahedral hole) and B (occupying octahedral hole) are two different transitional metal cations of 3d series with +2 and + 3 charge, respectively, have attracted considerable interest of the scientific community. The coupling of two different metal species of multiple valence states renders rich redox reactions in these spinel-mixed oxides. Manifestation of the enhanced performance by these ternary oxides than the oxides of their individual metal counterparts proved the synergistic effect between metal species [153]. Thus, these materials reveal interesting applications [154].

Yazdanbakhsh et al. [155] fabricated $ZnCr_2O_4$ nanospinel following sol-gel process and applied on Reactive blue 5 dye photodegradation. Experimental outcomes of photocatalytic color removal of azo-dye Reactive Blue 5 by $ZnCr_2O_4$ disclose that the decolorization process can be attained by adsorption method. The photocatalytic degradation of dye via $ZnCr_2O_4$ spinel under UV treatment at pH $= 1$ has been proved as the optimum condition.

Graphene oxide and graphene oxide-based nanomaterials: Graphene is a two-dimensional nanomaterial with few layers densely packed in six-membered rings of sp^2 carbon atoms, which presents incredible properties, such as high surface area (with a theoretical value of specific surface area at $2630 \, m^2/g$), strong mechanical properties, electrical properties, and chemical stability. Accordingly, it has the potential of applications in nanoelectronic devices, sensors, and nanocomposite materials. The graphene composite nanomaterials possess dual properties, high surface area, and catalytic active site of the nanomaterial.

Dutta et al. [156] followed a facile method of synthesis of copper sulfide in djurleite phase and reduced graphene oxide djurleite copper sulfide composites—under the action of sulfur-containing biomolecule, acting both as reducing agent and sulfur-providing source. The composite revealed outstanding photocatalytic activity under visible light as well as excellent phenol-sensing property.

Sun et al. [157] studied the use of magnetite/reduced graphene oxide nanocomposites (MRGO) to remove Rhodamine B and Malachite green. MRGO displayed high removal competence for Rh B (over 91%) and MG (over 94%). The removal efficiencies of Rh B and MG were still over 80% after five cycles. The high performance of MRGO suggested that this can be used effectively for the removal of dye pollutants.

3 Comparison between different treatment processes

To get an understanding of the efficiencies of different dye removal procedures, we need to go for a comparative study of various treatment techniques toward degradation of same dye or dye effluents under almost same experimental conditions. Table 10 summarizes a few literature reports depicting comparative study (between two or more techniques) for dye color and COD removal from wastewater, applying different treatment processes.

4 Conclusion and future perspective

Here in this chapter we have discussed the origin of synthetic dyes and their market application in innumerable ways. However, the enormous use of dyes and its untreated disposal imparts a threat to the environment and even to human beings in multiple ways. This posed a throttle to the concerned industries (textile industry in particular) to urgently come up with an efficient effluent treatment methodology. The need to reuse the treated dye wastewater has become important. This chapter had summarized the different wastewater treatment methods in terms of biological, chemical, and physical means.

Table 10 Comparative study for removal of various dyes or dye effluent using different treatment processes.

Name of the dye	Biological treatment technique	Physical treatment technique	Chemical treatment technique	Combinatorial treatment technique	Nanotechnology-based treatment technique	References
Reactive dye with azo functional group 1. Reactive Black 5 (RB5) 2. Reactive Red 231 (RR231)	**Enzymatic treatment:** Laccase from ascomycete *M. thermophile* selected Condition: dye solution = 0.1 g/L, at pH 5, at 50°C with 4 U/mL of [enzyme], time 24–48 h. Efficiency: 53%–61% (hydrolyzed dye) 88%–91% (unhydrolyzed dye) for RB5 88%–93% for RR231 dye.	**Coagulation:** Condition: 1000 mg/L *Moringa oleifera* oilseed residues suspension stirred at rt. for 2 h, 0.1 g/L dye. Efficiency: 90%–94% for RR231and RB5 slightly lower efficiency	**Electrochemical treatment:** Condition: Cathodes-Ti, Anodes-Ti covered by PtOx, flow rate = 20 L/h, [dye] = 0.1 g/L, [NaCl] = 20 g/L; current density = 24 mA/cm^2 Efficiency: Color removal 95%–96% for RB5 and 98%–100% for RR231 dye	–	–	[158]
Color effluent from Organized Industrial District (OID)	–	**Coagulation:** Condition: effluent = 500 mL, at pH 12.2 with a mixture MgCl$_2$ + Ca(OH)$_2$ dosage = 240 mg/L Efficiency: 87.4% of color removal	**Fenton and Fenton-like oxidation processes:** Conditions: FeSO$_4$·7H$_2$O, or FeCl$_3$·6H$_2$O (50–300 mg/L dosage), pH 2–6, H$_2$O$_2$ Efficiency: 93%, 80%, respectively, of color removal **Ozonation:** Conditions: ozone dosage 8 g/L·min, pH 12, contact time 3 min Efficiency: 93.6% color removal	–	–	[159]

Dye waste effluents of cotton, synthetic and woven fabric processing	**Aerobic biodegradation:** Condition: F/M ratio is 0.7, pH within 6.5–7.5. Efficiency: COD removal of 57%–67% observed.	**Coagulation:** Condition: 500 mL of dye waste, pH in range 4–12, chemical coagulant are $FeCl_3.6H_2O$, $FeSO_4.7H_2O$ and alum (~100–1250 mg/L). Anionic polyelectrolyte was added to increase flocculation. Efficiency: 45%–77% color and 48%–68% COD removal.	**Ozonation:** Condition: ozone dosage 0.8 g/L, for 30 min. Efficiency: 78%–100% of color removal and 5%–18% of COD removal seen **Combination of Ozone oxidation and biological process:** Efficiency: 62%–82% of COD removal. **Combined of Ozone oxidation and coagulation process:** Efficiency: 59%–71% of COD removal.	– [160]
Magenta MB	–	**Homogeneous Fenton's process:** Condition: [Initial dye] = 50 mg/L, $[H_2O_2] = 0.38$ M, $[FeSO_4] = 400$ mg/L, pH of 3, contact time = 15 min. Efficiency: Dye removal 90.9%.	**Heterogeneous Fenton's process:** Condition: [Initial dye] = 50 mg/L, $[H_2O_2] = 0.26$ M, $[Fe_3O_4$ nanocatalyst] = 600 mg/L, pH of 2, contact time = 60 min. Efficiency: Dye removal 86.8%	– [161]
Dye effluents from industry	–	**Photochemical process:** Condition: 100 mL of effluents, 1.0 mol L^{-1} $FeCl_3$ solution, Irradiation by UV-A light ($\lambda_{max} = 366$ nm, dissolved oxygen concentration >80%, temperature 20 ± 2°C. Efficiency: Even after 300 min of irradiation no significant change is observed in COD.	Condition: 100 mL of effluents, 2.5 g L^{-1} TiO_2 nanoparticle solution, Irradiation by UV-A light. Efficiency: Cationic dye degradation 37%–86%, but 14% COD removal after 300 min irradiation. Anionic dyes COD removal from effluents after 300 min irradiation was 63% and 11% for Acid Red 88 and Direct Black 22 dyes, respectively.	– [162]

The purpose of this chapter is to discuss the different methodologies, optimum removal conditions, and the removal efficiency (percentage of dye removal) of the contaminants. The combined use of different treatment methods has also been shown advantageous in a few cases of dealing with textile dye effluent. However, this literature survey demonstrates that nanotechnology-based treatment method is proved to be a potential candidate as the nanomaterials can be reused several times, which brings down the cost of running a treatment plant. Still further research work is required to ascertain a single plan, such that it becomes a solution to all the existing drawbacks that the current water treatment technologies are facing.

Acknowledgment

The Department of Science and Technology under the SERB (N-PDF) sponsored project (No. PDF/2017/000390), Government of India, is gratefully acknowledged.

References

[1] K. Othmer, Encycl. Chem. Technol. 7 (5) (2004).

[2] J.N. Murrel, The Theory of the Electronic Spectra of Organic Molecules, John Wiley & Sons, NJ, 1973.

[3] A. Pandey, P. Singh, L. Iyengar, Bacterial decolorization and degradation of azo dyes, Int. Biodeterior. Biodegrad. 59 (2) (2007) 73–84.

[4] M.S. Adb-Al-Kareem, H.M. Taha, Decolorization of malachite green and methylene blue by two microalgal species, Int. J. Chem. Environ. Eng. 3 (2012) 297–302.

[5] S. Garfield, Mauve: How One Man Invented a Color that Changed the World, W. W. Norton & Company, NY, 2002.

[6] Z. Aksu, G. Karabayir, Comparison of biosorption properties of different kinds of fungi for the removal of Gryfalan black RL metal-complex dye, Bioresour. Technol. 99 (2008) 7730–7741.

[7] Z. Aksu, S. Tezer, Equilibrium and kinetic modeling of biosorption of Remazol black B by Rhizopusarrhizus in a batch system: effect of temperature, Process Biochem. 36 (2000) 431–439.

[8] P.J.T. Morris, A.S. Travis, A history of the international dyestuff industry, Am. Dyest Rep. 81 (11) (1992).

[9] M.R. Khan, A.S.W. Kurny, F. Gulshan, Parameters affecting the photocatalytic degradation of dyes using TiO_2: a review, Appl Water Sci 7 (2017) 1569–1578.

[10] A. Akbari, J.C. Remigy, P. Aptel, Treatment of textile dye effluent using a polyamide-based nanofiltration membrane, Chem. Eng. Process. 41 (2002) 601–609.

[11] G. Mishra, M. Tripati, A critical review of the treatment for the decolorization of textile effulent, Colourage 40 (1993) 35–38.

[12] H. Alhassani, R. Muhammad, S. Ashraf, Efficient microbial degradation of toluidine blue dye by Brevibacillus sp, Dyes Pigments 75 (2007) 395–400.

[13] Society of Dyers and Colourists, Colour Index, Vol. 4, The Society of Dyers and Colourists, Bradford, UK, 1971, 3.

[14] K.A. Adegoke, O.S. Bello, Dye sequestration using agricultural wastes as adsorbents, Water Resour. Ind. 12 (2015) 8–24.

[15] V.K. Gupta, Suhas, application of low-cost adsorbents for dye removal—a review, J. Environ. Manag. 90 (8) (2009) 2313–2342.

[16] C.R. Holkar, A.J. Jadhav, D.V. Pinjari, N.M. Mahamuni, A.B. Pandit, A critical review on textile wastewater treatments: possible approaches, J. Environ. Manag. 182 (2016) 351–366.

[17] A. Srinivasan, T. Viraraghavan, Decolorization of dye wastewaters bybiosorbents: a review, J. Environ. Manag. 91 (10) (2010) 1915–1929.

[18] K. Hunger (Ed.), Industrial Dyes Chemistry, Properties, Applications, Wiley-VCH Verlag GmbH & Co. KGaA, Weinheim, 2003.

[19] C. Sandoval, G. Molina, P.V. Jentzsch, J. Pérez, F. Muñoz, Photocatalytic degradation of azo dyes over semiconductors supported on polyethylene terephthalate and polystyrene substrates, J. Adv. Oxid. Technol. 20 (2) (2017) 20170006.

[20] J. Madhavan, F. Grieser, M. Ashokkumar, Degradation of Orange-G by advanced oxidation processes, Ultrason. Sonochem. 17 (2010) 338–343.

[21] L. He, H.S. Freeman, L. Lu, S. Zhang, Spectroscopic study of anthraquinone dye/amphiphile systems in binary aqueous/organic solvent mixtures, Dyes Pigments 91 (2011) 389–395.

[22] N. Mathur, P. Bhatnagarp, P. Sharma, Review of the mutagenicity of textile dye products, Universal J. Environ. Res. Technol. 2 (2) (2012) 1–18.

[23] W.Y. Li, F.F. Chen, S.L. Wang, Binding of reactive brilliant red to human serum albumin: insights into the molecular toxicity of sulfonic azo dyes, Protein Pept. Lett. 17 (5) (2010) 621–629.

[24] S. Srivastava, R. Sinha, D. Roy, Toxicological effects of malachite green, Aquat. Toxicol. 66 (3) (2004) 319–329.

[25] M. Sudha, A. Saranya, G. Selvakumar, Microbial degradation of azo dyes: a review, Int. J. Curr. Microbiol. Appl. Sci. 3 (2) (2014) 670–690.

[26] K.P. Gopinath, S. Murugesan, J. Abraham, Bacillus sp. mutant for improved biodegradation of Congo red: random mutagenesis approach, Bioresour. Technol. 100 (24) (2009) 6295–6300.

[27] H.B. Mansour, Y. Ayed-Ajmi, R. Mosrati, Acid violet 7 and its biodegradation products induce chromosome aberrations, lipid peroxidation, and cholinesterase inhibition in mouse bonemarrow, Environ. Sci. Pollut. Res. 17 (7) (2010) 1371–1378.

[28] A.B. dos Santos, F.J. Cervantes, J.B.V. Lier, Review paper on current technologies for decolourisation of textile wastewaters: perspectives for anaerobic biotechnology, Bioresour. Technol. 98 (12) (2007) 2369–2385.

[29] N. Ballav, R. Das, S. Giri, A.M. Muliwa, K. Pillay, A. Maity, L-cysteine doped polypyrrole (Ppy@L-cyst): a super adsorbent for the rapid removal of Hg^{+2} and efficient catalytic activity of the spent adsorbent for reuse, Chem. Eng. J. 345 (2018) 621–630.

[30] K.D. Mojsov, D. Andronikov, A. Janevski, A. Kuzelov, S. Gaber, The application of enzymes for the removal of dyes from textile effluents, Adv. Technol. 5 (1) (2016) 81–86.

[31] A. Srinivasan, T. Viraraghavan, Decolorization of dye wastewaters by biosorbents: a review, J. Environ. Manag. 91 (10) (2010) 1915–1929.

[32] N. Das, D. Charumathi, Remediation of synthetic dyes from wastewater using yeast a review, Ind. J. Biotechnol. 11 (4) (2012) 369–380.

[33] M.Z. Alam, M.J.H. Khan, N.A. Kabbashi, S.M.A. Sayem, Development of an effective biosorbent by fungal immobilization technique for removal of dyes, Waste Biomass Valoriz. 9 (4) (2017) 681–690.

[34] Z. Aksu, S. Tezer, Biosorption of reactive dyes on the green alga *Chlorella vulgaris*, Process Biochem. 40 (3–4) (2005) 1347–1361.

[35] B. Rani, V. Kumar, J. Singh, S. Bisht, P. Teotia, S. Sharma, R. Kela, Bioremediation of dyes by fungi isolated from contaminated dye effluent sites for bio-usability, Braz. J. Microbiol. 45 (3) (2014) 1055–1063.

[36] T. Joshi, L. Iyengar, K. Singh, S. Garg, Isolation, identification and application of novel bacterial consortium TJ-1 for the decolorization of structurally different azo dyes, Bioresour. Technol. 99 (15) (2008) 7115–7121.

[37] L. Ayed, A. Mahdhi, A. Cheref, A. Bakhrouf, Decolorization and degradation of azo dye methyl redby an isolated *Sphingomonas paucimoboilis*: biotoxicity and metabolites characterization, Desalination 274 (1–3) (2011) 272–277.

[38] D. Charumathi, N. Das, Bioaccumulation of syntheticdyes by *Candida tropicalis* growing in sugarcane bagasseextract medium, Adv. Biol. Res. 4 (4) (2010) 233–240.

[39] Q. Yang, M. Yang, K. Pritsch, A. Yediler, A. Hagn, M. Schloter, A. Kettrup, Decolorization of synthetic dyes andproduction of manganese-dependent peroxidase by newfungal isolates, Biotechnol. Lett. 25 (2003) 709–713.

[40] R.P. Singh, P.K. Singh, R.L. Singh, Bacterial decolorization of textile azo dye acid orange by *Staphylococcus hominis* RMLRT03, Toxicol. Int. 21 (2) (2014) 160–166.

[41] M.S. Khehra, H.S. Saini, D.K. Sharma, B.S. Chadha, S.S. Chimni, Biodegradation of azo dye C.I. Acid Red 88 by ananoxic-aerobic sequential bioreactor, Dyes Pigments 70 (1) (2006) 1–7.

[42] K. Murugesan, I.H. Yang, Y.M. Kim, J.R. Jeon, Y.S. Chang, Enhanced transformation of malachite green by laccase of Ganodermalucidum in the presence of natural phenolic compounds, Appl. Microbiol. Biotechnol. 82 (2) (2009) 341–350.

[43] H. Horitsu, M. Takada, E. Idaka, M. Tomoyeda, T. Ogawa, Degradation of p-aminoazobenzene by *Bacillus subtilis*, Eur. J. Appl. Microbiol. Biotechnol. 4 (1977) 217–224.

[44] M. Tekere, A.Y. Mswaka, R. Zvauya, J.S. Read, Growth dye degradation and lignolytic activity studies on Zimbabwean white rot fungi, Enzym. Microb. Technol. 28 (4–5) (2001) 420–426.

[45] J.A. Libra, M. Borchent, S. Banit, Competition strategiesfor the decolorization of a textile reactive dye with thewhite rot fungi *Trametes versicolor* under non sterileconditions, Biotechnol. Bioeng. 82 (6) (2003) 736–744.

[46] V. Christian, R. Shrivastava, D. Shukla, H.A. Modi, B.R.M. Vyas, Degradation of xenobiotic compounds by lignindegrading white rot fungi: enzymology and mechanism involved, Ind. J. Exp. Biol. 43 (4) (2005) 301–312.

[47] Z. Aksu, Application of biosorption for the removal of organic pollutants: a review, Process Biochem. 40 (3–4) (2005) 997–1026.

[48] L. Jinqi, L. Houtian, Degradation of azo dyes by algae, Environ. Pollut. 75 (3) (1992) 273–278.

[49] M.M. Kamel, R.M. El-Shishtawy, B.M. Yussef, H. Mashaly, Ultrasonic assisteddyeing III. Dyeing of wool with lac as natural dye, Dyes Pigm. 65 (2) (2005) 103–110.

[50] H.H. Omar, Algal decolorization and degradation ofmonoazo and diazo dyes, Pak. J. Biol. Sci. 11 (10) (2008) 1310–1316.

[51] M.M. Martorell, H.F. Pajot, L.I. de Figueroa, Dye-decolorizing yeasts isolated from Las Yungas rainforest. Dye assimilation and removal used as selection criteria, Int. Biodeter. Biodegr. 66 (1) (2012) 25–32.

[52] C. Allegre, M. Maisseu, F. Charbit, P. Moulin, Coagulation-flocculationdecantationof dye house effluents: concentrated effluents, J. Hazard. Mater. 116 (1–2) (2004) 57–64.

[53] R.G. Saratale, G.D. Saratale, J.S. Chang, S.P. Govindwar, Decolorization and biodegradation of reactive dyes and dye wastewater by a developed bacterial consortium, Biodegradation 21 (6) (2010) 999–1015.

[54] R.G. Saratale, G.D. Saratale, D.C. Kalyani, J.S. Chang, S.P. Govindwar, Enhanced decolorization and biodegradation of textile azo dye Scarlet R by using developed microbial consortium-GR, Bioresour. Technol. 100 (9) (2009) 2493–2500.

[55] S.R. Dave, T.L. Patel, D.R. Tipre, Bacterial Degradation of AZO Dye Containing Wastes, Springer International Publishing, 2015, pp. 57–83.

[56] K. Wuhrmann, K.L. Mechsner, T.H. Kappeler, Investigation on rate—determining factors in the microbial reduction of azo dyes, Appl. Microbiol. Biotechnol. 9 (4) (1980) 325–338.

[57] M. Solís, A. Solís, H.I. Pérez, et al., Microbial decolouration of azo dyes: a review, Process Biochem. 47 (12) (2012) 1723–1748.

[58] E. Acuner, F.B. Dilek, Treatment of tectilon yellow 2G by *Chlorella vulgaris*, Process Biochem. 39 (5) (2004) 623–631.

[59] R.L. Singh, P.K. Singh, R.P. Singh, Enzymatic decolorization and degradation of azo dyes—a review, Int. Biodeter. Biodegr. 104 (2015) 21–31.

[60] A. Pandey, P. Singh, L. Iyengar, Bacterial decolorization and degradation of azo dyes, Int. Biodeterior. Biodegradation 59 (2) (2007) 73–84.

[61] T. Hadibarata, L.A. Adnan, A.R.M. Yusoff, A. Yuniarto, Rubiyanto, M.M.F.A. Zubir, A.B. Khudhair, Z.C. Teh, M.A. Naser, Microbial decolorization of an azo dye Reactive Black 5 using white rotfungus *Pleurotus eryngii* F032, Water Air Soil Poll. 224 (2013) 1595–1604.

[62] D. Wesenberg, I. Kyriakides, S.N. Agathos, White-rot fungi and their enzymes for the treatment of industrial dye effluents, Biotech. Adv. 22 (1–2) (2003) 161–187.

[63] R. Mehta, P. Singhal, H. Singh, et al., Insight into thermophiles and their wide-spectrum applications, Biotech. 6 (1) (2016) 1–9.

[64] M. Imran, F. Negm, S. Hussain, et al., Characterization and purification of membrane-bound azoreductase from azo dye degrading *Shewanella* sp. strain IFN4, CLEAN Soil Air Water 44 (11) (2016) 1523–1530.

[65] S. Galai, H. Korri-Youssoufi, M.N. Marzouki, Characterization of yellow bacterial laccase SmLac/role of redox mediators in azo dye decolorization, J. Chem. Technol. Biotechnol. 89 (11) (2014) 1741–1750.

[66] M.M. Al-Ansari, B. Saha, S. Mazloum, K.E. Taylor, J.K. Bewtra, N. Biswas, Soybean Peroxidase Applications in Wastewater Treatment, 6th ed., Nova Science Publishers, 2011.

[67] P.K. Singh, R.L. Singh, Bio-removal of azo dyes: a review, Int. J. Appl. Sci. Biotechnol. 5 (2) (2017) 108–126.

[68] A.D. Watharkar, R.V. Khandare, A.A. Kamble, A.Y. Mulla, S.P. Govindwar, J.P. Jadhav, Phytoremediation potential of *Petunia grandiflora* Juss., an ornamental plant to degrade a disperse, disulfonated triphenylmetnae textile dye Brilliant Blue G, Environ. Sci. Pollut. Res. Int. 20 (2) (2013) 939–949.

[69] G.S. Ghodake, A.A. Telke, J.P. Jadhav, S.P. Govindwar, Potential of *Brassica juncea* in order to treat textile effluent contaminated sites, Int. J. Phytoremediation 11 (4) (2009) 297–312.

[70] A. Mukimin, H. Vistanty, N. Zen, Oxidation of textile wastewater using cylinder Ti/β-PbO$_2$ electrode in electrocatalytic tube reactor, Chem. Eng. J. 259 (1) (2015) 430–437.

[71] M.F. Sevimli, H.Z. Sarikaya, Ozone treatment of textile effluents and dyes: effect of applied ozone dose, pH and dye concentration, J. Chem. Technol. Biotechnol. 77 (7) (2002) 842–850.

[72] E.E. Ebrahiem, M.N. Al-Maghrabi, A.R. Mobarki, Removal of organic pollutants from industrial wastewater by applying photofenton oxidation technology, Arab. J. Chem. 10 (2) (2017) S1674–S1679.

[73] A. Mukimin, K. Wijaya, A. Kuncaka, Oxidation of remazolbrilliant Bluer (RB.19) with in situ electrogenerated active chlorine using Ti/PbO$_2$ electrode, Sep. Purif. Technol. 95 (19) (2012) 1–9.

[74] G. Li, N. Wang, B. Liu, X. Zhang, Decolorization of azo dye Orange II by ferrate(VI)-hypochlorite liquid mixture, potassium ferrate(VI) and potassium permanganate, Desalination 249 (3) (2009) 936–941.

[75] S. Cotillas, J. Llanos, P. Canizares, D. Clematis, G. Cerisola, M.A. Rodrigo, M. Panizza, Removal of procion red MX-5B dye from wastewater by conductive-diamond electrochemical oxidation, Electrochim. Acta 263 (2018) 1–7.

[76] M. Amini, M. Arami, N.M. Mahmoodi, A. Akbari, Dye removal from colored textile wastewater using acrylic grafted nanomembrane, Desalination 267 (1) (2011) 107–113.

[77] N. Ertugay, F.N. Acar, Removal of COD and color from direct blue 71 azo dye wastewater by Fenton's oxidation: kinetic study, Arab. J. Chem. 10 (1) (2017) S1158–S1163.

[78] J. Hoigne, Mechanisms, Rates and selectivities of oxidations of organic compounds initiated by ozonation of water, in: Handbook of Ozone Technology and Applications, Vol. 1, Ann Arbor Science Publisher, 1982.

[79] F.H. Oliveira, M.E. Osugi, F.M.M. Paschoal, D. Profeti, P. Olivi, M.V.B. Zanoni, Electrochemical oxidation of an acid dye by active chlorine generated using Ti/Sn(1–x)Ir x O2 electrodes, J. Appl. Electrochem. 37 (5) (2007) 583–592.

[80] C. Gottschalk, J.A. Libra, A. Saupe, Ozonation of Water and Waste Water, Wiley-VCH, 2000.

[81] E. Forgacs, T. Cserháti, G. Oros, Removal of synthetic dyes from wastewaters: a review, Environ. Int. 30 (7) (2004) 953–971.

[82] M. Aleksić, H. Kušić, N. Koprivanac, D. Leszczynska, A.L. Božić, Heterogeneous Fenton type processes for the degradation of organic dye pollutant in water – the application of zeolite assisted AOPs, Desalination 257 (1–3) (2010) 22–29.

[83] Z. Eren, N.H. Ince, Sonolytic and sonocatalytic degradation of azo dyes by low and high frequency ultrasound, J. Hazard. Mater. 177 (1–3) (2010) 1019–1024.

[84] S. Wang, C.W. Ng, W. Wang, Q. Li, L. Li, A comparative study on the adsorption of acid and reactive dyes on multiwall carbon nanotubes in single and binary dye systems, J. Chem. Eng. Data 57 (5) (2012) 1563–1569.

[85] M. Panizza, G. Cerisola, Electrocatalytic materials for the electrochemical oxidation of synthetic dyes, Appl. Catal. B 75 (1–2) (2007) 95–101.

[86] A. Wang, Y.-Y. Li, J. Ru, The mechanism and application of the electro-Fenton process for azo dye Acid Red 14 degradation using an activated carbon fibre felt cathode, J. Chem. Technol. Biotechnol. 85 (11) (2010) 1463–1470.

[87] N.F. Cardoso, R.B. Pinto, E.C. Lima, T. Calvete, C.V. Amavisca, B. Royer, M.L. Cunha, T.H.M. Fernandes, I.S. Pinto, Removal of remazol black B textile dye from aqueous solution by adsorption, Desalination 269 (1–3) (2011) 92–103.

[88] Y.-L. Song, J.-T. Li, H. Chen, Removal of acid Brown 348 dye from aqueous solution by ultrasound irradiated exfoliated graphite, Indian J. Chem. Technol. 15 (5) (2008) 443–448.

[89] E. Akceylan, M. Bahadir, M. Yilmaz, Removal efficiency of a calix[4]arene-based polymer for water-soluble carcinogenic direct azo dyes and aromatic amines, J. Hazard. Mater. 162 (2–3) (2009) 960–966.

[90] N.M.H. Al-Nakib, Reverse osmosis polyamide membrane for the removal of blue and yellow dye from waste water, Iraqi J. Chem. Petrol. Eng. 14 (2) (2013) 49–55.

[91] S.S. Moghaddam, M.R.A. Moghaddam, M. Arami, Coagulation/flocculation process for dye removal using sludge from water treatment plant: optimization through response surface methodology, J. Hazard. Mater. 175 (1–3) (2010) 651–657.

[92] A.H. Hassani, R. Mirzayee, S. Nasseri, M. Borghei, M. Gholami, B. Torabifar, Nanofiltration process on dye removal from simulated textile wastewater, Int. J. Environ. Sci. Technol. 5 (3) (2008) 401–408.

[93] S. Kaur, S. Rani, R.K. Mahajan, Adsorption kinetics for the removal of hazardous dye congo red by biowaste materials as adsorbents. J. Chem. 2013 (2013), 628582. https://doi.org/10.1155/2013/628582.

[94] V.K. Garg, R. Gupta, A. Bala Yadav, R. Kumar, Dye removal from aqueous solution by adsorption on treated sawdust, Bioresour. Technol. 89 (2) (2003) 121–124.

[95] B. Royer, N.F. Cardoso, E.C. Lima, T.R. Macedo, C. Airoldi, A useful organofunctionalized layered silicate for textile dye removal, J. Hazard. Mater. 181 (1–3) (2010) 366–374.

[96] G.L. Dotto, V.M. Esquerdo, M.L.G. Vieira, L.A.A. Pinto, Optimization and kinetic analysis of food dyes biosorption by *Spirulina platensis*, Colloids Surf. B Biointerfaces 91 (2012) 234–241.

[97] B.P. Cho, T. Yang, L.R. Blankenship, J.D. Moody, M. Churchwell, F.A. Bebland, S.J. Culp, Synthesis and characterization of N-demethylated metabolites of malachite green and leuco malachite green, Chem. Res. Toxicol. 16 (3) (2003) 285–294.

[98] E. Daneshvar, M. Kousha, M.S. Sohrabi, A. Khataee, A. Converti, Biosorption of three acid dyes by the brown macroalga *Stoechospermum marginatum*: isotherm, kinetic and thermodynamic studies, Chem. Eng. J. 195–196 (2012) 297–306.

[99] G. Crini, Non-conventional low-cost adsorbents for dye removal: a review, Bioresour. Technol. 97 (9) (2006) 1061–1085.

[100] S.M.A.G. Ulson de Souza, E. Forgiarini, A.A. Ulson de Souza, Toxicity of textile dyes and their degradation by the enzyme horseradish peroxidase (HRP), J. Hazard. Mater. 147 (3) (2007) 1073–1078.

[101] C.Y. Suzuki, Media optimization for laccase production by *Trichoderma harzianum* ZF-2 using response surface methodology, J. Microbiol. Biotechnol. 23 (12) (2004) 1757–1764.

[102] P. Bhattacharya, S. Dutta, S. Ghosh, S. Vedajnananda, S. Bandyopadhyay, Crossflow microfiltration using ceramic membrane for treatment of sulphur black effluent from garment processing industry, Desalination 261 (1–2) (2010) 67–72.

[103] J. Cheria, A. Bakhrouf, Triphenylmethanes, malachite green and crystal violet dyes decolorization by *Sphingomonas paucimobilis*, Ann. Microbiol. 59 (1) (2009) 57–61.

[104] I. Eichlerova, L. Homoika, F. Nerud, Ability of industrial dyes decolorization and ligninolytic enzymes production by different Pleurotus species with special attention on *Pleurotus calyptratus*, strain CCBAS 461, Process Biochem. 41 (4) (2006) 941–946.

[105] M. Joshi, R. Bansal, R. Purwar, Colour removal from textile effluents, Indian J. Fibre Text. Res. 29 (2) (2004) 239–259.

[106] A. Ahmad, S. Hamidah, M. Setapar, C.-S. Chuong, A. Khatoon, W.A. Wani, R. Kumar, M. Rafatullah, Recent advances in new generation dye removal technologies: novel search for approaches to reprocess wastewater, RSC Adv. 5 (39) (2015) 30801–30818.

[107] F. Harrelkas, A. Azizi, A. Yaacoubi, A. Benhammou, M.N. Pons, Treatment of textile dye effluents using coagulation flocculation coupled with membrane processes or adsorption on powdered activated carbon, Desalination 235 (1–3) (2009) 330–339.

[108] D. Cui, Y.-Q. Guo, H.-Y. Cheng, B. Liang, F.-Y. Kong, H.-S. Lee, A.-J. Wang, Azo dye removal in a membrane-free up-flow biocatalyzed electrolysis reactor coupled with an aerobic bio-contact oxidation reactor, J. Hazard. Mater. 239–240 (2012) 257–264.

[109] V. Buscio, M.G. Jiménez, M. Vilaseca, V. López-Grimau, M. Crespi, C. Gutiérrez-Bouzán, Reuse of textile dyeing effluents treated with coupled nanofiltration and electrochemical processes, Materials 9 (6) (2016) 490.

[110] Z. Shkavro, V. Kochkodan, R. Ognyanova, T. Budinova, V. Goncharuk, Combined ultrafiltration-adsorption water purification of the cationic violet dye, J. Water Chem. Technol. 32 (2) (2010) 101–106.

[111] N. Azbar, T. Yonar, K. Kestioglu, Comparison of various advanced oxidation processes and chemical treatment methods for COD and color removal from a polyester and acetate fiber dyeing effluent, Chemosphere 55 (1) (2004) 35–43.

[112] H. Masoumbeigi, A. Rezaee, Removal of methylene blue (MB) dye from synthetic wastewater using UV/H_2O_2 advanced oxidation process, J. Health Policy Sustain. Health 2 (1) (2015) 160–166.

[113] M.F. Abid, M.A. Zablouk, A.M. Abid-Alameer, Experimental study of dye removal from industrial wastewater by membrane technologies of reverse osmosis and nanofiltration, Iran. J. Environ. Health Sci. Eng. 9 (1) (2012) 17.

[114] J. Moon, C.Y. Yun, K.-W. Chung, M.-S. Kang, J. Yi, Photocatalytic activation of TiO_2 under visible light using acid red 44, Catal. Today 87 (1–4) (2003) 77–86.

[115] R. Suárez-Parra, I. Hernández-Pérez, M.E. Rincón, S. López-Ayala, M.C. Roldán-Ahumada, Visible light-induced degradation of blue textile azo dye on TiO_2/CdO–ZnO coupled nanoporous films, Sol. Energy Mater. Sol. Cells 76 (2) (2003) 189–199.

[116] H. Wang, C. Xie, W. Zhang, S. Cai, Z. Yang, Y. Gui, Comparison of dye degradation efficiency using ZnO powders with various size scales, J. Hazard. Mater. 141 (3) (2007) 645–652.

[117] Z. Ali, M.N. Chaudhry, S.T. Hussain, S.A. Batool, S.M. Abbas, N. Ahmad, N. Ali, N.A. Niaz, Copper and cobalt co-modified nitrogen doped titania nano photocatalysts for degradation of erichrome black-T, Dig. J. Nanomater. Bios. 8 (3) (2013) 1271–1280.

[118] R. Wei, T. Zhou, J. Hu, J. Li, Glutatione modified ultrathin SnS2 nanosheets with highly photocatalytic activity for wastewater treatment, Mater. Res. Express 1 (2) (2014) 1–14.

[119] N.M. Mahmoodi, Photocatalytic ozonation of dyes using copper ferrite nanoparticle prepared by co-precipitation method, Desalination 279 (1–3) (2011) 332–337.

[120] M. Ahmad, E. Ahmed, Z.L. Hong, W. Ahmed, A. Elhissi, N.R. Khalid, Photocatalytic, sonocatalytic and sonophotocatalytic degradation of rhodamine B using ZnO/CNTs composites photocatalysts, Ultrason. Sonochem. 21 (2) (2014) 761–773.

[121] G. Elango, S.M. Roopan, Efficacy of SnO_2 nanoparticles toward photocatalytic degradation of methylene blue dye, J. Photochem. Photobiol. B 155 (2016) 34–38.

[122] S. Li, S. Hu, W. Jiang, Y. Liu, Y. Zhou, J. Liu, Z. Wang, Facile synthesis of cerium oxide nanoparticles decorated flower-like bismuth molybdate for enhanced photocatalytic activity toward organic pollutant degradation, J. Colloid Interface Sci. 530 (2018) 171–178.

[123] M. Roy, M.K. Naskar, Alkali metal ion induced cube shaped mesoporous hematite particles for improved magnetic properties and efficient degradation of water pollutants, Phys. Chem. Chem. Phys. 18 (30) (2016) 20528–20541.

[124] M. Roy, S. Ghosh, M.K. Naskar, Synthesis of morphology controllable porous Co_3O_4 nanostructures with tunable textural properties and their catalytic application, Dalton Trans. 43 (26) (2014) 10248–10257.

[125] Z.-C. Wu, Y. Zhang, T.-X. Tao, L. Zhang, H. Fong, Silver nanoparticles on amidoxime fibers for photocatalytic degradation of organic dyes in waste water, Appl. Surf. Sci. 257 (3) (2010) 1092–1097.

[126] R. Arunachalam, S. Dhanasingh, B. Kalimuthu, M. Uthirappan, C. Rose, A.B. Mandal, Phytosynthesis of silver nanoparticles using *Coccinia grandis* leaf extract and its application in the photocatalytic degradation, Colloids Surf. B: Biointerfaces 94 (2012) 226–230.

[127] V.S. Suvith, D. Philip, Catalytic degradation of methylene blue using biosynthesized gold and silver nanoparticles, Spectrochim. Acta A 118 (2014) 526–532.

[128] S. Bhakya, S. Muthukrishnan, M. Sukumaran, M. Muthukumar, T. Senthil, M.V. Rao, Catalytic degradation of organic dyes using synthesized silver nanoparticles: a green approach, J. Bioremed. Biodeg. 6 (5) (2015) 1000312.

[129] N. Rahman, Z. Abedin, M.A. Hossain, Rapid degradation of azo dyes using nano-scale zero valent iron, Am. J. Environ. Sci. 10 (2) (2014) 157–163.

[130] X. Liu, D. Wang, Y. Li, Synthesis and catalytic properties of bimetallic nanomaterials with various architectures, NanoToday 7 (5) (2012) 448–466.

[131] J.A. Rodriguez, D.W. Goodman, The nature of the metal-metal bond in bimetallic surfaces, Science 257 (5072) (1992) 897–903.

[132] J. Tanori, N. Duxin, C. Petit, I. Lisiecki, P. Veillet, M.P. Pileni, Synthesis of nanosize metallic and alloyed particles in ordered phases, Colloid Polym. Sci. 273 (1995) 886–892.

[133] Q. Ge, Y. Huang, F. Qiu, S. Li, Bifunctional catalysts for conversion of synthesis gas to dimethyl ether, Appl. Catal. A 167 (1) (1998) 23–30.

[134] W. Du, Q. Wang, D. Saxner, N.A. Deskins, D. Su, J.E. Krzanowski, A.I. Frenkel, X. Teng, Highly active iridium/iridium–tin/tin oxide heterogeneous nanoparticles as alternative electrocatalysts for the ethanol oxidation reaction, J. Am. Chem. Soc. 133 (38) (2011) 15172–15183.

[135] A.D. Bokare, R.C. Chikate, C.V. Rode, K.M. Paknikar, Iron-nickel bimetallic nanoparticles for reductive degradation of azo dye Orange G in aqueous solution, Appl. Catal. B. 79 (3) (2008) 270–278.

[136] J. Wang, C. Liu, I. Hussain, C. Li, J. Li, X. Sun, J. Shen, W. Han, L. Wang, Iron–copper bimetallic nanoparticles supported on hollow mesoporous silica spheres: the effect of Fe/ Cu ratio on heterogeneous Fenton degradation of a dye, RSC Adv. 6 (59) (2016) 54623–54635.

[137] E. Alvarez-Ayuso, A. García-Sánchez, X. Querol, Purification of metal electroplating waste waters using zeolites, Water Res. 37 (20) (2003) 4855–4862.

[138] M.N. Chong, Z.Y. Tneu, P.E. Poh, B. Jin, R. Aryal, Synthesis, characterisation and application of TiO_2–zeolite nanocomposites for the advanced treatment of industrial dye wastewater, J. Taiwan Inst. Chem. Eng. 50 (2015) 288–296.

[139] K. Nakajima, T. Fukui, H. Kato, M. Kitano, J.N. Kondo, S. Hayashi, M. Hara, Structure and acid catalysis of mesoporous $Nb_2O_5 \cdot nH_2O$, Chem. Mater. 22 (11) (2010) 3332–3339.

[140] M. Moliner, J.M. Serra, A. Corma, E. Argente, S. Valero, V. Botti, Application of artificial neural networks to high-throughput synthesis of zeolites, Microporous Mesoporous Mater. 78 (1) (2005) 73–81.

[141] S. Dutta, A. Ghosh, H. Kabir, R.N. Saha, Facile one pot synthesis of zinc oxide nanorods and statistical evaluation for photocatalytic degradation of a diazo dye, Water Sci. Technol. 74 (3) (2016) 698–713.

[142] M.S. Islam, Y. Kusumoto, J. Kurawaki, M. Abdulla-AL-Mamun, H. Manaka, A comparative study on heat dissipation, morphological and magnetic properties of hyperthermia suitable nanoparticles prepared by co-precipitation and hydrothermal methods, Bull. Mater. Sci. 35 (7) (2012) 1047–1053.

[143] A.S. Adekunle, K.I. Ozoemena, Voltammetric and impedimetric properties of nano-scaled γ-Fe_2O_3 catalysts supported on multi-walled carbon nanotubes: catalytic detection of dopamine, Int. J. Electrochem. Sci. 5 (12) (2010) 1726–1742.

[144] A. Afkhami, M. Saber-Tehrani, H. Bagheri, Modified maghemite nanoparticles as an efficient adsorbent for removing some cationic dyes from aqueous solution, Desalination 263 (1–3) (2010) 240–248.

[145] D. Maity, D.C. Agrawal, Synthesis of iron oxide nanoparticles under oxidizing environment and their stabilization in aqueous and non-aqueous media, J. Magn. Magn. Mater. 308 (1) (2007) 46–55.

[146] S. Yang, Y. Xu, Y. Sun, G. Zhang, D. Gao, Size-controlled synthesis, magnetic property, and photocatalytic property of uniform α-Fe_2O_3 nanoparticles via a facile additive-free hydrothermal route, CrystEngComm 14 (23) (2012) 7915–7921.

[147] H.G. Cha, C.W. Kim, Y.H. Kim, M.H. Jung, E.S. Ji, B.K. Das, J.C. Kim, Y.S. Kang, Preparation and characterization of α-Fe_2O_3 nanorod-thin film by metal–organic chemical vapor deposition, Thin Solid Film 517 (5) (2009) 1853–1856.

[148] C. Liang, H. Liu, J. Zhou, X. Peng, H. Zhang, One-step synthesis of spherical γ-Fe_2O_3 nanopowders and the evaluation of their photocatalytic activity for Orange I degradation. J. Chem. 2015 (2015), 791829. https://doi.org/10.1155/2015/791829.

[149] M. Stylidi, D.I. Kondarides, X.E. Verykios, Mechanistic and kinetic study of solar-light induced photocatalytic degradation of acid Orange 7 in aqueous TiO_2 suspensions, Int. J. Photoenergy 5 (2) (2003) 59–67.

[150] R. Sankar, P. Manikandan, V. Malarvizhi, T. Fathima, K.S. Shivashangari, V. Ravikumar, Green synthesis of colloidal copper oxide nanoparticles using *Carica papaya* and its application in photocatalytic dye degradation, Spectrochim. Acta Part A 121 (2014) 746–750.

[151] B. Nagappa, G.T. Chandrappa, Mesoporous nanocrystalline magnesium oxide for environmental remediation, Microporous Mesoporous Mater. 106 (1–3) (2007) 212–218.

[152] J.P. Dhal, M. Sethi, B.G. Mishra, G. Hota, MgO nanomaterials with different morphologies and their sorption capacity for removal of toxic dyes, Mater. Lett. 141 (2015) 267–271.

[153] D. Fang, J. Xie, D. Mei, Y. Zhang, F. He, X. Liu, Y. Li, Effect of $CuMn_2O_4$ spinel in Cu–Mn oxide catalysts on selective catalytic reduction of NO_x with NH_3 at low temperature, RSC Adv. 4 (49) (2014) 25540–25551.

[154] L. Kuang, F. Ji, X. Pan, D. Wang, X. Chen, D. Jiang, Y. Zhang, B. Ding, Mesoporous $MnCo_2O_{4.5}$ nanoneedle arrays electrode for high-performance asymmetric supercapacitor application, Chem. Eng. J. 315 (2017) 491–499.

[155] M. Yazdanbakhsh, K. Iman, E.K. Goharshadi, A. Youssefi, Fabrication of nanospinel $ZnCr_2O_4$ using sol-gel method and its application on removal of azo dye from aqueous solution, J. Hazard. Mater. 184 (1–3) (2010) 684–689.

[156] S. Dutta, S. Biswas, R.C. Maji, R.N. Saha, Environmentally Sustainable fabrication of $Cu_{1.94}S$-rGO composite for dual environmental application: visible-light-active photocatalyst and room-temperature phenol sensor, ACS Sustain. Chem. Eng. 6 (1) (2018) 835–845.

[157] H. Sun, L. Cao, L. Lu, Magnetite/reduced graphene oxide nanocomposites: One step solvothermal synthesis and use as a novel platform for removal of dye pollutants, Nano Res. 4 (2011) 550–562.

[158] V. López-Grimau, M. Vilaseca, C. Gutiérrez-Bouzán, Comparison of different wastewater treatments for colour removal of reactive dye baths, Desalin. Water Treat. (2015) 1–8.

[159] M.Y. Kilica, T. Yonara, S. Tekera, K. Kestioğlu, Comparing treatment methods that remove color from the effluent of an organized industrial district (OID), Desalin. Water Treat. (2014) 1–10.

[160] T. Turan-Ertas, Biological and physical-chemical treatment of textile dyeing wastewater for color and COD removal, Ozone Sci. Eng. 23 (2001) 199–206.

[161] S. Xavier, R. Gandhimathi, P.V. Nidheesh, S.T. Ramesh, Comparison of homogeneous and heterogeneous Fenton processes for the removal of reactive dye magenta MB from aqueous solution, Desalin. Water Treat. (2013) 1–10.

[162] E. Adamek, W. Baran, J. Ziemiańska, A. Sobczak, The comparison of photocatalytic degradation and decolorization processes of dyeing effluents, Int. J. Photoenergy 2013 (2013) 1–11.

7

An intelligent estimation model for water quality parameters assessment at Periyakulam Lake, South India

T.T. Dhivyaprabha[a], P. Subashini[a], M. Krishnaveni[a], N. Santhi[b], Ramesh Sivanpillai[c], and G. Jayashree[d]

[a]DEPARTMENT OF COMPUTER SCIENCE, AVINASHILINGAM INSTITUTE FOR HOME SCIENCE AND HIGHER EDUCATION FOR WOMEN, COIMBATORE, INDIA [b]DEPARTMENT OF BIOINFORMATICS, AVINASHILINGAM INSTITUTE FOR HOME SCIENCE AND HIGHER EDUCATION FOR WOMEN, COIMBATORE, INDIA [c]DEPARTMENT OF BOTANY, UNIVERSITY OF WYOMING, LARAMIE, WY, UNITED STATES [d]NORTHEASTERN UNIVERSITY, BOSTON, MA, UNITED STATES

1 Introduction

Water is an indispensable renewable resource for the survival of living organisms on Earth. Water is essential to carry out daily activities that involve drinking, agriculture, industry, cleaning, recreation, animal husbandry, and producing electricity for domestic and commercial use in our day-to-day lives. All over the world, among the total water resources on Earth, less than 3% of inland waters occur in which less than 1% is fresh water; the remaining is groundwater that occurs below the surface or is locked in the ice caps [1]. Freshwater lakes are critically important for the drinking water supply, survival of aquatic species, cultivation, terrestrial habitat, manufacturing, recharging the groundwater, climate regulation, soil conservation, flood mitigation, nutrient recycling, aquaculture, and reducing poverty (inland fisheries provides fishing and fish farming job opportunities, and food security to a tremendous amount of the population around the world). In India, water resources are categorized into brackish water, streams and canals, derelict water bodies, lakes and ponds, reservoirs, and tanks. Table 1 depicts the area of water bodies' level in India [2]. It is speculated that more than 50% of inland water bodies were lost during the 20th century. Freshwater bodies are rapidly lost due to several reasons such as vegetation clearance, drainage effluents, urbanization, overharvesting, species hunting, pollution, eutrophication, and global climate change.

Freshwater bodies always support ecological and economic activities in every region, especially India that has a large number of rivers, lakes, and wetlands. The surface water resources in India are slowly degrading where the water is in unacceptable quality [3,4]. In particular, lake water is highly polluted by discharge of industrial effluents, agricultural

Intelligent Environmental Data Monitoring for Pollution Management. https://doi.org/10.1016/B978-0-12-819671-7.00007-5

Table 1   Inland water resources in India.

Rivers & Canals (length in km)	195,210
Other water bodies (area in Mha)	
Reservoirs	2.91
Tanks & ponds	2.14
Flood plain lakes & derelict water bodies	0.80
Brackish water	1.24

Source: Water and related statistics reported by Central Water Commission, India, 2013.

wastes, and unexpected climate changes that lead to deterioration of water quality [5]. Hence, the assessment of water bodies and finding the harmful effects is helpful to sustaining these valuable resources. Coimbatore City, also called Manchester of South India, is the 16th largest urban area in India [6]. During the 8th and 9th centuries, this city was ruled by Kongu Chola kings who constructed 30 wetlands in the surrounding region of the Noyyal River. It was widely used for irrigation, flood control, domestic purposes, and recharging groundwater [7]. Over a period of time, these wetlands were slowly destroyed due to garbage dumping, sewage effluent, urbanization, deforestation, and encroachment. Currently, Coimbatore has eight major lakes, namely, Singanallur, Valankulam, Ukkadam Periyakulam, Selvampathy, Krishnampathi, Selvachinthamani, Narasampathy, and Kumaraswami Lakes [8]. These lakes are well interconnected and facilitate as a water source for irrigating agricultural land as well as forming a flood buffer during the rainy season [9]. Among these, Ukkadam Periyakulam Lake is a primary one, where approximately 400 families are actively involved in fish farming and fishing businesses (oral information given by Periyakulam lake fisherman to Mr. G.K. Arumugam). Moreover, it has 36 genera of zooplankton with increased eutrophication; it also acts as a shelter for various species of birds and other aquatic organism. The lake has been contaminated with effluents from municipal sewage released from the city, dumping of domestic garbage and industrial wastage by inhabitants, and also by the spread of water hyacinth (Eichhorniacrassipes). Siruthuli, a nonprofit public movement (NGO), joined with municipal corporation bodies and people to desilt, remove water hyacinth, and perform continuous assessment of water quality on this lake [10]. Municipal corporation bodies and NGOs show keen interest in cleaning Coimbatore lakes and restoring the existing remaining water bodies. Therefore, remedial measures/strategies need to be formulated to keep Coimbatore lakes pollution-free for economic development, recharging groundwater, due to concern with ecological components like fauna and flora, and environmental conservative aspects.

Ukkadam Periyakulam Lake is one of the primary lakes situated in Ukkadam, Coimbatore City, South India. This lake encompasses a rich set of biodiversity and acts as a shelter for a wide variety of bird species. This lake attracts a huge population of birds, specifically, little grebes, painted storks, and purple moorhen, which are observed in the late month of summer (March), pre-monsoon period (November), and winter season (December, January, and

February). It provides local employment involve fishing and fish farming businesses that are actively carried out by fishermen living in the surrounding regions. This lake supports ecosystem and ecology by offering various measures such as flood mitigation, fishing, species habitat, recharging groundwater, and storing rainwater. Indeed, monitoring the physicochemical parameters is quite important and prominently beneficial to protect wetland quality, prevent soil erosion, and retain nutrients in Ukkadam lake [5]. It is considered as a primary source for irrigation and fishing, and can be used for domestic purposes by means of applying water treatment methods. Henceforth, aforementioned features of this lake are motivated to carry out this research by keenly observing the physicochemical parameters for a period of 4 years (2015–2018). This study has been conducted in four different sampling sites on a temporal, spatial, and seasonal basis. Water quality parameters are dynamic, irregular, and nonlinear in nature. It is frequently subject to change corresponding to temporal as well as spatial location basis. Henceforth, assessments of water quality parameters are essential for the identification of water quality degradation, and then implementing pollution control policies to further improve water quality standards. In this chapter, a novel estimation model based on Kalman filter integrated with synergistic fibroblast optimization (SFO) algorithm has been developed for forecasting physicochemical properties in the corresponding sampling sites in Periyakulam Lake. Monitoring and estimating the water quality parameters could be greatly helpful to Coimbatore Municipal Corporation and NGOs to formulate remedial measures/water quality management programs to rejuvenate water quality standards in this lake.

Several research works were carried out in the assessment of water quality in lakes. Rachna Bhateria and Disha Jain et al. surveyed various kinds of problems that impose threats to the survival of aquatic species in lakes. Lake water quality was assessed by physical and chemical properties, namely, pH, total dissolved solids (TDS), electrical conductivity (EC), biochemical oxygen demand (BOD), and total suspended solids (TSS). It illustrated that heavy metals, namely magnesium, phosphorus, calcium, sulphate, sodium, and potassium, contaminate the water, which causes severe problems to habitats existing in the surrounding environment. Harmful algal blooms, eutrophication, climate change, encroachment, discharge of domestic sewage, and industrial effluents cause hazards to the ecosystem as well as human habitants. It suggested that bioremediation techniques need to be adopted to improve lake water quality and prohibit all the activities that cause pollution [11]. Sarah Hembrow and Kathryn Taffs applied a palaeoecological approach for the detection of ecological variations that degrade the water quality in Lake McKenzie located in Fraser Island, Australia. Sediments were extracted from frozen water samples to measure parameters such as pH, EC, dissolved oxygen (DO), and heavy metals, namely total phosphorus (TP), orthophosphate (PO_4), ammonia (NH4), nitrite (NO2), and nitrogen oxide (NOx) in the prefixed five zones. Investigational studies reveal that a high variability of climate conditions had a severe effect on the nutrient degradation in this lake. Therefore, long-term monitoring of water quality is required, and reforms planning strategies are to be formulated to enhance recreational activities and conserve the ecosystem of this lake [12]. Water quality analysis among the 28 freshwater bodies and reservoirs in China were conducted by Xiaojie Meng et al. Water quality was tested based on spatial and temporal distribution of lakes from 2005 to 2010 to identify the correlation between water quality and socioeconomic development. Analysis of results shows that water quality rapidly deteriorated in the lakes located among the densely inhabited and industrial areas, and lakes in the rest of the areas do not exhibit significant changes in water quality [13]. The characteristics of water quality were assessed in Keenjhar Lake, and its waterway connected to Kotri and Dhabeji treatment plant in Sindh Pakistan was investigated by Muhammad Afzal Farooq et al. Water samples were acquired from the 14 predetermined sites in Kotri and Dhabeji treatment plants, and 22 water quality parameters were analyzed. It revealed that industrial effluents were discharged into the canal that severely affected the water quality, and it was not safe for human consumption [14]. Kiran Kumar et al. did a case study on water quality analysis of Avaragere Lake, India. In total, 22 water quality parameters and heavy metals, namely nitrate, chloride, chromium, zinc, magnesium, phosphate, sodium, and alkalinity, were analyzed in the water samples, which were periodically acquired, monthly twice from February 2013 to April 2013. This study concluded that alkalinity and acidity level were slightly high but found to be within the tolerable limits of Bureau of Indian Standards (BIS) drinking water standards. It indicates that water quality was acceptable, and it could be used for consumption and domestic and agricultural needs. However, it required appropriate water treatment methods to improve the water quality standards and also to protect the lake ecosystem from pollutants [15]. Salim Aijaz Bhat and Ashok Pandit performed water quality analysis at five sites in Wular Lake, Kashmir. In total, 27 physicochemical parameters and heavy metals were analyzed to evaluate spatiotemporal variations from 2011 to 2013 and identify various pollution sources in four sampling seasons. Discriminant analysis (DA) technique was applied to measure water quality index at five sites. Investigation demonstrates that water quality standards were acceptable, and it can be utilized for drinking, and agriculture and domestic purposes [16]. Goldin Quadros et al. investigated the wetlands of Coimbatore city in different perspectives. It involves a diverse set of analyses such as physicochemical parameters, heavy metal contamination, phytoplankton, fishing, diversity of birds, and amphibian and reptile species. The summarized report illustrated that water quality was tremendously deteriorated in all the Coimbatore lakes. Immediate actions or remedial measures need to be taken by various stakeholders in a collaborative manner such as prohibiting sewage dumping and industrial waste, encroachment of activities of lakes, construction of bunds for bird shelters, implementation of water treatment strategies, and formulating control policies for fishing and fish farming activities to rejuvenate lakes and conserve the aquatic ecosystem in the wetlands of Coimbatore [17].

Studies on promising literature works emphasize the importance of conserving the ecosystem, protecting ecological species, and restoring the structure and function of the natural ecosystem. The development of a predictive model for estimating the physicochemical parameters is needed for conservation of freshwater bodies. It could be useful to formulate a control policy and implement preventive measures to preserve the ecosystem.

The remaining section of this chapter is structured as follows. Section 2 explains the study area and materials used in the data collection site. The statistical method utilized to identify the correlation between water quality parameters is presented in Section 3. The proposed predictive model for the estimation of water quality parameters is described in Section 4. Section 5 deals with experimental results and discussions. Conclusion and future works are summarized in Section 6.

2 Materials and methods

Ukkadam Periyakulam Lake, also known as Ukkadam Big Tank (10°58′54′ N 76°57′17′ E), has a wide spread area over 1.295 km^2 (0.500 sq. mi), average depth of 5.82 m (19.1 ft), and is situated in Coimbatore City, Tamil Nadu, India. This lake has seven inlets (one from the Noyyal River, one from Selvachinthamani Lake, and five sewage inlets) and five outlets (it feeds water to Valankulam Lake, which is located on the east side of the lake through a Weir, and four Sluices) [7]. Water quality is degraded dramatically by weathering of bedrock minerals, atmospheric processes of evapotranspiration, deposition of dust and salt by wind, by the natural leaching of organic matter and nutrients from soil, hydrological factors, and biological processes within the aquatic species, climate change, seasonal variation, algae blooms, fish culture, water level, and pollutant discharge that can alter the physical and chemical composition in Periyakulam Lake. In this research, groundwater samples are acquired to investigate the water quality by analyzing the physicochemical properties from the year 2015 to 2018 in Ukkadam Periyakulam Lake. Fig. 1 shows the four sampling points of Ukkadam Periyakulam Lake, namely, Center location (C1); Corner location (E1)—near center point; Corner location (W1)—Perur Bye Pass Road; and Corner location (E2)—Palakkad Road, which were selected as data collection zones for this study. In general, the physical properties of water include water temperature (WT), EC, TSS, transparency or turbidity, TDS, odor, color, taste, and chemical properties that involve pH, BOD, chemical oxygen demand (COD), DO, total hardness (TH), heavy metals, nitrate, orthophosphates, pesticides, and surfactants monitored to measure the characteristics of water quality [18]. Secchi disk transparency is strongly related with turbidity, which is the low-cost method used for measuring water clarity in lakes and streams [19,20]. Among these, Water Resources Control Board advised the following: Salinity (ppm), DO (mg/L), water temperature (°C), pH, EC (μs/m), TDS (ppm), turbidity (NTU), and atmospheric temperature (°C) are viewed as the primary physicochemical parameters for the existence of aquatic species in water. Hence, the aforementioned physicochemical parameters have been chosen for experimental analysis in this study [21–23]. Water quality parameters were regularly assessed from four sampling sites, namely, center point (C1) and three spatial points (E1), (W1), and (E2) every week from October 2015 to December 2018 and tested using portable water quality

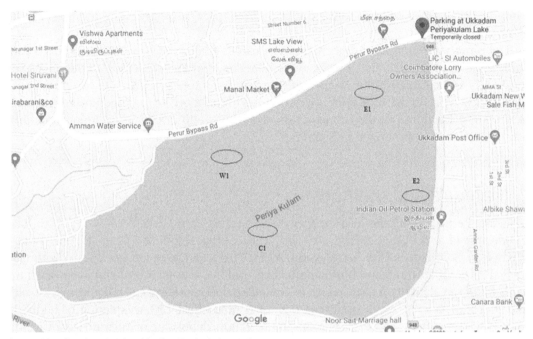

FIG. 1 Sampling location in Ukkadam Periyakulam Lake.

measurement equipment (Systronics digital model: Oxygen meter: Make.Lutron PDO519; Salinity tester: HANNA HI 98203; Portable water analyzer HANNA HI 9813-5). Tables 2–5 represent the sample dataset collected in C1, E1, E2, and W1 sites.

3 Statistical analysis for water quality assessment

Suitable control measures are essential for forecasting the growing demand for water in future. Numerous research works have categorized the water quality by various physico-chemical parameters. Descriptive methods of statistical analysis, such as correlation and regression analysis are part of the exploratory analyses necessary to develop estimation models on lake water quality. To develop a predictive model for estimating the physico-chemical parameters on a spatioseasonal basis, selection of the significant parameters is a critical task. In this study, Pearson correlation coefficient method was implemented to calculate the association existing between water quality parameters as recommended by Sami Daraigan et al. [24]. The range of correlation coefficients lies between −1 to +1. It specifies the degree of association that exists between water quality parameters and, hence, it is utilized for the development of forecasting model. The mathematical representation of correlation coefficient (R) that determines the relationship between water quality variables is given in the following section. Numerical results of correlation coefficient (R) for the prefixed water quality parameters are presented in Tables 6–9.

Table 2 Water quality parameters of samples collected at C1 site.

S. No.	Date (DD/MM/YY)	Time (AM)	Comment	Salinity (ppt)	DO (mg/L)	WT (°C)	pH	EC (µs/cm)	TDS (mg/L)	AT (°C)	SDT (cm)
1	08/11/15	9.30	Sunny	1.38	7.13	28.50	9.50	1950	1328	28.00	10.00
2	15/11/15	10.20	Rainy	1.39	7.71	26.20	9.50	1470	1073	25.00	16.00
3	22/11/15	11.00	Rainy	1.41	7.36	29.10	10.00	1520	1106	30.00	17.30
4	06/12/15	10.55	Sunny	1.40	7.67	26.50	9.70	1340	978	26.10	16.20
5	13/12/15	11.34	Sunny	1.45	7.42	28.60	9.90	1380	1006	27.90	17.10
6	20/12/15	10.10	Sunny	1.48	7.56	27.40	10.20	1480	1081	26.70	13.60
7	27/12/15	3.50	Sunny	1.38	7.56	27.40	9.50	1950	1240	28.00	12.70
8	03/01/16	12.50	Sunny	1.29	7.22	30.40	10.30	1650	1185	27.40	12.20
9	10/01/16	11.10	Sunny	1.38	7.60	27.10	10.50	1630	1199	26.40	13.90
10	16/01/17	10.30	Sunny	1.64	7.40	28.80	9.70	1640	1202	25.80	11.30
11	28/12/18	1.56	Sunny	1.17	7.06	31.90	11.40	2060	1422	30.70	13.22
Mean				1.39	7.42	28.35	10.01	1642.73	1165.45	27.45	14.00

Table 3 Water quality parameters of samples collected at E1 site.

S. No.	Date (DD/MM/YY)	Time (AM)	Comment	Salinity (ppt)	DO (mg/L)	WT (°C)	pH	EC (µs/cm)	TDS (mg/L)	AT (°C)	SDT (cm)
1	08/11/15	9.50	Sunny	1.71	7.49	28.00	8.80	0930	662	26.20	16.30
2	15/11/15	9.30	Rainy	1.33	7.71	26.20	9.40	1560	1131	25.50	19.00
3	22/11/15	10.15	Rainy	1.42	7.87	25.00	10.00	1480	1076	29.40	16.80
4	06/12/15	10.25	Sunny	1.27	7.67	26.50	9.90	1480	1080	25.00	19.50
5	13/12/15	10.50	Sunny	1.36	7.39	28.90	10.20	1510	1107	28.50	14.00
6	20/12/15	9.40	Sunny	1.39	7.53	27.70	09.20	1530	1122	27.70	17.20
7	27/12/15	3.30	Sunny	1.28	7.43	28.50	09.70	1460	1020	26.50	16.80
8	03/01/16	12.30	Sunny	1.22	7.47	28.20	09.50	1480	1083	27.10	15.00
9	10/01/16	10.30	Sunny	1.29	7.60	27.10	10.50	1580	1205	26.70	14.50
10	16/01/16	10.15	Sunny	1.29	7.56	27.40	10.90	1640	1204	23.20	14.90
11	17/02/17	10.08	Sunny	1.27	7.54	27.60	09.10	1790	742	26.30	45.70
12	24/12/18	2.54	Sunny	1.25	7.16	31.00	11.30	2080	1495	28.80	15.20
Mean				1.34	7.53	27.67	9.87	1543.33	1077.3	26.74	18.70

In these tables, red (dark gray in print version) value indicates strong correlation ($R > 0.7$), yellow (light gray in print version) denotes moderate correlation ($R > 0.2$), and green (dark gray in print version) indicates weak correlation ($R > 0.1$), respectively.

$$R = \frac{n\sum (x_i y_i) - \left(\sum x_i\right) \cdot \left(\sum y_i\right)}{\sqrt{\left[n\sum x_i^2 - \left(\sum x_i\right)^2\right]\left[n\sum y_i^2 - \left(\sum y_i\right)^2\right]}} \tag{1}$$

Table 4 Water quality parameters of samples collected at E2 site.

S. No.	Date (DD/MM/YY)	Time (AM)	Comment	Salinity (ppm)	DO (mg/L)	WT (°C)	pH	EC (µs/cm)	TDS (mg/L)	AT (°C)	SDT (cm)
1	08/11/15	9.06	Sunny	1.37	7.5	27.90	9.40	1880	1357	27.70	28.00
2	15/11/15	10.15	Rainy	1.44	7.72	26.10	9.30	1620	1177	25.40	28.00
3	22/11/15	11.15	Rainy	1.40	7.48	28.10	9.20	1500	1095	30.00	22.00
4	06/12/15	10.45	Sunny	1.46	7.59	27.20	9.70	1410	1030	26.40	22.70
5	13/12/15	11.20	Sunny	1.53	7.40	28.80	10.00	1470	1073	28.60	15.20
6	20/12/15	10.15	Sunny	1.48	7.62	26.90	10.20	1520	1111	26.10	18.20
7	27/12/15	4.10	Sunny	1.36	7.81	25.40	9.40	1580	1345	27.50	19.00
8	01/03/16	12.45	Sunny	1.35	7.32	29.50	10.20	1700	1270	26.80	14.80
9	01/10/16	10.55	Sunny	1.41	7.60	27.10	10.40	1620	1193	26.70	15.40
10	16/01/16	10.25	Sunny	1.38	7.41	28.70	9.80	1640	1200	25.30	16.50
11	23/01/16	10.30	Sunny	1.4	7.56	27.40	9.90	1480	1202	27.30	14.60
12	31/01/16	10.20	Sunny	1.33	7.54	27.60	8.50	1670	1228	24.80	27.30
13	01/02/16	9.35	Sunny	1.25	7.55	27.50	10.30	1670	1227	26.60	100.00
14	17/02/17	9.15	Sunny	1.44	7.38	29.00	10.20	1750	750	27.40	30.48
15	10/12/18	2.15	Sunny	1.27	7.13	31.30	11.30	2100	1417	28.00	27.94
Mean				1.39	7.50	27.9	9.85	1640.67	1178.33	27.00	26.67

Table 5 Water quality parameters of samples collected at W1 site.

S. No.	Date (DD/MM/YY)	Time (AM)	Comment	Salinity (ppt)	DO (mg/L)	WT (°C)	pH	EC (µs/cm)	TDS (mg/L)	AT (°C)	SDT (cm)
1	08/11/15	8.30	Sunny	1.25	7.43	28.50	9.70	1760	1225	27.70	8.00
2	15/11/15	10.45	Rainy	1.40	7.76	25.80	8.80	1330	965	26.40	12.00
3	22/11/15	11.40	Rainy	1.40	7.79	25.60	9.20	1350	987	25.80	14.00
4	06/12/15	11.20	Sunny	1.38	7.65	26.70	8.90	1460	1068	26.10	14.00
5	13/12/15	11.55	Sunny	1.45	7.34	29.30	9.30	1530	1128	28.60	12.20
6	20/12/15	10.40	Sunny	1.50	7.53	27.70	9.30	1530	1122	26.60	15.50
7	27/12/15	4.30	Sunny	1.71	7.49	28.00	9.90	0960	1120	28.50	14.80
8	03/01/16	1.10	Sunny	1.35	7.23	30.30	9.80	1890	1325	30.40	11.50
9	10/01/16	11.30	Sunny	1.38	7.23	30.30	9.80	1780	1302	30.60	9.80
10	16/01/16	10.45	Sunny	1.36	7.48	28.10	9.60	1690	1235	26.30	6.60
14	17/02/17	9.20	Sunny	1.35	7.48	28.10	9.80	1770	768	27.60	30.50
15	10/12/18	2.35	Sunny	1.27	7.15	31.10	11.30	1990	1366	28.10	22.90
Mean				1.4	7.46	28.29	9.61	1586.67	1134.25	27.72	14.30

where

  $x = x$-variable values,
  $y = y$-variable values,
  $n =$ number of data points,
  $i = i$th variable, and
  $\Sigma =$ summation.

Table 6 Correlation coefficient of trace elements at C1 site.

Parameter	Saline	Dissolved oxygen	Water temperature	pH	Electrical conductivity	Total dissolved solids	Atmospheric temperature	Secchi disk transparency
Saline	1.00							
Dissolved oxygen	−0.03	1.00						
Water temperature	−0.26	−0.22	1.00					
pH	−0.16	0.31	0.57	1.00				
Electrical conductivity	−0.63	−0.06	0.61	0.42	1.00			
Total dissolved solids	−0.62	0.49	0.03	0.05	0.44	1.00		
Atmospheric temperature	0.09	0.19	0.30	0.22	0.03	−0.06	1.00	
Secchi disk transparency	−0.56	0.12	−0.04	0.13	0.19	0.05	−0.09	1.00

Red (dark gray in print version) value indicates strong correlation ($R > 0.7$), *yellow* (light gray in print version) denotes moderate correlation ($R > 0.2$), and *green* (dark gray in print version) indicates weak correlation ($R > 0.1$), respectively.

From the study area, it is observed that water was colorless, odorless, and free from slight turbidity. It was inferred from the dataset that the temperature of the water is in the range 21–30°C. An increase in the indication of temperature may enhance the development of micro-organisms, which cause corrosion problems [25]. The pH values are obtained with the mean value of 8.1, and the minimum and maximum was 10 and 10.2, respectively. The range indicates that the limit was slightly high for water that has a domestic purpose. pH value above 8.0 would be harmful in the treatment of drinking water with chlorine [25]. The DO level of the lake water lies between 25.3% and 43.2%, which may vary for the suitability of fishing. There was no positive correlation between salinity and Secchi disk transparency with any of the observed parameters. Correlation analysis of the water quality parameters shows a strong correlation between temperature correlated very strongly with EC and TDS. The most significant strong correlation existed between EC and TDS, as well as TDS also exhibiting a strong correlation with EC. It also depends upon the increase in temperature. From the inference of EC and TDS values, it portrays that this lake has less concentration of salt and water suitable for human consumption and can be used for crop production. An R-squared test is applied to find the coefficient of determination between two parameters [18]. It is utilized to analyze how the observed data fit into the fitted regression line using Eq. (2).

Table 7 Correlation coefficient of trace elements at E1 site.

Parameter	Saline	Dissolved oxygen	Water temperature	pH	Electrical conductivity	Total dissolved solids	Atmospheric temperature	Secchi disk transparency
Saline	1.00							
Dissolved oxygen	0.16	1.00						
Water temperature	−0.47	−0.05	1.00					
pH	−0.61	−0.15	0.68	1.00				
Electrical conductivity	−0.48	0.22	0.53	0.32	1.00			
Total dissolved solids	−0.47	0.13	0.62	0.46	0.96	1.00		
Atmospheric temperature	−0.56	0.03	0.75	0.54	0.53	0.51	1.00	
Secchi disk transparency	−0.01	−0.43	−0.28	−0.02	−0.70	−0.72	0.03	1.00

Red (dark gray in print version) value indicates strong correlation ($R > 0.7$), *yellow* (light gray in print version) denotes moderate correlation ($R > 0.2$), and *green* (dark gray in print version) indicates weak correlation ($R > 0.1$), respectively.

$$R^2 = 1 - \frac{\left[Sum_{(i=1\,to\,n)}\left\{ w_i (y_i - f_i)^2 \right\} \right]}{\left[Sum_{(i=1\,to\,n)}\left\{ w_i (y_i - f_{av})^2 \right\} \right]} \tag{2}$$

where

f_i = estimated value from the fit,
f_{av} = average value of the observed data,
y_i = observed data value, and
w_i = weight applied to data value.

The range of R^2 test value lies between 0 and 1, and R^2 is high if the value is closer to 1. Otherwise, the relativity between two variables are low, and R^2 is closer to 0. $Y = Ax + B$ is the linear relation followed to find the best fit line. Figs. 2–4 gives a graph based on regression analysis equations for saline and EC, saline and TDS, and lastly TDS and EC.

Table 8 Correlation coefficient of trace elements at E2 site.

Parameter	Saline	Dissolved oxygen	Water temperature	pH	Electrical conductivity	Total dissolved solids	Atmospheric temperature	Secchi disk transparency
Saline	1.00							
Dissolved oxygen	0.23	1.00						
Water temperature	−0.16	0.00	1.00					
pH	−0.48	0.24	0.34	1.00				
Electrical conductivity	−0.77	−0.39	0.37	0.63	1.00			
Total dissolved solids	−0.57	0.36	0.31	0.84	0.68	1.00		
Atmospheric temperature	0.09	0.09	0.23	0.06	0.17	0.20	1.00	
Secchi disk transparency	−0.15	−0.91	−0.12	−0.40	0.26	−0.52	−0.13	1.00

Red (dark gray in print version) value indicates strong correlation (*R* > 0.7), *yellow* (light gray in print version) denotes moderate correlation (*R* > 0.2), and *green* (dark gray in print version) indicates weak correlation (*R* > 0.1), respectively.

4 Proposed methodology

Analysis on the linear correlation of data indicates that the highest significance exists between EC and TDS parameters, which are taken as the primary variables for the development of forecasting model. The objective of the work is to design and develop a Kalman estimation model based on SFO algorithm for forecasting the water quality parameters, which includes TDS and EC in the four sites, namely, C1, E1, E2, and W1 corresponding to South India seasons such as Monsoon, Autumn, Winter, and Prevernal. The structure of the Kalman estimation model based on SFO is portrayed in Fig. 5 and is described as follows.

4.1 Feature extraction

Dimensionality reduction is considered the appropriate method that needs to be defined for the removal of noisy and irrelevant features. It is classified into two types, namely, feature extraction and feature selection. Feature extraction mapping features into a

Table 9   Correlation coefficient of trace elements at W1 site.

Parameter	Saline	Dissolved oxygen	Water temperature	pH	Electrical conductivity	Total dissolved solids	Atmospheric temperature	Secchi disk transparency
Saline	1.00							
Dissolved oxygen	0.26	1.00						
Water temperature	−0.26	0.39	1.00					
pH	−0.23	0.02	0.77	1.00				
Electrical conductivity	0.16	−0.09	0.68	0.53	1.00			
Total dissolved solids	−0.21	0.68	0.72	0.53	0.44	1.00		
Atmospheric temperature	0.01	0.55	0.81	0.44	0.37	0.52	1.00	
Secchi disk transparency	−0.03	−0.59	0.08	0.39	0.16	−0.49	−0.06	1.00

Red (dark gray in print version) value indicates strong correlation ($R > 0.7$), *yellow* (light gray in print version) denotes moderate correlation ($R > 0.2$), and *green* (dark gray in print version) indicates weak correlation ($R > 0.1$), respectively.

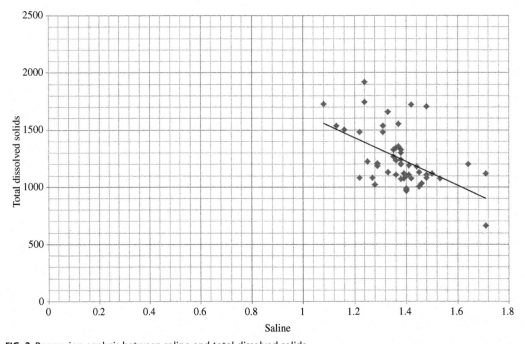

FIG. 2 Regression analysis between saline and total dissolved solids.

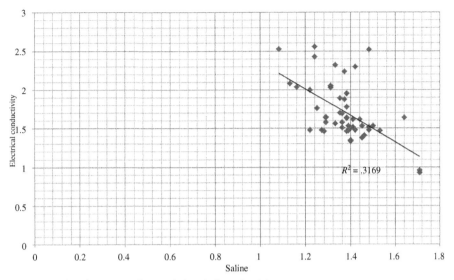

FIG. 3 Regression analysis between saline and electrical conductivity.

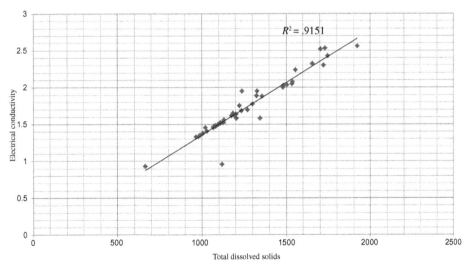

FIG. 4 Regression analysis between total dissolved solids and electrical conductivity.

new feature space with lower dimensionality and the new features are integrated with original features space. The most popular feature extraction techniques are, namely, Principle Component Analysis (PCA), Linear Discriminant Analysis (LDA), and Canonical Correlation Analysis (CCA). Feature selection technique selects a feature subset from

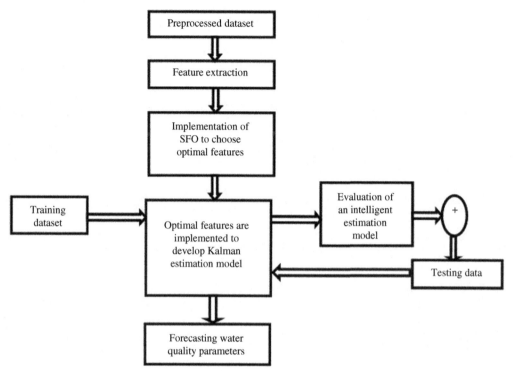

FIG. 5 Process flow of SFO-based Kalman estimation model.

feature space, which reduces redundancy and improves the relevance of target features. Both feature extraction and feature selection techniques are significantly useful to improve learning ability and minimize redundancy, computational complexity, and memory space required for the construction of an estimation model. In this work, TDS and EC features do not need any transformation and are required to maintain the original features for building an effective model. So, feature selection technique can be a suitable method for constructing an optimal range of TDS and EC parameters.

A study on the common feature selection techniques such as Information Gain, Relief, Fisher Score, and Lasso illustrate that Information Gain select features in a univariate way, and, hence, it cannot perform well with redundant features. Lasso technique is an efficient one, when the features satisfy a nontrivial condition. As water dataset encompass scattered and redundant TDS and EC values, Information Gain and Lasso techniques could not be applied to this water dataset efficiently. But Fisher score evaluates features individually, and it assigns similar instances to the same class and dissimilar instances to different classes. Henceforth, Fisher score is considered as an appropriate method for this work. It is utilized to choose fittest TDS and EC features subset and maps into the target class label, namely C1, E1, E2, and W1 sites. The obtained results are presented in Tables 10 and 11. The permissible limit of EC and TDS ranges for the inland surface water bodies

Table 10   Fisher Score value for TDS.

Site	Fisher score	Mean	Variance
C1	1.30E-10	1270.00	103,840.3
E1	2.01E-07	1242.10	16,678.54
E2	5.37E-08	1211.06	58,992.73
W1	4.42E-07	1353.67	38,192.97

Table 11   Fisher Score value for EC.

Site	Fisher score	Mean	Variance
C1	1.21E-08	1787.27	154,421.8
E1	2.38E-08	1712.22	40,444.44
E2	9.71E-08	1591.25	158,469.64
W1	4.63E-08	1788.08	76,638.63

recommended by Central Pollution Control Board (CPCB) and Indian Standards Institute (ISI-IS: 2296-1982) have been considered to select features subset relevant to target variables [21,22]. The mathematical expression of Fisher score method is represented in Eq. (3).

$$S_i = \frac{\sum_{k=1}^{K} n_j \left(\mu_{ij} - \mu_i \right)^2}{\sum_{k=1}^{K} n_j \rho_{ij}^2} \qquad (3)$$

where

S_i = Fisher Score of ith feature,
μ_{ij} = mean of ith feature in the jth class,
ρ_{ij} = variance of ith feature in the jth class,
n_j = number of instances in the jth class,
μ_i = mean of ith feature,
K = total number of features,
k = kth feature, and
Σ = summation.

4.2 Feature selection

The procedural steps of SFO algorithm employed for choosing optimal feature vector space is deliberate given in this section [26].

4.2.1 Procedure

Step 1: The discrete set of cells (n size) with random generation of position (x^n) and velocity (v^n) are initialized in the two-dimensional problem space. The random selection of

TDS and EC features are represented as collagen deposition defined in extracellular matrix (ecm). The parameters of diffusion coefficient $(\rho) = 0.8$ and cell speed $(s = \frac{s}{k_{ro}L})$ are predetermined.

Step 2: Calculate the mathematical expression of Mean Square Error (MSE) used as an objective function.

$$MSE = \frac{1}{M}\sum_{i=1}^{M}[\hat{\theta} - \theta]^2 \tag{4}$$

where

$\hat{\theta}$ = expected observation,
Θ = actual observation,
M = total number of data instances, and
Σ = summation.

Step 3: The optimal TDS and EC features obtained at every iteration are compared with fittest features obtained at each step (Cbest). If cell gives better TDS and EC feature (cbest$_i$) than previous features (cbest$_{i-1}$), hence cbest$_i$ has taken as the fittest value. Otherwise, no update operations have been done.

```
if (cbest_i < Cbest)
    Cbest = cbest_i;
else
    Cbest = cbest_{i-1};
```

Step 4: The velocity and position of cell is updated at every iteration using the following Eqs. (5), (6).

$$v^{i(t+1)} = v^i(t) + (1 - \rho)*c\big(f^i(t), t\big) + \rho*\frac{f^i(t - \tau)}{\|f^i(t - \tau)\|} \tag{5}$$

where

t = current iteration t,
τ = time lag,
v^i = velocity of ith cell,
f^i = ith fibroblast, and
c = function denotes candidate solution (collagen deposition) defined in the extracellular matrix, which is stochastically selected by ith fibroblast (f^i).

$$x^{i(t+1)} = x^i(t) + \text{cell\_speed}*\frac{v^i(t)}{\|v^i(t)\|} \tag{6}$$

where

$$cellspeed = \frac{s}{k_{ro}L}, \; s = 15 \, \mu m \, h^{-1},$$
$$K_{ro} = 10^3 k_{ro=10^3} \mu/min,$$
$$L = 10, \text{ and}$$
$$x^i = \text{position of } i\text{th cell.}$$

Step 5: The optimal TDS and EC feature subset obtained by cell at every iteration is updated in the ecm.

Step 6: Steps 2 to 5 are repeated until either prefixed number of iterations are attained to 1000 cycles or the fittest TDS and EC features are obtained. SFO code is repeatedly executed five times, and the fine-tuned TDS and EC features have been chosen based on average values of best outcomes presented in Table 12.

Step 7: The resultant fittest TDS and EC features (C_{best}) solution utilized in Kalman filter for estimating TDS and EC parameters in the four sites, namely, C1, E1, E2, and W1 correspond to seasons.

4.3 Development of an intelligent estimation model

Kalman filter is an efficient method for estimating the future state variables in the non-linear prediction system. Studies on Kalman filtering technique signify that it is considered as an optimal estimator, which is well suited for the development of a predictive model while the small numbers of a dataset are available. Kalman estimator is used to predict future data with respect to noisy values. It empirically estimates the prediction variables based on historical data. TDS value of the water samples is very high, ranging from 662 to 1495 ppm. It reveals that TDS exhibit very strong correlation with EC ($R^2 = .925$ and $R^2 = .458$) in E1 and E2 sites. This is because EC variation showed the situation of inorganic pollutants with water. TDS and EC are the continuous target variables of the water dataset. TDS and EC parameters are greatly influenced by various factors including temperature, sewage effluents, and biological activity in this lake. Indeed, Stochastic Kalman filter is a suitable method to design a prediction system for a nondeterministic characteristic of water dataset subjected to high uncertainty. Consider a stochastic process

Table 12 Sample of optimal EC and TDS parameters chosen by SFO.

Parameters	No. of cycles	Site	Minima	Maxima
TDS	5	C1	719, 1250, 1150, 1240	1563, 1472, 1391, 1586
		E1	866,1259, 1506, 357, 1627	1428, 1557, 1785, 1805
		E2	612, 729, 1265, 1450	1135, 1570, 1770, 1169
		W1	1376, 551, 1291, 1825	1510, 1755, 1747, 1911
EC	5	C1	914, 1212, 1220, 1236	1520, 1751, 1822, 1445
		E1	1270, 1240, 1320, 1280	1750, 1870, 1540,1536
		E2	920, 1604, 916, 1030	1160, 1786, 1280, 1260
		W1	981, 923, 1175, 1212	1102, 1210, 1549, 1781

$$\hat{x} = A_x + w \tag{7}$$

$$x(t_0) = x_0 \tag{8}$$

$$y = C_x + v \tag{9}$$

where

x = previous state variable,
$\hat{x}$ = next state variable,
$w(t)$ and $v(t)$ = noisy values,
x_0 = random variable, and
A_x and C_x = constant factors.

The optimal TDS and EC parameters are given as previous state variables (priori estimate) in the Kalman estimator to estimate the future state variables (posteriori estimate) in the corresponding four sites, namely C1, E1, E2, and W1. Based on the initial state condition, Kalman filter has computed the estimated measurement or prediction (future state) using the following Filtering Riccati equations.

$$\varepsilon\left[w(t)w^T(\tau)\right] = Q(t)\delta(t - \tau) \tag{10}$$

$$\varepsilon\left[v(t)v^T(\tau)\right] = R(t)\delta(t - \tau) \tag{11}$$

$$\varepsilon\left[v_i(t)w_j(\tau)\right] = 0 \tag{12}$$

$$\varepsilon\left[x_0 x_0^T\right] = S \tag{13}$$

where

$\varepsilon(x)$ is the expected value (mean) of the random variable x,
$v(t)$ is unrelated to $w(t)$ whenever $\tau \neq t$,
$v(t)$ is not related to $w(\tau)$ at all,
$Q(t)$ = expected size of $w(t)$ and $w^T(t)$,
$R(t)$ = expected size of $v(t)$ and $v^T(t)$,
S = expected size of $x_0 x_0^T$.

5 Experimental observations and discussions

The experimental study was carried out in MATLAB (R2018b) run on Intel (R) Core (TM) i7-4790 CPU executing 3.60 GHz with 8 GB RAM. Windows 7 professional 64-bit machine OS environment, and the examined results were visualized in R tool. Sum of Squared Error (SSE) is applied to calculate the overall measurement of prediction error given in Eq. (14), and the experimental results are portrayed in Tables 13 and 14.

$$SSE = \sum (y - \hat{y})^2 \tag{14}$$

where

y = actual value,
$\hat{y}$ = predicted value.

Table 13 Analysis of SSE results for total dissolved solids.

Site	Kalman	SFO + Kalman
C1	61,637	364
E1	57,947	74
E2	555,683	529
W1	180,879	4356

Table 14 Analysis of SSE results for electrical conductivity.

Site	Kalman	SFO + Kalman
C1	18,900	1090
E1	2800	100
E2	210,200	6400
W1	215,689	0

The mathematical expression employed to measure the improvement of SSE values obtained by conventional Kalman method and SFO-based Kalman method is given in Eq. (15) [27].

$$\text{Success rate} = \frac{(V_1 - V_2)}{V_1} * 100 \tag{15}$$

where

V_1 = SSE value measured by Kalman estimator,
V_2 = SSE value measured by SFO based Kalman estimator.

In the SFO algorithm, collaborative and self-learning characteristics of fibroblast enable one to find optimal features in the feature vector space. The fittest features chosen by SFO are implemented to regulate the state estimator of Kalman filtering method. SFO-based Kalman estimation model inherit fine-tuned features used in the state estimator of Kalman method to predict TDS and EC parameters in the corresponding four sites, namely C1, E1, E2, and W1. The experimental results shown in Tables 15 and 16 confirmed that SFO algorithm significantly calibrated the estimation accuracy of Kalman filter, which gives better results than a conventional approach.

Analysis on the accuracy of Kalman filtering method and SFO-based Kalman filtering method are also extended to cover South Indian seasons such as Autumn (October 15 to December 14), Winter (December 15 to February 14), Prevernal (February 15 to April 14), and Monsoon (April 15 to October 14) [28]. The investigational results presented in

Table 16   Improvement of result attained for electrical conductivity.

Site	Kalman filtering method	SFO based Kalman filtering method (SSE)	Success rate (%) (Increase (+)/Decrease (−))
C1	18,900	1090	+ 94.23
E1	2800	100	+ 96.43
E2	210,200	6400	+ 96.96
W1	215,689	0	+ 100.00

Table 15   Improvement of result for total dissolved solids.

Site	Kalman filtering method (SSE)	SFO based Kalman filtering method (SSE)	Success rate (%) (Increase (+)/Decrease (−))
C1	61,637	364	+ 99.41
E1	57,947	74	+ 99.87
E2	555,683	529	+ 99.90
W1	180,879	4356	+ 97.59

Tables 17–22 had demonstrated that the fine-tuning features chosen by SFO algorithm enhanced the estimator of Kalman method that achieves better predictive results than a conventional approach. The observed TDS and EC values for the aforementioned seasons and its corresponding estimated values of both Kalman method and SFO-based Kalman method are portrayed from Figs. 6–37. The SSE obtained for the two parameters, namely TDS and EC, in the four sites such as C1, E1, E2, and W1 shows that there is a wide difference found between *actual values* and *estimated values*. TDS and EC values significantly varied at regular intervals due to various factors that include climate change, seasonal variation, pollutants, physicochemical characteristics, and aqua culture. There are multiple unknown sewage sources (inlets) connected to the regions of W1 and E1 sites. Therefore, the level of turbidity is very high and leads to reduced DO levels in these regions. Massive growth of algal blooms and eutrophication level is highly observed in these regions. Comparatively, C1 and E2 sites are too far away from inhabitant's access point, and these regions are greatly focused on fishing. So, DO level is quite high, and turbidity level is slightly low in these zones.

SFO algorithm was employed in Kalman filter to develop a predictive model for the estimation of time series data. An optimal estimation model was evaluated using the real-time water samples collected from Ukkadam Periyakulam lake for forecasting water quality parameters on a temporal basis. The performance of SFO-based Kalman filter was compared with other well-known global optimization algorithms to conduct a better investigation of this study [27,29,30]. Generally, estimating the time-varying data is a quite challenging task, because time series data are irregular, dynamic, and nonlinear. Analysis of experimental results illustrated that collaborative and self-adaptive nature enables SFO algorithm to choose optimal features (TDS and EC parameters) that do not easily get stuck in a quick convergence problem. But, other comparative metaheuristic algorithms,

Table 17   Seasonal analysis in C1 site based on SSE.

Parameter	Autumn		Winter		Prevernal		Monsoon	
	Kalman	SFO + Kalman	Kalman	SFO + Kalman	Kalman	SFO + Kalman	Kalman	SFO + Kalman
TDS	112,009	234	277,515	170	138,004	26	128,021	52
EC	148,600	916	80,200	145	2056	1	5129	67

Table 18   Seasonal analysis in E1 site based on SSE.

Parameter	Autumn		Winter		Prevernal		Monsoon	
	Kalman	SFO + Kalman	Kalman	SFO + Kalman	Kalman	SFO + Kalman	Kalman	SFO + Kalman
TDS	40,178	9	312,154	25	503,653	5	234,771	72
EC	64	49	566,600	0	135,600	0	4510	84

Table 19   Seasonal analysis in E2 site based on SSE.

Parameter	Autumn		Winter		Prevernal		Monsoon	
	Kalman	SFO + Kalman	Kalman	SFO + Kalman	Kalman	SFO + Kalman	Kalman	SFO + Kalman
TDS	7785	10	1,314,530	83	511,445	19	147,871	37
EC	780	65	2658	194	1769	0	7732	98

Table 20   Seasonal analysis in W1 site based on SSE.

Parameter	Autumn		Winter		Prevernal		Monsoon	
	Kalman	SFO + Kalman	Kalman	SFO + Kalman	Kalman	SFO + Kalman	Kalman	SFO + Kalman
TDS	734,459	51	5262	17	99,193	0	43,176	35
EC	60,100	0	3109	16	3681	9	5421	76

namely, Particle Swarm Optimization (PSO), Differential Evolution (DE), and Cuckoo Search (CS) were not able to exhibit significant performance due to which were affected in a stagnation problem. These optimal features chosen by SFO algorithm were incorporated into Kalman filter to estimate the future state variables of time series data (TDS and EC parameters). It shows that the Kalman estimation model based on SFO was highly effective, reliable, and exhibits robust performance in predicting time-varying data when compared with other metaheuristic algorithms [27]. Therefore, the efficiency of improved Kalman model is further being validated by incorporating the optimal features selected by SFO algorithm on a seasonal basis in this work.

Table 21   Seasonal variation of TDS based on SSE.

Season	Site	Kalman filtering method	SFO based Kalman filtering method	Success rate (%) (Increase (+)/Decrease (−))
Autumn	C1	112,009	234	+ 99.79
	E1	40,178	9	+ 99.98
	E2	7785	10	+ 99.87
	W1	734,459	51	+ 99.99
Winter	C1	277,515	170	+ 99.94
	E1	312,154	25	+ 99.99
	E2	1,314,530	83	+ 99.99
	W1	5262	17	+ 99.68
Prevernal	C1	138,004	26	+ 99.98
	E1	503,653	5	+ 99.99
	E2	511,445	19	+ 99.99
	W1	99,193	0	+ 100.00
	C1	128,021	52	+ 99.95
Monsoon	E1	234,771	72	+ 99.96
	E2	147,871	37	+ 99.97
	W1	43,176	35	+ 99.91

Table 22   Seasonal variation of EC based on SSE.

Season	Site	Kalman filtering method	SFO based Kalman filtering method	Success rate (%) (Increase (+)/Decrease (−))
Autumn	C1	148,600	916	+ 99.38
	E1	64	49	+ 23.43
	E2	780	65	+ 91.66
	W1	60,100	0	+ 100.00
Winter	C1	80,200	145	+ 99.82
	E1	230,856	0	+ 100.00
	E2	2658	194	+ 92.70
	W1	3109	16	+ 99.49
Prevernal	C1	2056	1	+ 99.95
	E1	135,600	0	+ 100.00
	E2	1769	0	+ 100.00
	W1	3681	9	+ 99.76
Monsoon	C1	5129	67	+ 98.75
	E1	4510	84	+ 98.13
	E2	7732	98	+ 98.73
	W1	5421	76	+ 98.59

As per CPCB and ISI standards (ISI-IS: 2296-1982), the classification of inland surface waters is depicted in Table 23. As per the standards prescribed by CPCB and ISI, the tolerance limit of TDS for Class A, Class C, Class D, and Class E are 500, 1500, 300, and 2100, respectively, and EC for Class D and Class E are 1000 and 2250, respectively.

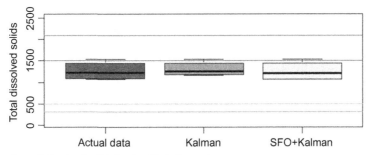

FIG. 6 Estimation of total dissolved solids in Site C1 during autumn.

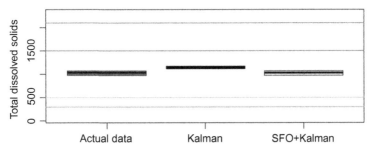

FIG. 7 Estimation of total dissolved solids in Site E1 during autumn.

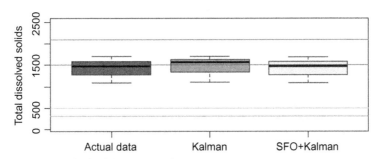

FIG. 8 Estimation of total dissolved solids in Site E2 during autumn.

Examined results show that water quality is well suited for irrigation and industrial purposes, and it could be useful for drinking purposes if followed by a conventional water treatment method. It emphasizes that regular monitoring of the water quality parameters and adopting remedial measures could be beneficial to improve the water quality standards in this lake. It is also illustrated that the estimation model is able to recognize the pattern of water quality parameters that delivers promising outcomes (TDS and EC features) in Ukkadam Periyakulam Lake, Coimbatore.

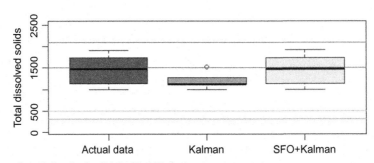

FIG. 9 Estimation of total dissolved solids in Site W1 during autumn.

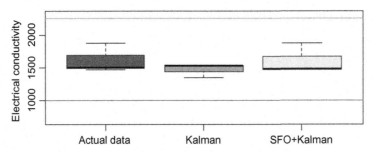

FIG. 10 Estimation of electrical conductivity in Site C1 during autumn.

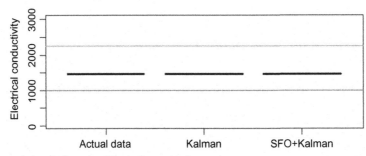

FIG. 11 Estimation of electrical conductivity in Site E1 during autumn.

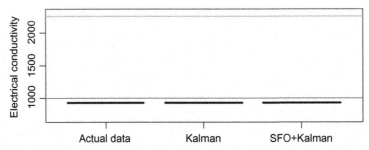

FIG. 12 Estimation of electrical conductivity in Site E2 during autumn.

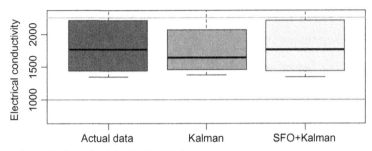

FIG. 13 Estimation of electrical conductivity in Site W1 during autumn.

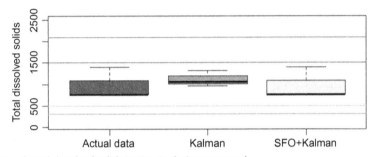

FIG. 14 Estimation of total dissolved solids in Site C1 during prevernal.

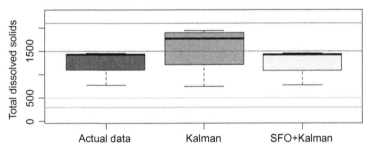

FIG. 15 Estimation of total dissolved solids in Site E1 during prevernal.

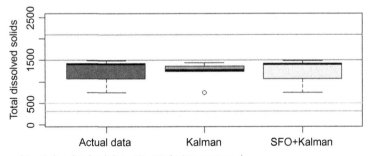

FIG. 16 Estimation of total dissolved solids in Site E2 during prevernal.

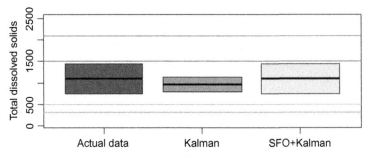

FIG. 17 Estimation of total dissolved solids in Site W1 during prevernal.

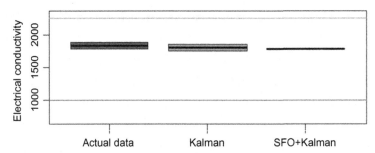

FIG. 18 Estimation of electrical conductivity in Site C1 during prevernal.

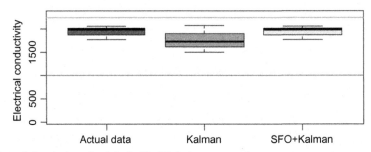

FIG. 19 Estimation of electrical conductivity in Site E1 during prevernal.

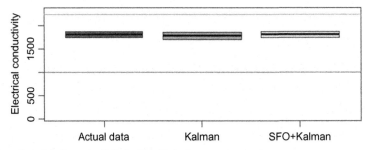

FIG. 20 Estimation of electrical conductivity in Site E2 during prevernal.

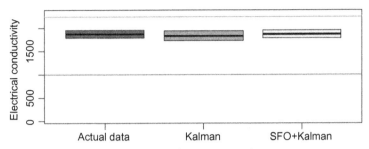

FIG. 21 Estimation of electrical conductivity in Site W1 during prevernal.

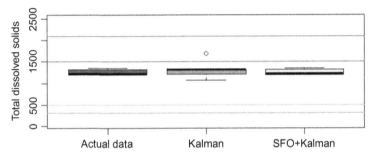

FIG. 22 Estimation of total dissolved solids in Site C1 during winter.

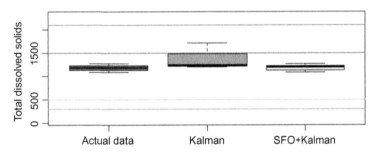

FIG. 23 Estimation of total dissolved solids in Site E1 during winter.

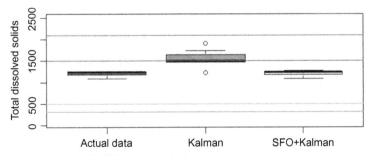

FIG. 24 Estimation of total dissolved solids in Site E2 during winter.

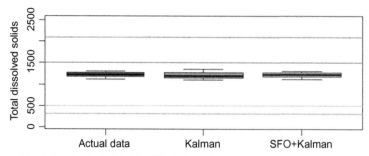

FIG. 25 Estimation of total dissolved solids in Site W1 during winter.

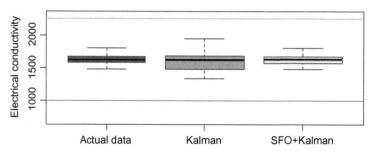

FIG. 26 Estimation of electrical conductivity in Site C1 during winter.

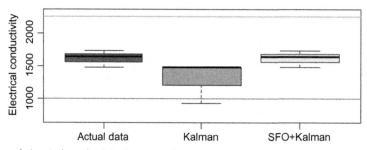

FIG. 27 Estimation of electrical conductivity in Site E1 during winter.

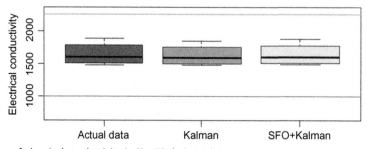

FIG. 28 Estimation of electrical conductivity in Site E2 during winter.

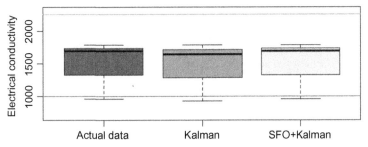

FIG. 29 Estimation of electrical conductivity in Site W1 during winter.

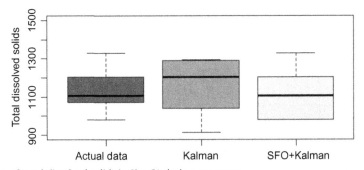

FIG. 30 Estimation of total dissolved solids in Site C1 during monsoon.

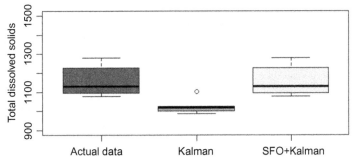

FIG. 31 Estimation of total dissolved solids in Site E1 during monsoon.

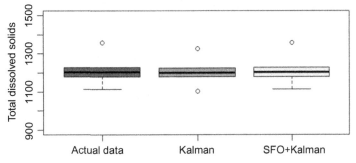

FIG. 32 Estimation of total dissolved solids in Site E2 during monsoon.

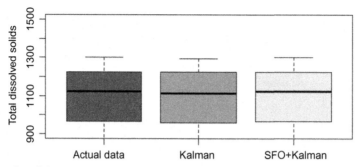

FIG. 33 Estimation of total dissolved solids in Site W1 during monsoon.

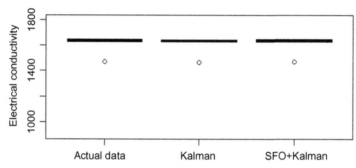

FIG. 34 Estimation of electrical conductivity in Site C1 during monsoon.

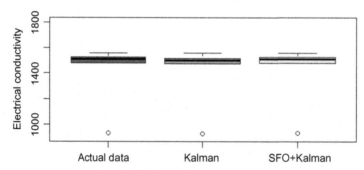

FIG. 35 Estimation of electrical conductivity in Site E1 during monsoon.

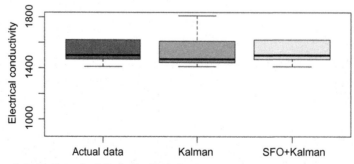

FIG. 36 Estimation of electrical conductivity in Site E2 during monsoon.

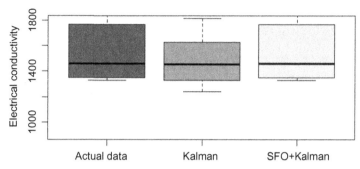

FIG. 37 Estimation of electrical conductivity in Site W1 during monsoon.

Table 23   Categorization of water.

Classification	Type of use
Class A	Drinking water source without conventional treatment but after disinfection
Class B	Outdoor bathing
Class C	Drinking water source with conventional treatment followed by disinfection.
Class D	Fish culture and wild life propagation
Class E	Irrigation, industrial cooling or controlled waste disposal

Source: Central Pollution Control Board (CPCB).

Investigational studies reveal that water quality standards of this lake have extremely deteriorated by numerous factors, namely, domestic garbage dumping, sewage disposal, and effluents from the surrounding textile dyeing industries and municipal market. The outcomes observed provide quantitative information on water quality that is significantly beneficial to various stakeholders to formulate control policies and preventive measures that conserve the existence of aquatic species, ecosystem, and rejuvenate the lake from multiple source of pollutants, preserve inland water bodies, promote socioeconomic development, sustain biodiversity, enhance underground water availability, and rapidly decline water scarcity problem in Coimbatore City.

5.1 Remedial measures suggested improving the benefits of wetlands

- Immediate steps have to be taken to prohibit garbage dumps/sewage disposal/industrial effluents into the lake
- Ban on the several unknown sewage inlets connected to this lake
- Action need to be taken for controlling the lake encroachment activities by local people who are living in the surrounding areas
- Promote activities to enhance fishing and fish farming
- Joining hands with NGOs, academicians, research organizations, and municipal corporations should formulate plans/strategies for cleaning algal blooms, removing water hyacinth, and rejuvenating ecological characteristics of the lake

- Water treatment plants need to be installed for water recycling that could be beneficial to various domestic purposes
- Construction of bunds in the lake to attract bird species

6 Conclusion

Monitoring and assessment of water quality parameters in freshwater bodies are essential to reduce the deterioration of water quality and conserve the aquatic ecosystem. The statistical analysis is the most recommended technique to identify the linear correlation between water quality parameters. The technique had been applied to validate the variations in water quality of Ukkadam Periyakulam Lake, Coimbatore. It identified that TDS and EC exhibit strong correlation among the water quality parameters. The extracted TDS and EC features are utilized to construct an estimation model. It reveals that TDS and EC parameters widely vary, which would affect the performance of conventional Kalman estimation model. SFO algorithm is utilized to find optimal TDS and EC features that have been incorporated into Kalman filter for the estimation of time series data. Investigation of results confirmed that SFO is considered an efficient computing paradigm, which greatly calibrated the estimation accuracy of Kalman filter. This study can be further extended to increase the number of sampling sites and observation of heavy metal contamination for monitoring and assessment of water quality standards of this lake in the future.

References

[1] Inland Waters Biodiversity – Why is it Important?, 2015, https://www.cbd.int/waters/importance/default.shtml. (Accessed 7 November 2015).

[2] D.P. Mondal, Water and related statistics, in: Central Water Commission Report, 2013, pp. 1–212.

[3] J.M. Mauskar, Status of surface water quality in India. Part I, in: Central Pollution Control Board Report, 2007, pp. 1–262.

[4] Setting right water quality in India, 2015, http://www.news18.com/news/india/setting-right-water-quality-in-india-737239.html/. (Accessed 16 February 2015).

[5] C. Rachna, K.A. Nishadh, P.A. Azeez, Monitoring water quality of Coimbatore wetlands, Tamil Nadu, India. Environ. Monit. Assess. 169 (1–4) (2010) 671–676, https://doi.org/10.1007/s10661-009-1206-0.

[6] Preface of Coimbatore city, 2015, https://en.wikipedia.org/wiki/Coimbatore/. (Accessed 1 February 2015).

[7] A.P.P. Jain, Environmental degradation of the Coimbatore wetlands in the Noyyal river basin, in: EIA Resource and Response Centre (ERC) Report, 2013, pp. 1–62.

[8] Coimbatore attempts to save water bodies, 2017, http://www.downtoearth.org.in/coverage/coimbatore-attempts-to-save-waterbodies-7045/. (Accessed 26 September 2016).

[9] Save Coimbatore Wetlands, 2015, http://www.coimbatorewetlands.org/?p=244/. (Accessed 5 February 2015).

[10] A lake comes to life, 2017, http://www.thehindu.com/features/magazine/a-lake-comes-to-life/article4817432.ece/. (Accessed 16 February 2016).

[11] R. Bhateria, D. Jain, Water quality assessment of lake water: a review. Sustain. Water Resour. Manag. 2 (2) (2016) 161–173, https://doi.org/10.1007/s40899-015-0014-7.

[12] C. Sarah Hembrow, H. Kathryn Taffs, Water quality changes in Lake Mckenzie, Fraser island, Australia: a palaeolimnological approach. Aust. Geogr. 43 (3) (2012) 291–302, https://doi.org/10.1080/00049182.2012.706207.

[13] X. Meng, Y. Zhang, X. Yu, J. Zhan, Y. Chai, A. Critto, Y. Li, J. Li, Analysis of the temporal and spatial distribution of lake and reservoir water quality in China and changes in its relationship with GDP from 2005 to 2010. Sustainability 7 (2) (2015) 2000–2027, https://doi.org/10.3390/su7022000.

[14] M.A. Farooq, A. Zubair, S.S. Shaukat, M.U. Zafar, W. Ahmad, Water quality characteristics of Keenjhar Lake, Sindh Pakistan, World Appl. Sci. J. 27 (3) (2013) 297–301.

[15] H.S. Kiran Kumar, Y.B. Bharatharaj Etigi, Varun, C.G. Punith, Water quality analysis of avaragere lake – a case study, Int. Res. J. Eng. Technol. 5 (12) (2018) 526–532.

[16] S.A. Bhat, K. Ashok Pandit, Surface water quality assessment of wular lake, a Ramsar site in kashmir himalaya, using discriminant analysis and wqi. Hindawi J. Ecosyst. 2014 (2014) 1–18, https://doi.org/10.1155/2014/724728.

[17] G. Quadros, B. Hemambika, A. Julffia Begam, P.A. Azeez, Lakes of Coimbatore City, ENVIS Publication, 2014, 43.

[18] C. Gajendran, P. Thamarai, Study on statistical relationship between ground water quality parameters in nambiyar river basin, Tamil Nadu, India, Int. J. Pollut. Res. 27 (4) (2008) 679–683.

[19] C. Boyd, Secchi disk visibility: correct measurement, interpretation, Glob. Aquacult. Advoc. 7 (1) (2004) 66–67.

[20] Water quality, 2016, https://https://en.wikipedia.org/wiki/Water_quality/. (Accessed 16 February 2016).

[21] CPCB, Tolerance Limit for Surface WATER MINARS/27, Guidelines for Water Quality Monitoring, Central Pollution Control Board, New Delhi, 2008, pp. 1–35.

[22] ISI, Tolerance Limit for Inland Surface Water Subject to Various Purpose, Tolerance and Classification, Indian Standards Institute, New Delhi, 1991, pp. 1–9.

[23] S. Mishra, M.P. Sharma, A. Kumar, Assessment of water quality in Surha lake based on physiochemical parameters, India, J. Mater. Environ. Sci. 6 (9) (2015) 2446–2452.

[24] G. Sami Daraigan, S. Ahmed Wahdain, S. Ahmed Ba-Mosa, H. Manal Obid, Linear correlation analysis study of drinking water quality data for al-Mukalla city hadhramout, Yemen, Int. J. Environ. Sci. 1 (7) (2011) 1699–1708.

[25] UNICEF, Handbook on Water Quality, United Nations Children's Fund (UNICEF), New York, USA, 2008, 179.

[26] P. Subashini, T.T. Dhivyaprabha, M. Krishnaveni, Synergistic fibroblast optimization. in: S. Dash, K. Vijayakumar, B. Panigrahi, S. Das (Eds.), Artificial Intelligence and Evolutionary Computations in Engineering Systems, Advances in Intelligent Systems and Computing, Vol. 517, Springer, Singapore, 2017, pp. 285–294, https://doi.org/10.1007/978-981-10-3174-8_25.

[27] T.T. Dhivyaprabha, P. Subashini, M. Krishnaveni, N. Santhi, R. Sivanpillai, G. Jayashree, A novel synergistic fibroblast optimization based kalman estimation for forecasting time-series, Evol. Syst. 10 (2) (2019) 205–220.

[28] Climate of India, https://en.wikipedia.org/wiki/Climate_of_India/, 2017. (Accessed 17 July 2017).

[29] S. Chatterjee, S. Sarkar, N. Dey, S.A. Ashour, S. Sen, A.E. Hassanien, Application of cuckoo search in water quality prediction using artificial neural network. Int. J. Computat. Intel. Stud. 6 (2/3) (2017) 229–244, https://doi.org/10.1504/IJCISTUDIES.2017.089054.

[30] S. Chatterjee, S. Sarkar, N. Dey, S. Sen, Non-dominated sorting genetic algorithm-II-induced neural-supported prediction of water quality with stability analysis. J. Inf. Knowl. Manag. 17 (02) (2018) 1850017(1)–1850017(21), https://doi.org/10.1142/S0219649218500168.

8

Recent trends in air quality prediction: An artificial intelligence perspective

Ibrahim Kok[a], Metehan Guzel[b], and Suat Ozdemir[c]

[a]DEPARTMENT OF COMPUTER SCIENCE, GAZI UNIVERSITY, ANKARA, TURKEY [b]DEPARTMENT OF COMPUTER ENGINEERING, GAZI UNIVERSITY, ANKARA, TURKEY [c]DEPARTMENT OF COMPUTER ENGINEERING, HACETTEPE UNIVERSITY, ANKARA, TURKEY

1 Introduction

Air pollution is a crucial social and environmental problem that has a large number of detrimental impacts on human health, ecosystem, and climate [1]. Millions of people suffer from diseases related to air pollution every year [2]. For this reason, air pollution is currently taking over the attention of all countries around the world [3]. To overcome this problem, countries have to take the necessary precautions by estimating air pollution with advanced methods [4]. Therefore, many studies have mainly focused on air pollutant and air quality prediction models in recent years [5].

Traditionally, the approaches used in air quality prediction can be classified as follows:

- *Deterministic prediction* is performed based on pollutant sources, real-time emission quantities, and chemical reactions between exhaust gases under the planetary boundary layer [6]. Deterministic models do not require large historical data and require precise knowledge of parameters, which is an expensive requirement in the real world and not always accurate it must be. In addition, deterministic models are extremely complex and need very high computational power. Advantage of deterministic prediction is to be able to extend a model in a spatial manner.
- *Statistical prediction* requires large historical data under different meteorological conditions. Using regression and machine learning methods, complex relations between pollutants and meteorological conditions can be addressed. But, statistical models work on singular stations and do not have a spatial resolution.
- *Neural network-based prediction* is performed based on neural network models. Neural network models are frequently used and potentially good tools to model nonlinear relations.

Deep learning (DL) is a subset of machine learning (ML) methodologies that imitate the structure and function of human brain. In traditional learning, feature engineering is needed to represent data as abstracted and high-level feature form. Such learning models are domain-specific and require extra effort. However, feature engineering is performed automatically without extra effort in DL [7]. In this aspect, DL models demonstrated great success in many application domains, such as computer vision, natural language processing, robotics, and many more [8]. As a result, in recent years, DL models are applied to trending and critical domains such as autonomous systems, environmental science, and bioinformatics [9,10].

In this chapter, we discuss recent artificial intelligence-based air quality prediction methods in the literature by categorizing them into two subgroups as neural network and DL-based approaches. In this context, we first give preliminary information about air quality, air quality index, particulate matter, and air pollutant gases. Then, we present a comprehensive analysis of neural network-based prediction models and deep learning-based prediction models in the area of air quality prediction.

2 Preliminary information

Air quality index (AQI) is used to report daily air quality. AQI is calculated from major air pollutants and determines if the air is clean or unhealthy. There are two types of pollutants are specified in AQI context [11] (see Table 1).

- **Particulate matter (PM):** Particulate matter is classified according to its diameter. PM_{10} and $PM_{2.5}$ represent particle with a diameter less than 10 and 2.5 micron, respectively [12]. PM emissions arise from vehicles [13], residual heating and industrial processes [6]. PM has severe effects on human health. PM can absorb numerous toxic materials (heavy metals, carbonaceous materials, etc.) and introduce them to the human body. High PM levels can cause cardiovascular and respiratory diseases, immune disorders, and even lung cancer [13,14].
- **Pollutant gasses:** In general, five pollutant gasses are identified and used for AQ assessment. These are ozone (O_3), nitrogen dioxide (NO_2), sulfur dioxide (SO_2), ammonia (NH_3), and carbon monoxide (CO). Among these, NO_2 and O_3 are identified as a criterion for all air quality indexes. SO_2 and CO are also common but do not include in all indexes. Similar to PM, pollutant gases also cause serious health problems [15–17]. In addition, these pollutants are precursors to PM [18].

Table 1 Usage of pollutant for AQ assessment [19].

| Country | Particulate matter | | Air pollutants | | | | | |
	$PM_{2.5}$	PM_{10}	NO_2	O_3	SO_2	CO	NH_3	Pb
Canada	x		x	x				
Hong Kong	x	X	x	x	x			
China	x	X	x	x	x	x		
India	x	X	x	x	x	x	x	x
Mexico	x	X	x	x	x	x		
Singapore	x	X	x	x	x	x		
UK	x	X	x	x	x			
US	x	X	x	x	x	x		
Europe	x	X	x	x				

Neural network- and deep learning-based air quality prediction methods learn from experience. A model is trained with historical data. Training process uncovers the relation between inputs and outputs, and embeds acquired knowledge into the prediction model. Then, model is used for prediction of unseen data.

2.1 Input types for air quality prediction models

In the scope of this chapter, outputs of mentioned works are either AQI or pollutant concentrations. As input, there are three sets of variables used for air quality prediction.

- In the process of prediction of pollutant concentrations and air quality, the most commonly used indicators are historical pollutant concentrations (PC). Temporal, spatial, and interpollutant correlations are used to predict air quality/air pollution and pollutant concentrations.
- Another set of factors that may be used for prediction is meteorological conditions. Emissions cause pollutants and meteorological variables transfer and diffuse them [16]. For this reason, numerous research [20,21] use *meteorological variables (MV)* as a part of prediction inputs. The most used meteorological variables are humidity, temperature, solar radiation, atmospheric pressure, wind speed, wind direction, precipitation, etc.
- *Time-wise variables (TV)* are another indicator of air quality. In cold seasons, because of heating systems total emission of pollutants increases significantly. Also, traffic emission increase on weekdays and rush hours because of increased traffic. It is obvious that the usage of TV can enhance prediction accuracy. The most used TVs are DoW (day of the week), DoY (day of the year), and HoD (hour of the day).

2.2 Structures of air quality prediction models

The structure of prediction models also differs. There are three main approaches in the literature.

- *Single-learner (SL) approach:* In SL approach, data are given to a singular model that performs prediction.
- *Ensemble learner (EL) approach:* EL approach is the concept of using numerous weak learners to perform a single machine learning task. Especially at the complex machine learning tasks, EL performs significantly better than SL [22]. In the EL approach, firstly data are split into a number of pieces. Then, predictions are performed separately for each piece and finally combined into a single prediction. Figurative illustration of EL approach is given in Fig. 1.
- *Hybrid learner (HL):* HL is pretty similar to EL approach. Numerous models are employed for a single task. However, unlike EL, in HL different types of prediction models are used.

2.3 Assisting methods for air quality prediction models

Prediction methods generally accompanied with feature selection and optimization methods to increase accuracy. Therefore, before going into details of neural network–based prediction methods, assisting algorithms will be explained briefly.

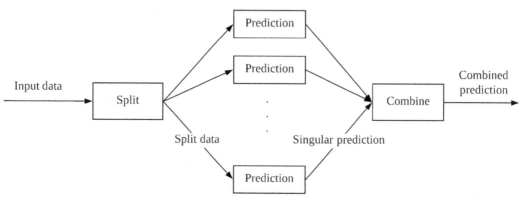

FIG. 1 Ensemble learner approach.

2.3.1 Feature selection methods for air quality prediction models

Numerous research employs feature selection algorithms before the prediction process. There are two reasons behind this. Firstly, by reducing the number of features the computational complexity of model decreases significantly. Also, irrelevant features can affect the accuracy of prediction models in a negative way. Therefore, eliminating irrelevant features increases the success of prediction models. The most used methods are as following:

- **Principal content analysis (PCA)** detects correlated features and combines them into new features called PCs. PCA generates uncorrelated PCs [23,24]. *Principal Content Regression (PCR) is* a regression model built upon PCA and in the literature, is used as a feature selection algorithm [13].
- **Classification and regression trees (CART)** partitions input space into a binary tree [25]. When partitioning, the aim is maximizing the variance between data, therefore maximizing the knowledge gained from each split. Recently, a rather complex variation of CART, *Random Forest (RF)*, takes the spotlight. RF is composed of multiple CARTs, generally used in a way that combines feature selection and prediction tasks [5].
- **Stepwise regression (SR)**is composed of two methods, namely, forward selection and backwards elimination [25]. Forward selection starts with no predictor and adds predictors based on their correlation with target. Backward elimination starts with all features and iteratively removes the least correlated one [26].

In addition to the mentioned methods *Gray Model (GM)*, *RM (Rolling Mechanism)*, *Fuzzy Comprehensive Evaluation (FCE)*, *Fuzzy Rough Sets (FRS)*, and *Multiple Linear Regression (MLR)* are also present in the literature. For more information, reader may refer to corresponding papers. Table 2 shows research that employs feature selection in air quality prediction process.

Table 2   Feature selection methods for air quality prediction models.

Reference	Feature selection method(s)
[13]	Multiple Linear Regression (MLR), Principal Content Regression (PCR)
[26]	Stepwise Regression (SR)
[27]	Gray Model (GM), Rolling Mechanism (RM),
[25]	Classification and Regression Trees (CART), Principal Content Analysis (PCA), Stepwise Regression (SR)
[17]	Fuzzy Comprehensive Evaluation (FCE)
[12]	Principal Content Analysis (PCA)
[2]	Fuzzy Rough Sets (FRS)
[20]	Gray Model (GM)
[28]	Random Forest (RF)

2.3.2 Optimization methods for air quality prediction models

Optimization of NN structure enhances prediction accuracy. In addition, Extreme Learning Machines (ELM) and Elman Neural Networks (ENN) involve some structural randomness. If this randomness is not removed, prediction models based on ELM and ENN tend to fall into local minimums or overfitting. To overcome this risk, optimization algorithms are used. Review of literature shows us diversity at optimization algorithms. For more information, reader may refer to corresponding papers. Some of the algorithms are listed here and researches that use them are given in Table 3.

- **Cuckoo search (CS)** algorithm mimics cuckoo birds' reproduction [32]. Cuckoo birds leave their eggs to alien nests. If host finds an egg, either throw it out or builds up a new nest. Otherwise, it feeds baby cuckoo bird like one of its own. Inspired from this method, an evolutionary optimization CS proposed. In the algorithm, each cuckoo bird leaves their egg to a nest every generation. Nests with high-quality eggs will be transferred to the next generations. Hosts identify these eggs according to a probabilistic function. In case of identification, egg will be tossed away or a new nest will be built.
- **Imperialist competitive algorithm (ICA)** simulates competition between empires to assimilate and conquer colonies [33]. In the algorithm, every individual of population is a country. Some of them are colonies and others are empires. Empires try to take

Table 3   Optimization methods for air quality prediction models.

Reference	Optimization Method(s)
[29]	Differential Evolution (DE)
[17]	Cuckoo Search (CS), Differential Evolution (DE)
[30]	Cuckoo Search (CS)
[2]	Imperialist Competitive Algorithm (ICA)
[31]	Multi objective optimization algorithm MSSA

control of states and the last standing empire represents the optimal solution to the problem.

- **_Differential evolution (DE)_** is a simple but powerful evolutionary algorithm [34]. DE is a population-based approach similar to genetic algorithm; evolutionary mechanisms like mutation and crossover are employed to find optimal solutions. In comparison, DE is more stable and able to highlight strengths of populations [35]. Also, implementation of DE to real-world problems is easier.

2.3.3 Decomposition methods for air quality prediction models

As mentioned, the first and one of the most crucial steps of ensemble learner approach is the decomposition of the data. Review of the literature reveals usage of three main approaches, which are given as follows. For more information, reader may refer to corresponding papers.

- **_Empirical mode decomposition (EMD)_** method decomposes a signal into multiple signals that are named as intrinsic mode functions (IMF) [22]. EMD is used to especially when data are nonlinear and unstable [2]. Review of the literature reveals that, EMD and its variations, namely, _Ensemble EMD (EEMD)_, _Complete EEMD (CEEMD)_ [36], _Fast EEMD (FEEMD)_ [37], _CEEMD with Adaptive Noise (CEEMDAN)_ [38], and _Improved CEEMDAN (ICEEMDAN)_ [39] are the most used set of methods for neural network-based air quality prediction models.
- **_Variational mode decomposition (VMD)_** is a nonrecursive decomposition model where modes are extracted concurrently [40]. VMD claimed to outperform EMD in manners of noise robustness and tone extraction.
- **_Wavelet transformation (WT)_** decomposes time series to summation of shifted wavelets in different scales [41].

3 Neural network-based prediction models

Artificial Neural Networks(ANNs) are a set of bioinspired methods that resemble the structure of the brain. Like a brain, an NN is composed of a large number of neurons and connections in between. Each neuron is a processor connected to other neurons with weighted connections. NNs operate in a parallel manner and learn from experience. NNs have a number of advantages that make it a good candidate for prediction tasks. The advantages of NNs are given as following [42].

- NNs can discover and formulate both linear and nonlinear problems effectively.
- NNs can learn from experience. NNs create a mapping between input and output space. For mapping, connections between neurons are optimized with the objective of minimizing the error function.
- NNs can perform tasks on unseen samples using the knowledge gained from training dataset.
- NNs have high adaptability because of their training methods. NNs can change weights in real time to reflect changes occurred in data samples.

Because of the given advantages, NNs are frequently used for AQ prediction tasks. In this section, different types of NNs and their applicability to AQ prediction domain will be discussed.

3.1 Multilayer perceptron

Multilayer Perceptron (MLP) is one of the most popular NN types. MLP is a feed-forward NN that is composed of an input layer, an output layer, and a number of hidden layers. Number of the nodes in the input layer depends on input number and output layer is composed of a single node [25]. Number of hidden layers and number of the nodes in the hidden layers are changeable. Connections of nodes are weighted. The structure of an MLP is given in Fig. 2.

The advantage of MLP is the ability to increase the number of hidden layers. By adding more layers, more complex linear/nonlinear relations can be uncovered. For learning, backpropagation method is used [12].

In Feng [6], an MLP-based ensemble prediction method is utilized to predict daily $PM_{2.5}$ levels in two day advance. 13 different air pollution monitoring stations are used to develop the model. By using WD, $PM_{2.5}$ signal is decomposed into multiple subsignals. A prediction model is applied to each decomposed signal and predictions are joined to acquire a final prediction. The proposed model uses $PM_{2.5}$ concentration of past, $PM_{2.5}$ levels of neighbor nodes, and meteorological variables.

In Cabaneros et al. [25], an MLP-based NO_2 prediction model is proposed. The proposed method uses pollutant concentrations and meteorological variables to perform

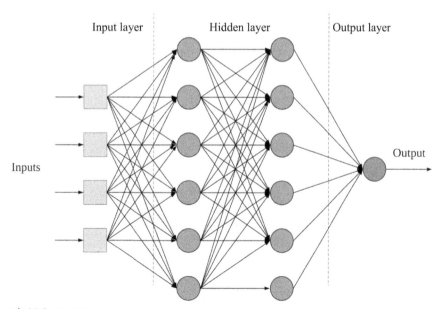

FIG. 2 Sample MLP structure.

prediction. The total number of predictors is 18. To reduce this number, three different feature selection methods, namely, Stepwise Regression, Principal Content Analysis, and CART, are employed. Test results show that feature selection methods select different predictors. Overall, all of the methods enhance MLP's prediction accuracy and decrease computational complexity. Obtained results indicate that PCA usually outperforms the other methods.

In addition to the outdoor, air quality prediction in the metro stations is also studied. In Perez and Gramsch [43], MLP-based particulate matter concentration prediction model is proposed. In the paper, timely pollutant concentrations, meteorological variables, spatial variables, and ventilation factors of the stations are used.

Another MLP-based prediction method is proposed in [12]. The method uses pollutant concentrations, meteorological variables, and spatial correlations between air quality stations to predict hourly particulate matter ($PM_{2.5}$, PM_{10}) concentrations. PCA is employed to select features and K-means is used to cluster data groups. Usage of clustering and feature selection enhances prediction accuracy.

In Gogikar et al. [21], Wavelet-based MLP is employed to predict daily PM concentrations. Their work takes seasonality and meteorological variables into consideration. Previous PM concentrations are transformed with discrete wavelet transform (DWT) and used as inputs of the MLP model. The paper reveals that usage of wavelet transformed pollutant concentrations instead of originals enhances prediction accuracy. The structures of proposed MLP-based prediction models are summarized in Table 4.

Table 4 Summary of MLP methods.

Reference	Prediction target	Method(s)	MLP structure
[6]	$PM_{2.5}$	MLP	10-8-1
[25]	NO_2	MLP	18-39-1
		MLP + CART	7-15-1
		MLP + PCA	11-23-1
		MLP + SR	5-11-1
[43]	$PM_{2.5}$	MLP	13-8-1
[12]	$PM_{2.5}$ (Station A)	MLP	9-2-1
		MLP_1	10-2-1
	PM_{10} (Station A)	MLP	11-4-1
		MLP_1	12-2-1
	$PM_{2.5}$ (Station B)	MLP	9-2-1
		MLP_1	10-2-1
	PM_{10} (Station B)	MLP	11-2-1
		MLP_1	12-2-1
[21]	$PM_{2.5}$, PM_{10}	MLP	6-8-2
		WMLP	8-8-2

3.2 Extreme learning machines

Extreme Learning Machine (ELM) is a feed-forward network that does not require updating of internode weights. At initialization, weights between input and hidden neurons are assigned randomly and ELM achieves an optimal solution by adjusting the number of hidden neurons [29]. Because of this approach, training time of ELM becomes significantly shorter than backpropagation NNs. But in return, ELMs are tending to fall into local minimums or overfitting. To overcome this problem, the weights of the ELM must be appropriately optimized. In addition to fast learning, ELM has good generalization ability [2]. To further improve this ability, an ensemble learning method is frequently used with ELM [28].

In Wang [15], an ensemble ELM is employed to predict air quality index. CEEMD and VMD are used for decomposition and Differential Evolution (DE) algorithm optimizes weights of ELM. Firstly, CEEMD decomposes time series into IMFs. Low-frequency IMFs are immediately used for prediction. High-frequency ones are decomposed a second time with VMD and then used for prediction. Prediction is performed by DE optimized ELM and then combined into a single prediction. Test results show that two-layer decomposition strategy surpasses singular implementations of both VMD and CEEMD. In addition, DE optimization significantly enhances the prediction accuracy of ELM.

A similar approach is followed in [2]. In this work, FRZ is employed for feature selection, improved complete ensemble empirical mode decomposition(ICEEMDAN) is used to decompose time series, and ELM is used for pollutant concentration prediction. Optimization of ELM is achieved with a metaheuristic algorithm called ICA. To evaluate method, pollutant concentrations acquired from six cities are used. Meteorological and seasonal variables are not used; only pollutant concentrations are used for prediction. Test results justify the usage of the proposed method.

In Luo [30], an ensemble PM_{10} prediction model is proposed. By using PM_{10} time-series data, predictions are performed. The proposed approach has two phases. In the first phase, original data are decomposed using FEEMD and initial forecast series are acquired using ELM. Then, initial forecast series are decomposed using VMD and ELM employed to acquire error forecast sequence. As last task, initial forecast and error forecast series are summed to generate final prediction. Cuckoo Search (CS) is employed to optimize weights of ELM at both phases. Proposed method is tested on data acquired from two cities of China. Results indicate that CS improves the accuracy of ELM and error correction phase improves the precision of the model significantly.

A rather innovative approach is proposed at [18]. In work, an Adaptive Neuro-Fuzzy Inference System (ANFIS) is utilized to reduce the randomness of ELM. $PM_{2.5}$, PM_{10}, CO, and NO predictions are performed using the proposed ANFIS-WELM algorithm. Results reveal that usage of ANFIS optimization enhances ELM prediction accuracy but slows down algorithm 20–25 times. A summary of this subsection is given in Table 5.

Table 5 Summary of ELM methods.

Reference	Model type	Decomposition method	Optimization method
[29]	Ensemble	CEEMD, VMD	DE
[2]	Ensemble	ICEEMDAN	ICA
[30]	Ensemble	FEEMD, VMD	CS
[18]	Hybrid	–	ANFIS

3.3 Elman neural networks

Elman Neural Network (ENN) is composed of three main layers, namely, input layer, output layer and a single hidden layer. In addition, ENN structure has an extra context layer that enhances usability at time series mining tasks. Similar to ELM, weights and thresholds of ENN are assigned randomly at initialization. Therefore, optimization of the model is a requirement also at ENN [31]. The structure of an ENN is given in Fig. 3.

ENN has two deep learning variations that are trendy topics, namely, Recurrent NN (RNN) and Long Short-Term Memory (LSTM). As more advanced variances of ENN, RNN and LSTM inherit ENN's ability to process time series and dominates the literature. More information about RNN and LSTM will be given in deep learning section.

In Hao and Tian [31], a novel data decomposition method MCEEMDAN and a new ENN-based prediction method proposed. A hybrid of CEEMD and VMD, MCEEMDAN uses VMD to decompose high-frequency IMFs of CEEMDAN. A multiobjective optimization algorithm MSSA is used to optimize ENN. Then, singular predictions of each IMF are combined into a single one. The proposed approach is evaluated using data acquired from three stations. Test results indicate higher-quality prediction compared to other methods.

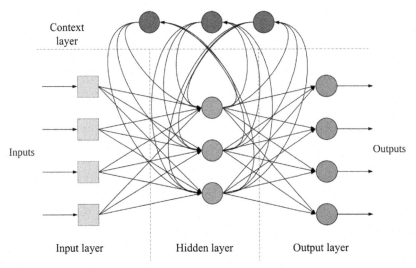

FIG. 3 Elman neural network structure.

In Yang and Wang [17], an ENN-based ensemble model is proposed to predict concentrations of six pollutants (SO_2, NO_2, CO, O_3, $PM_{2.5}$, and PM_{10}). CEEMD is employed to decompose and denoise the data. ENN is employed for prediction on IMFs and optimized by an optimization algorithm based on CS and DE, namely, MCSDE. The proposed approach is tested with hourly data acquired from two different cities.

3.4 Radial basis function neural networks

Radial Basis Function Neural Network (RBFNN) is a feed-forward NN that utilizes radial basis function (RBF) instead of the logistic function. Conventional FF mechanism is good for classification and decision-making tasks, but in the usage of the continuous value of RBF outperforms usage logistic function. Radial function is implemented in the neurons of hidden layer to approximate a nonlinear function [44]. RBFNN is composed of three layers, an input layer, a single hidden layer, and an output layer. A sample RBFFNN structure is given in Fig. 4. In Ha [44], RBFNN is employed to estimate spatial distribution of ozone. For this purpose, daily nitrogen oxide (NO_x) and volatile organic component (VOC) emissions, topography, coordinates of the prediction site and temperature are used. Proposed model uses Gaussian-based equation to train the network.

Generalized regression neural networks (GRRN) is a variation of RBFNN [13]. It is a simple method that learns with a single pass and has a high generalization ability [42]. A small input data set can be used to make an accurate prediction with GRRN. Between input and output, GRNN includes pattern and summation layers. The neurons in the pattern layer can memorize relations between input neurons and proper responses to give. Pattern layer gives GRNN ability to process time series better than RBFNN [13]. A sample GRRN structure is given in Fig. 5.

In Zhou [13], an ensemble-based GRNN prediction model is proposed for $PM_{2.5}$ prediction. Firstly, MLR and PCR are used to find the best input parameters. Then inputs

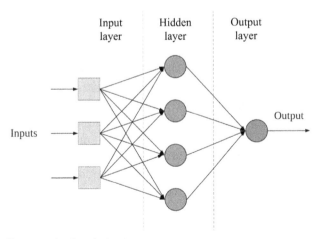

FIG. 4 Radial basis function neural network structure.

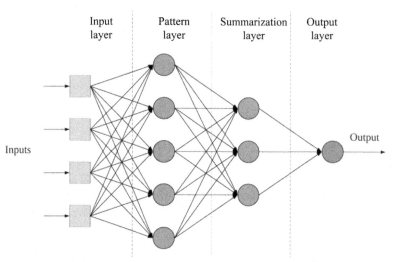

FIG. 5 GRNN structure.

composed of pollutant concentrations and meteorological variables are decomposed using EEMD. Acquired IMFs are fed to GRRN models and predictions are acquired for each IMF separately. Then, as final step predictions are put together.

3.5 Hybrid approaches

In addition to the singular and ensemble prediction approaches, hybrid approach that employs different type of prediction models for a single prediction task are also present in the literature. It must be noted that the mentioned works in this section have hybridized prediction models. Addition of feature selection or optimization models to improve prediction model is accepted as hybridization in the literature. But in this subsection, researches that employ that kind of hybridization are excluded.

In Wang [15], ANN/SVR-based hybrid prediction method for PM_{10} and SO_2 is proposed. The aim of the proposal is to enhance ANN and SVR's prediction capabilities using Taylor Expansion forecasting model. Approach tested and application on both ANN and SVR are proved to increase prediction accuracy compared to singular approach.

In Zhu [20], support vector regression and GRNN are used for daily $PM_{2.5}$ prediction. The proposed method has both ensemble and hybrid characteristics. Two kinds of predictions are performed. The first one is performed using previous $PM_{2.5}$ concentrations. Concentrations are decomposed using CEEMD and predictions are performed for each IMF with a Support Vector Regression (SVR) model. Then, predictions are combined, and a temporal prediction is acquired. In the second phase of prediction, GCA is used for feature selection on meteorological variables and pollutant concentrations. Selected variables are used for prediction using a GRNN model. Predictions of SVR and GRNN are combined into final prediction. Particle Swarm Optimization and Gravitational Search Algorithm (PSOGSA) are used for the optimization of both SVR and GRNN prediction models.

Summary of this section is given in Table 6. In addition to the previously mentioned research, a few extra ones are added to the table.

4 Deep learning models for air quality prediction

In this subsection, we give a brief overview of deep learning. Then, we present deep learning models and recent studies about deep learning-based air quality prediction in detail.

4.1 Deep learning

Deep Learning (DL) is inspired from the function and learning process of human brains. DL is an application of artificial neural networks that is composed of multilayer and multi-neuron [48]. DL performs supervised and unsupervised learning techniques such as regression, classification, and clustering [49]. Recently, DL has gained widespread popularity due to the advancement of computing technologies, training, and learning algorithms [50]. DL models have produced promising results in the field of AI, including computer vision, visual object recognition, visual data processing, natural language processing, speech recognition, object detection [51]. DL models are also applied in other domains such as autonomous systems, robotics, recommendation systems, medical diagnostics, cybersecurity, neuroscience, climate and weather [52,53]. Herein, we mainly focus on several DL models that are widely used for air quality prediction. These DL models are Convolution Neural Networks (CNN), Recurrent Neural Network (RNN), Long Short-Term Memory (LSTM), Gated Recurrent Units (GRU), Deep Belief Networks (DBN), and Auto Encoders (AEs).

4.1.1 Convolutional neural networks (CNN)

CNN contain many hierarchical layers as shown in Fig. 6. These layers are input layer, convolutional layer (Conv), pooling layer, fully connected layer (FC), and output layer. Conv layer is responsible for filtering large multidimensional matrices into feature maps. The pooling layers provide reduced feature maps in order to decrease parameter count and computation time. They also prevent overfitting by providing an abstracted feature representation. In FC layer, each neuron is connected to every neuron in the next layer to learn the final features for classification [7].

Soh et al. [54] propose a general predictive model for air quality prediction called ST-DNN. Authors use combined neural network architecture using ANN, CNN, and LSTM to combine air quality correlations according to spatiotemporal dependency. ST-DNN aims to forecast $PM_{2.5}$ for up to 48 h by taking current and previous hours air pollutants ($PM_{2.5}$, PM_{10}) and meteorological data. ST-DNN is trained with two real-world datasets, which include the information from 76 locations and 23 cites in Taiwan and Beijing. Results show that ST-DNN outperformed the state-of-the-art models.

Table 6 Summary of the section.

References	Prediction model structure	Feature selection algorithm(s)	Ensemble algorithm	Prediction algorithm	Opt. algorithm	Target	Input types
[13]	EL	MLR, PCR	EMD	GRRN	–	$PM_{2.5}$	PC, MV
[6]	EL	–	WD	MLP	–	$PM_{2.5}$	PC, MV, TV
[27]	SL	GM, RM	–	FFNN		$PM_{2.5}, PM_{10}$	PC, MV
[44]	SL	–	–	RBFNN	–	O_3	PC, MV
[26]	SL	SR	–	ANN	–	PM_{10}	PC, MV
[15]	HL	–	–	ANN	–	PM_{10}, SO_2	PC, MV
[16]	EL	–	WD	BPNN	–	$PM_{10}, NO_2,$ SO_2	PC, MV
[14]	SL	–	–	FFNN	–	PM_{10}	PC, MV
[43]	SL	–	–	MLP	–	$PM_{2.5}, PM_{10}$	PC
[25]	SL	CART, PCA, SR	–	MLP	–	NO_2	PC, MV, TV
[45]	SL	–	–	MLP, RBFNN, GRNN	–	O_3	PC, MV
[46]	SL	–	–	BPNN	–	$PM_{2.5}, PM_{10},$ $CO, NO_2, O_3,$ SO_2	PC
[15]	EL	–	EMD, VMD	ELM	DE	AQI	PC
[17]	EL	FCE	EMD	ENN	CS, DE	$PM_{2.5}, PM_{10},$ $CO, NO_2, O_3,$ SO_2	PC
[12]	SL	PCA	–	MLP	–	$PM_{2.5}, PM_{10}$	PC, MV, TV
[18]	HL	–	–	ELM		$PM_{2.5}, PM10,$ CO, NO	PC
[2]	EL	FRS	EMD	ELM	ICA	$PM_{2.5}, PM_{10},$ $CO, NO_2, O_3,$ SO_2	PC
[30]	EL	–	EMD, VMD	ELM	CS	PM_{10}	PC
[47]	SL	–	–	ANN	–	PM_{10}	PC
[20]	EL, HL	GM	EMD	GRRN	PSO, GSA	PM_{10}	PC, MV
[21]	SL	–	–	MLP	–	$PM_{2.5}, PM_{10}$	PC, MV
[31]	EL	–	–	ENN	MSSA	AQI, $PM_{2.5},$ $PM_{10}, CO,$ NO_2, O_3, SO_2	PC
[28]	EL	RF	–	ELM	–	$PM_{2.5}$	PC

Wen et al. [55] propose a spatiotemporal convolutional LSTM extended network called C-LSTME. C-LSTME takes the historical air pollutant ($PM_{2.5}$) concentrations of both the present and k-nearest stations as inputs into the model. Then, the model predicts $PM_{2.5}$ concentration for future hours. In this model, high-level spatiotemporal features are

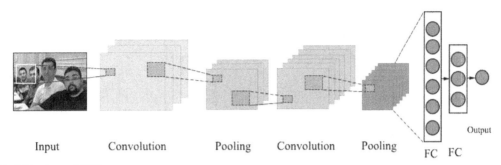

FIG. 6 Structure of CNN.

learned by combining both CNN and LSTM. C-LSTME is tested with meteorological data and aerosol data, which are gathered at 1233 monitoring stations in Beijing and China. The overall results illustrate that C-LSTME has achieved better performance than traditional state-of-the-art models.

4.1.2 Recurrent neural network (RNN), long short-term memory (LSTM), and gated recurrent units (GRU)

Traditional DL models do not take the sequential and time series data into account. They are not also convenient to learn features for time series and sequential data. Therefore, RNN, LSTM, and GRU are sequential learning models. They have been developed to extract the patterns in sequences of data with time series or sequential characteristics, such as speech, texts, handwriting, or sensor data. RNNs are used in the applications of natural language processing, speech recognition, and machine translation [8,51]. The structure of a basic RNN model is shown in Fig. 7.

Long short-term memory (LSTM) is an extension of RNN. It was proposed to solve the vanishing and exploding gradient problems in conventional RNNs [56]. There are memory cells that consist of an input, forget, and output gate in LSTM instead of the standard neurons in RNN. These gates control information flow between the memory cells. A schematic structure of LSTM memory cells [57] is given in Fig. 8.

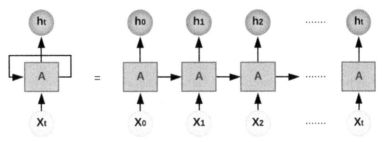

FIG. 7 Structure of basic RNN.

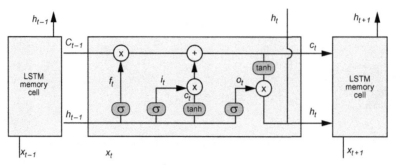

FIG. 8 Structure of LSTM.

Gated recurrent unit (GRU) is a simplified version of LSTM network. GRU contains only two gates: update gate and reset gate [58]. GRU also aims to tackle the vanishing gradient problem, which comes with RNN.

Current studies based on RNN, LSTM, and GRU models for air quality prediction are presented.

Septiawan and Endah [59] proposed Elman RNN, Jordan RNN, and Hybrid RNN network architecture to predict air pollutant concentration. In this study, O_3, PM_{10}, $PM_{2.5}$, and SO_2 pollutants were used to predict NO_2 value. The proposed models trained with data obtained from the city of London between January 2008 and March 2018. The authors present comparative analysis results of all models. Experimental results show that Jordan RNN network better performance than the other two networks.

Kuo et al. [60] apply GRU-based RNN network for air quality prediction. The proposed RNN network uses air pollutants and meteorological data as inputs, which consist of CO, NO_x, NO_2, O_3, PM_{10}, $PM_{2.5}$, humidity, temperature, wind direction, and wind speed to predict the NO_2 pollutant. Gaussian process was used to determine the best model parameters. Experiments were performed according to the network results generated with the best parameters. The results show that GRU-based RNN provides better results than basic RNN and backpropagation NN.

Liu et al. [61] proposed two-stage particular matter prediction system, which consists of multiple NNs that are capable of processing spatiotemporal relations. Authors employ attention-based RNN network architecture similar to the encoder-decoder system. The proposed system takes in 72 h of historical PM_{25}, PM_{10}, SO_2 data as inputs and predicts up to 24 h of $PM_{2.5}$. The proposed system is compared with various LSTM models and Unorganized Machines (UM) models. The experiment results show that the proposed models show better performance than multiple RNN networks and traditional methods.

Liu et al. [62] propose an n-step air quality predictor (AAQP), which is an attention-based seq2seq model. AAQP uses air pollutant data (SO_2, CO, NO_2, O_3, $PM_{2.5}$, PM_{10}) and meteorological data of the previous 24 h as input to predict the $PM_{2.5}$ values of future 24 h. AAQP is compared with the variants of LSTM and GRU models. The authors state that the proposed n-step model significantly decreased training time and error accumulation compared with the original seq2seq attention model.

Athira et al. [63] propose an air quality forecasting framework that uses RNN, LSTM, and GRU. The authors obtain optimal network topology by testing the models according to different layer numbers. Then, the authors used AirNet data that include temperature, humidity, wind components, precipitation, and total cloud cover data to develop the proposed models. RNN, LSTM and GRU networks within the framework predict the future PM_{10} value by taking the data mentioned here.

Bai et al. [4] propose an ensemble learning strategy-based long short-term memory network called E-LSTM. Authors developed E-LSTM by using three steps. First, they utilized ensemble empirical mode decomposition (EEMD) to extract multimodal features. Second, LSTM approach is used to learn multimodal features. Finally, they used inverse EEMD computation for multimodal feature estimated integration. The proposed network used $PM_{2.5}$ concentrations and meteorological data as inputs to predict $PM_{2.5}$ concentrations. Bejing environmental monitoring data were used for training and testing of E-LSTM network. Authors used the MAPE, RMSE, and CC metrics to evaluate the performance of the prediction models. Overall results show that E-LSTM achieves better prediction performance than single LSTM and classical NN.

Li et al. [64] propose a long short-term memory-based extended model called LSTME. LSTME predicts the multiscale air pollutants by using air pollutants, meteorological and timestamp data as inputs. LSTME is compared with the spatiotemporal deep learning (STDL) model, the time-delay neural network model (TDNN), ARMA, SVR, and traditional LSTM. Experiment results show that LSTME provides better prediction performance than SVR, ARMA and TDNN, and traditional LSTM models. Besides, it is stated that the use of meteorological and timestamp data can notably improve prediction performance.

Zhou et al. [65] propose a deep multioutput LSTM model called DM-LSTM that combines three DL models for multistep air quality prediction. In the model construction stage, DM-LSTM is trained with 580 inputs variables to show the effectiveness of DM-LSTM. Input variables consist of air pollutants, air quality, meteorological, and station time lags data. The model has performed the multistep prediction of $PM_{2.5}$, PM_{10}, and NO_x concentrations by taking all these inputs. In the training process, minibatch strategy, dropout, and L2 regularization are used for reducing error accumulation and propagation of the model. The results demonstrate that DM-LSTM can notably improve the prediction accuracy of air quality concentrations.

Wang and Song [66] propose a deep spatiotemporal ensemble model called STE. STE comprises the ensemble method, spatial correlation, an LSTM network. STE takes historical air quality and meteorological data as inputs to forecast future air quality concentrations. STE predicts $PM_{2.5}$ concentration with prediction intervals varying from 6 to 48 h according to input data. Results show that STE outperforms the traditional regression methods (e.g., Linear Regression, Regression Tree) and machine learning methods (ANN, DNN).

Kök et al. [67] propose a new LSTM-based air quality prediction models. The proposed models take historical air pollutant data (O_3, NO_2) and predict the future O_3, NO_2 concentrations. The proposed models are designed to predict future n-step concentration for each air pollutant. In this study, the predicted pollutant values are compared with the decision table created according to the reference values of the EPA to determine the state of air pollution. In this way, the authors aim a novel DL-based framework that can activate early warning systems. The proposed model performance are compared with SVR, which is the traditional nonlinear regression model. The overall results show that proposed LSTM model has a better performance than SVR.

Ahn et al. [68] propose GRU- and LSTM-based air quality prediction system. The proposed models in the system take temperature, humidity, light intensity, CO, fine dust, and volatile organic compounds (VOC) data as inputs and predict the same input variables. The models are tested under different hyperparameters and evaluated comparatively. GRU model shows significantly higher prediction performance than LSTM model.

4.1.3 Autoencoders (AEs) and stacked autoencoders (SAEs)

Autoencoder (AE) is a special type of ANN and has a two-stage structure, i.e., encoding stage and decoding stage (see Fig. 9). AEs create a compressed representation of data in the hidden layer by transforming input to output with the lowest possible distortion [69]. In the encoder stage, inputs are transformed to the hidden layer via encoding function, then the hidden representation is reconstructed back to initial input by using decoding function in the decoding stage. Generally, nonlinear mapping functions such as sigmoid function (sig), tangent function (tanh), and Rectified Linear units function (ReLu) are used in these two stages [8].

Stacked autoencoder (SAE) is a specific neural network that consist of multilayered sparse autoencoders (see Fig. 9). The reason for stacking is to enhance feature abstraction level by creating a deep structure with AEs [70]. In the literature, there are several autoencoder types such as variational autoencoder (VAE), denoising autoencoder (DAE), and sparse autoencoder (SAE). Recent studies related to AEs are presented here.

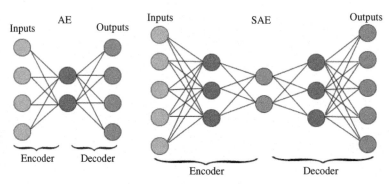

FIG. 9 Structures of autoencoders (AE) and stacked autoencoders (SAE).

Qi et al. [71] develop a DL-based air quality prediction approach called DAL. DAL combines the feature selection, interpolation, prediction, and analysis processes into a model. For this purpose, AEs and Spatiotemporal Semisupervised regression (STSR) are used in input and output layer of DAL, respectively. Real meteorological and air pollution data from Beijing city were used in the training and evaluation of the model. DAL was compared with many traditional methods (e.g., logistic regression, ARIMA, laplacian regression). Results demonstrate that DAL has superior performances than the compared methods in terms of interpolation and prediction.

Li et al. [72] propose a new spatiotemporal deep learning method (STDL) with two stages that consist of a SAE model and a logistic regression model, respectively. In the first stage, SAEs use to learn latent air quality features. Then, STDL is constructed according to the learned representation extracted by SAEs. In the second stage, a logistic regression model is used for real-value regression. The model uses the hourly $PM_{2.5}$ concentration of Beijing city data, which were collected 12 air quality monitoring stations. STDL aims to predict the air quality for all stations simultaneously and to show the temporal stability according to seasons. Results illustrate that STDL performs a superior performance in predicting air quality than spatiotemporal ANN, ARMA, and SVR models.

Li et al. [73] propose a sparse denoising autoencoding NN architecture (SSDAE) that uses spatiotemporal optimization strategy for air quality prediction. Authors used sparse denoising autoencoders to strengthen the scarcity and stability of the model. To evaluate the prediction accuracy, SSDAE is compared with SVM and BP neural network. Results show that SSDAE achieves better prediction performances than compared models.

Do et al. [74] propose a graph convolutional autoencoders-based variational model (AVGAE). AVGAE aims to incorporate the spatiotemporal correlations via a temporal smoothness constraint and graph convolutional operation. Authors solve air quality problems by formulating it as a graph-based matrix completion problem.

4.1.4 Restricted Boltzmann machines (RBM) and deep belief networks (DBN)

RBM is a two-layer stochastic recurrent neural network that can be used for dimensionality reduction, prediction, classification, and feature learning [53]. RBM consists of visible layers and hidden layers. In RBM, there is a full connection between units of visible and hidden layers, but no connections between the same layer units. *Deep Belief Networks (DBN)* are composed of stacked RBMs or Autoencoders. DBNs have the capability of extracting a hierarchical representation of the data in a "layer by layers" manner [48]. They also have the advantage of learning the generative models without imposing subjective filter selection. However, DBNs are sensitive to hyperparameter selection. Therefore, parameters such as momentum, weights, learning rate, number of epoch and layer should be well adjusted [75].

Harrou et al. [76] propose DBN-based anomaly detection scheme (DBN-OCSVM), which combines a DBN model and a one-class support vector machine (OCSVM). DBN-OCSVM aims to detect abnormal ozone pollution. It is tested with real data from Isere in France. The proposed scheme is compared with the deep-stacked autoencoders and RBM-based OCSVM and other clustering procedures. Results show that DBN-OCSVM methodology can effectively detect the abnormalities in ozone measurements.

Li et al. [77] propose a geo-intelligent deep learning model (Geoi-DBN) to predict ground-level $PM_{2.5}$ concentration. The input variables of Geoi-DBN are satellite-derived AOD, meteorological parameters, normalized difference vegetation index (NDVI), and spatiotemporal terms. The RBMs are used in hidden layer of the model to extract the essential features associated with $PM_{2.5}$. To evaluate the effectiveness of Geoi-DBN, authors compare Geoi-DBN with data-driven learning models such as GRNN and BPNN. The proposed model achieves good results but is not showing the expected performance due to some evidence of bias. The contribution of Geoi-DBN is to show that $PM_{2.5}$ can be estimated from satellite data.

Yi et al. [78] propose a deep neural network approach called *DeepAir*, which composes of a spatial transformation component and a deep fusion network. The model aims to predict the fine-grained air quality for Chinese cities every hour by fusing heterogeneous urban data such as air pollutants, weather data, and meteorological data. DeepAir was compared with 10 baseline methods. On short-term, long-term, and sudden changes prediction, DeepAir provides 2.4%, 12.2%, 63.2% accuracy improvements, respectively, compared to baseline models.

All the studies described here are summarized in Table 7.

Table 7 Summaries of the DL-based air quality prediction models.

Ref.	Used model(s)	Model input(s)	Model output(s)	Model evaluation/ performance metric(s)	Dataset location and duration
[54]	ANN, CNN, LSTM	$PM_{2.5}$, PM_{10}, Meteorological data (temperature, wind speed, wind direction, relative humidity)	$PM_{2.5}$	MAE	Taiwan (from January 2014 to September 2017), and Beijing (from May 2014 to Apr. 2015)
[55]	CNN, LSTM	Meteorological and Aerosol data	$PM_{2.5}$	RMSE, MAE, MAPE	Beijing and the whole of China from January 2016 to December 2017
[59]	Elman RNN, Jordan RNN, Hybrid RNN	PM_{10}, $PM_{2.5}$, SO_2, O_3	NO_2	MAPE	London (from January 2008 to March 2018)
[60]	RNN	$PM_{2.5}$, PM_{10}, CO, NO, NO_2, NO_x, O_3,, meteorological data (relative humidity temp, wind direction, wind speed)	NO_2	MSE	Taiwan (from January 1, 2017, to December 31, 2017)
[61]	RNN	$PM_{2.5}$,PM_{10},SO_2	$PM_{2.5}$	RMSE, MAPE	Northern China (from March 2014 to January 2016)

Table 7 Summaries of the DL-based air quality prediction models—cont'd

Ref.	Used model(s)	Model input(s)	Model output(s)	Model evaluation/ performance metric(s)	Dataset location and duration
[62]	RNN	$PM_{2.5}$, PM_{10}, SO_2, CO, NO_2, O_3, meteorological data (precipitation, humidity, temperature, wind force, and wind direction)	$PM_{2.5}$	MAE, R2	Beijing city data /China (from April 2017 to March 2018)
[63]	RNN, LSTM, GRU	Temperature, Humidity, U-Wind component, V-wind component, Precipitation, Total cloud cover	PM_{10}	RMSE, MAPE	AirNet dataset (from April 1, 2015, to September 1, 2017)
[4]	LSTM	$PM_{2.5}$ concentrations and meteorological data	$PM_{2.5}$	MAPE, RMSE, CC	Beijing city data/China
[64]	LSTM	$PM_{2.5}$, Timestamp, Meteorological data (temperature, humidity, wind speed, and visibility)	$PM_{2.5}$	RMSE, MAE, MAPE	Beijing city data /China (from January 2014 to May 2016)
[65]	LSTM	$PM_{2.5}$, PM_{10}, O_3, NO_x, NO_2, NO, SO_2, CO, time lags, monitoring stations, meteorological data (rainfall, temperature, wind speed, wind direction, and relative humidity)	$PM_{2.5}$, PM_{10}, NO_x	RMSE, Gbench	Taipei city data, Taiwan (from 2010 to 2016)
[66]	LSTM	PM_{10}, $PM_{2.5}$, O_3, CO, NO_2, SO_2, Meteorological data (temperature, humidity, wind speed, and wind direction)	$PM_{2.5}$	RMSE, MAE, Accuracy	Beijing city data, China (from May 2013 to April 2017)
[67]	LSTM	O_3, NO_2	O_3, NO_2	RMSE, Precision, Recall, F1-Measure, Accuracy	City Pulse dataset (Aarhus/Denmark, Brasov/Romania)
[68]	GRU, LSTM	Temperature, Humidity, Dust, Light, CO_2, VOC	CO_2, Dust, Temperature, Humidity, Light, VOC	Accuracy	Indoor data (from 22 February 2016 to 22 April 2016)
[71]	AE	$PM_{2.5}$, PM_{10}, SO_2, NO_2, CO, O_3, Meteorological data (temperature, humidity, barometric pressure, precipitation, wind direction, and wind strength)	$PM_{2.5}$	RMSE	Beijing city data/China
[72]	SAE	$PM_{2.5}$	$PM_{2.5}$	RMSE, MAE, MAPE	Beijing city (from January 2014 to May 2016)

Continued

Table 7 Summaries of the DL-based air quality prediction models—cont'd

Ref.	Used model(s)	Model input(s)	Model output(s)	Model evaluation/ performance metric(s)	Dataset location and duration
[73]	SDAE	Air quality and Meteorological data	Air quality	MAPE	Zhengzhou City data/China (from January 2015 to December 2017)
[74]	VAE	NO_2, $PM_{2.5}$	NO_2, $PM_{2.5}$	RMSE, MAE	Antwerp city/Belgium
[76]	DBN	O_3	O_3	AUC	Isere city data/France (January 2015, to March 2015)
[77]	DBN	Satellite-derived aerosol optical depth (AOD), meteorological parameters, normalized difference vegetation index (NDVI), and spatiotemporal terms	$PM_{2.5}$	RMSE, R2, MPE, RPE	(from January 2015 to December 2015)
[78]	DBN	$PM_{2.5}$, PM_{10}, O_3, SO_2, NO_2, CO, Meteorological data, weather data, Time Station ID, AOIs	Air quality	MAE, Accuracy	Seven Chinese cities data (from May 2014 to April 2017)

5 Conclusion

In this chapter, we have focused on recent artificial intelligence-based studies and developed models for air quality prediction. Selected studies are examined under two main sections. Firstly, we examine the studies involving neural network-based methods. Our findings in these studies can be summarized as follows:

- NN-based methods are frequently used for AQ prediction and prove to be accurate.
- MLP-based methods are generally used for approaches that use different types of features. For example, features composed of three sets of variables (pollutant concentrations, meteorological variables, and time-wise variables) are used for prediction.
- Especially at MLP-based methods where number of features is generally high, feature selection is a must.
- Encountered feature selection methods are PCA, CART, SR, RM, and GM. Among them, PCA and CART are the most popular ones.
- ELMs are fast and accurate methods. They require optimization algorithms to optimize thresholds and neural connections.
- Ensemble approach enhances EML's ability to generalize and frequently employed.
- Because of their ability to process time series, ENN and variations of ENN (RNN, LSTM) are accurate predictors for AQ.

- Similar to ELM, randomly generated threshold values and neural connection weights can cause ENN to overfit or fall into local minimums. To overcome this problem, the optimization of ENN must be done.
- In short, NN-based methods surpasses traditional prediction models in a number of ways. Firstly, on account of the training method of NNs, NN-based methods are more adaptive and flexible, has a higher degree of accuracy compared to the traditional methods. By learning from experience, NN-based methods have a relatively easier training process. Lastly, NNs ability to model linear and nonlinear relations in a single prediction model makes it possible to use NN-based models for temporal, spatial, and spatiotemporal prediction models.

Second, we examine the studies involving deep learning-based prediction methods. Similarly, our findings in these studies can be summarized as follows:

- Recent studies revealed the necessity of extracting compressed or latent representations of the data for air quality prediction.
- Traditional machine learning approaches seem to be inadequate in extracting hidden features and processing spatiotemporal and time series data.
- It has been shown in the examined studies that machine learning models called deep learning are quite efficient in analyzing time series and spatiotemporal data.
- Models based on CNN, RNN, LSTM, GRU, AE, and RBM and their combinations have been widely used in air quality prediction. Particularly, it is illustrated that RNN- and LSTM-based models are inherently more convenient to air quality prediction.
- In light of the obtained information promising, it is predicted that deep learning models will be widely used especially in the analysis and prediction of data showing spatiotemporal and time series data.

As seen from the findings and the literature review provided in this chapter, artificial intelligence-based air quality prediction methods have a great potential to predict air quality.

References

[1] E.E. Agency, Air Quality in Europe, 2016.

[2] C. Li, Z. Zhu, Research and application of a novel hybrid air quality early-warning system: a case study in China, Sci. Total Environ. 626 (2018) 1421–1438.

[3] K. Gu, J.F. Qiao, W.S. Lin, Recurrent air quality predictor based on meteorology- and pollution-related factors, IEEE Trans. Ind. Inform. 14 (9) (2018) 3946–3955.

[4] Y. Bai, et al., An ensemble long short-term memory neural network for hourly PM2.5 concentration forecasting, Chemosphere 222 (2019) 286–294.

[5] S.-Q. Dotse, et al., Application of computational intelligence techniques to forecast daily PM 10 exceedances in Brunei Darussalam, Atmos. Pollut. Res. 9 (2) (2018) 358–368.

[6] X. Feng, et al., Artificial neural networks forecasting of PM 2.5 pollution using air mass trajectory based geographic model and wavelet transformation, Atmos. Environ. 107 (2015) 118–128.

[7] B. Jan, et al., Deep learning in big data analytics: a comparative study, Comput. Electr. Eng. 75 (2019) 275–287.

[8] Q. Zhang, et al., A survey on deep learning for big data, Inf. Fusion 42 (2018) 146–157.

[9] Y.J. Jung, et al., Design of sensor data processing steps in an air pollution monitoring system, Sensors 11 (12) (2011) 11235–11250.

[10] M.M. Najafabadi, et al., Deep learning applications and challenges in big data analytics, J. Big Data 2 (1) (2015) 1.

[11] Agency, U.S.E.P, Air Quality Index: A Guide to Air Quality and Your Health, 2014.

[12] F. Franceschi, M. Cobo, M. Figueredo, Discovering relationships and forecasting PM10 and PM2.5 concentrations in Bogotá, Colombia, using artificial neural networks, principal component analysis, and k-means clustering, Atmos. Pollut. Res. 9 (5) (2018) 912–922.

[13] Q. Zhou, et al., A hybrid model for PM(2).(5) forecasting based on ensemble empirical mode decomposition and a general regression neural network, Sci. Total Environ. 496 (2014) 264–274.

[14] S.K. Hur, et al., Evaluating the predictability of PM10 grades in Seoul, Korea using a neural network model based on synoptic patterns, Environ. Pollut. 218 (2016) 1324–1333.

[15] P. Wang, et al., A novel hybrid forecasting model for PM(1)(0) and SO(2) daily concentrations, Sci. Total Environ. 505 (2015) 1202–1212.

[16] Y. Bai, et al., Air pollutants concentrations forecasting using back propagation neural network based on wavelet decomposition with meteorological conditions, Atmos. Pollut. Res. 7 (3) (2016) 557–566.

[17] Z. Yang, J. Wang, A new air quality monitoring and early warning system: air quality assessment and air pollutant concentration prediction, Environ. Res. 158 (2017) 105–117.

[18] Y. Li, et al., Research on air pollutant concentration prediction method based on self-adaptive neuro-fuzzy weighted extreme learning machine, Environ. Pollut. 241 (2018) 1115–1127.

[19] Wikipedia, Air Quality Index, Wikipedia, 2019.

[20] S. Zhu, et al., PM2.5 forecasting using SVR with PSOGSA algorithm based on CEEMD, GRNN and GCA considering meteorological factors, Atmos. Environ. 183 (2018) 20–32.

[21] P. Gogikar, B. Tyagi, A.K. Gorai, Seasonal prediction of particulate matter over the steel city of India using neural network models, Model. Earth Syst. Environ. 5 (1) (2018) 227–243.

[22] N.E. Huang, et al., The empirical mode decomposition and the Hilbert spectrum for nonlinear and non-stationary time series analysis, Proc. R. Soc. Lond. Ser. A: Math. Phys. Eng. Sci. 454 (1971) (1998) 903–995.

[23] D. Mishra, P. Goyal, Development of artificial intelligence based NO_2 forecasting models at Taj Mahal, Agra, Atmos. Pollut. Res. 6 (1) (2015) 99–106.

[24] W. Sun, J. Sun, Daily PM2.5 concentration prediction based on principal component analysis and LSSVM optimized by cuckoo search algorithm, J. Environ. Manag. 188 (2017) 144–152.

[25] S.M.S. Cabaneros, J.K.S. Calautit, B.R. Hughes, Hybrid artificial neural network models for effective prediction and mitigation of urban roadside NO_2 pollution, Energy Procedia 142 (2017) 3524–3530.

[26] A. Russo, et al., Neural network forecast of daily pollution concentration using optimal meteorological data at synoptic and local scales, Atmos. Pollut. Res. 6 (3) (2015) 540–549.

[27] M.L. Fu, et al., Prediction of particular matter concentrations by developed feed-forward neural network with rolling mechanism and gray model, Neural Comput. Applic. 26 (8) (2015) 1789–1797.

[28] Z. Shang, et al., A novel model for hourly PM2.5 concentration prediction based on CART and EELM, Sci. Total Environ. 651 (Pt 2) (2019) 3043–3052.

[29] D. Wang, et al., A novel hybrid model for air quality index forecasting based on two-phase decomposition technique and modified extreme learning machine, Sci. Total Environ. 580 (2017) 719–733.

[30] H. Luo, et al., Research and application of a novel hybrid decomposition-ensemble learning paradigm with error correction for daily PM 10 forecasting, Atmos. Res. 201 (2018) 34–45.

[31] Y. Hao, C. Tian, The study and application of a novel hybrid system for air quality early-warning, Appl. Soft Comput. 74 (2019) 729–746.

[32] X.S. Yang, S. Deb, Engineering optimisation by cuckoo search, Int. J. Math. Model. Numer. Optim. 1 (4) (2010) 330.

[33] E. Atashpaz-Gargari, C. Lucas, Imperialist competitive algorithm: an algorithm for optimization inspired by imperialistic competition, in: 2007 IEEE Congress on Evolutionary Computation, IEEE, 2007.

[34] R. Storn, K. Price, Differential evolution –a simple and efficient heuristic for global optimization over continuous spaces, J. Glob. Optim. 11 (4) (1997) 341–359.

[35] B. Hegerty, C.-C. Hung, K. Kasprak, A comparative study on differential evolution and genetic algorithms for some combinatorial problems, in: Proceedings of 8th Mexican International Conference on Artificial Intelligence, 2009.

[36] J.-R. Yeh, J.-S. Shieh, N.E. Huang, Complementary ensemble empirical mode decomposition: a novel noise enhanced data analysis method, Adv. Adapt. Data Anal. 02 (02) (2010) 135–156.

[37] Y.-H. Wang, et al., On the computational complexity of the empirical mode decomposition algorithm, Phys. A: Stat. Mech. Appl. 400 (2014) 159–167.

[38] M.E. Torres, et al., A complete ensemble empirical mode decomposition with adaptive noise, in: 2011 IEEE International Conference on Acoustics, Speech and Signal Processing (ICASSP), 2011.

[39] M.A. Colominas, G. Schlotthauer, M.E. Torres, Improved complete ensemble EMD: A suitable tool for biomedical signal processing, Biomed. Signal Process. Control 14 (2014) 19–29.

[40] K. Dragomiretskiy, D. Zosso, Variational mode decomposition, IEEE Trans. Signal Process. 62 (3) (2013) 531–544.

[41] S.G. Mallat, A theory for multiresolution signal decomposition: the wavelet representation, IEEE Trans. Pattern Anal. Mach. Intell. 7 (1989) 674–693.

[42] D.F. Specht, A general regression neural network, IEEE Trans. Neural Netw. 2 (6) (1991) 568–576.

[43] P. Perez, E. Gramsch, Forecasting hourly PM2.5 in Santiago de Chile with emphasis on night episodes, Atmos. Environ. 124 (2016) 22–27.

[44] Q.P. Ha, et al., Enhanced radial basis function neural networks for ozone level estimation, Neurocomputing 155 (2015) 62–70.

[45] N. Kumar, A. Middey, P.S. Rao, Prediction and examination of seasonal variation of ozone with meteorological parameter through artificial neural network at NEERI, Nagpur, India, Urban Clim. 20 (2017) 148–167.

[46] J.Z. Wang, et al., Developing an early-warning system for air quality prediction and assessment of cities in China, Expert Syst. Appl. 84 (2017) 102–116.

[47] S. Park, et al., Predicting PM10 concentration in Seoul metropolitan subway stations using artificial neural network (ANN), J. Hazard. Mater. 341 (2018) 75–82.

[48] M. Mohammadi, et al., Deep learning for IoT big data and streaming analytics: a survey, IEEE Commun. Surv. Tutor. 20 (4) (2018) 2923–2960.

[49] W.G. Hatcher, W. Yu, A survey of deep learning: platforms, applications and emerging research trends, IEEE Access 6 (2018) 24411–24432.

[50] Y.M. Guo, et al., Deep learning for visual understanding: a review, Neurocomputing 187 (2016) 27–48.

[51] Y. LeCun, Y. Bengio, G. Hinton, Deep learning, Nature 521 (7553) (2015) 436–444.

[52] D.F. Xie, L. Zhang, L. Bai, Deep learning in visual computing and signal processing, Appl. Comput. Intell. Soft Comput. 10 (2017) 1–13.

[53] S. Wang, J. Cao, P.S. Yu, Deep Learning for Spatio-Temporal Data Mining: A Survey, arXiv preprint arXiv:1906.04928, 2019.

[54] P.W. Soh, J.W. Chang, J.W. Huang, Adaptive deep learning-based air quality prediction model using the most relevant spatial-temporal relations, IEEE Access 6 (2018) 38186–38199.

[55] C. Wen, et al., A novel spatiotemporal convolutional long short-term neural network for air pollution prediction, Sci. Total Environ. 654 (2019) 1091–1099.

[56] S. Hochreiter, J. Schmidhuber, Long short-term memory, Neural Comput. 9 (8) (1997) 1735–1780.

[57] C. Olah, Understanding LSTM Networks, 2015.

[58] R. Dey, F.M. Salemt, Gate-variants of gated recurrent unit (GRU) neural networks, in: 2017 IEEE 60th International Midwest Symposium on Circuits and Systems (MWSCAS), IEEE, 2017.

[59] W.M. Septiawan, S.N. Endah, Suitable recurrent neural network for air quality prediction with back-propagation through time, in: 2018 2nd International Conference on Informatics and Computational Sciences (ICICoS), 2018.

[60] R.J. Kuo, B. Prasetyo, B.S. Wibowo, Deep learning-based approach for air quality forecasting by using recurrent neural network with gaussian process in Taiwan, in: 2019 IEEE 6th International Conference on Industrial Engineering and Applications (ICIEA), 2019.

[61] T. Liu, et al., Fine-grained air quality prediction using attention based neural network, in: 2018 International Joint Conference on Neural Networks (IJCNN), 2018.

[62] B. Liu, et al., A sequence-to-sequence air quality predictor based on the n-step recurrent prediction, IEEE Access 7 (2019) 43331–43345.

[63] V. Athira, et al., DeepAirNet: applying recurrent networks for air quality prediction, Procedia Comput. Sci. 132 (2018) 1394–1403.

[64] X. Li, et al., Long short-term memory neural network for air pollutant concentration predictions: method development and evaluation, Environ. Pollut. 231 (Pt 1) (2017) 997–1004.

[65] Y.L. Zhou, et al., Explore a deep learning multi-output neural network for regional multi-step-ahead air quality forecasts, J. Clean. Prod. 209 (2019) 134–145.

[66] J.S. Wang, G.J. Song, A deep spatial-temporal ensemble model for air quality prediction, Neurocomputing 314 (2018) 198–206.

[67] I. Kök, M.U. Şimşek, S. Özdemir, A deep learning model for air quality prediction in smart cities, in: 2017 IEEE International Conference on Big Data (Big Data), 2017.

[68] J. Ahn, et al., Indoor air quality analysis using deep learning with sensor data, Sensors 17 (11) (2017) 2476.

[69] P. Baldi, Autoencoders, unsupervised learning, and deep architectures, in: Proceedings of ICML Workshop on Unsupervised and Transfer Learning, 2012.

[70] S. Vieira, W.H.L. Pinaya, A. Mechelli, Using deep learning to investigate the neuroimaging correlates of psychiatric and neurological disorders: methods and applications, Neurosci. Biobehav. Rev. 74 (2017) 58–75.

[71] Z. Qi, et al., Deep air learning: Interpolation, prediction, and feature analysis of fine-grained air quality, IEEE Trans. Knowl. Data Eng. 30 (12) (2018) 2285–2297.

[72] X. Li, et al., Deep learning architecture for air quality predictions, Environ. Sci. Pollut. Res. 23 (22) (2016) 22408–22417.

[73] Y. Li, et al., Spatio-temporal-aware sparse denoising autoencoder neural network for air quality prediction, in: 2018 5th IEEE International Conference on Cloud Computing and Intelligence Systems (CCIS), 2018.

[74] T.H. Do, et al., Matrix completion with variational graph autoencoders: application in hyperlocal air quality inference, in: ICASSP 2019-2019 IEEE International Conference on Acoustics, Speech and Signal Processing (ICASSP), IEEE, 2019.

[75] L. Shao, et al., Performance evaluation of deep feature learning for RGB-D image/video classification, Inf. Sci. 385 (2017) 266–283.

[76] F. Harrou, et al., Detecting abnormal ozone measurements with a deep learning-based strategy, IEEE Sensors J. 18 (17) (2018) 7222–7232.

[77] T. Li, et al., Estimating ground-level PM2.5 by fusing satellite and station observations: a geo-intelligent deep learning approach, Geophys. Res. Lett. 44 (23) (2017) 11985–11993.

[78] X.W. Yi, et al., Deep distributed fusion network for air quality prediction, in: Kdd'18: Proceedings of the 24th Acm Sigkdd International Conference on Knowledge Discovery & Data Mining, 2018, pp. 965–973.

9

Optimization of absorption process for exclusion of carbaryl from aqueous environment using natural adsorbents

Soumya Chattoraj[a] and Kamalesh Sen[b]

[a]UNIVERSITY INSTITUTE OF TECHNOLOGY, GENERAL SCIENCE AND HUMANITIES, THE UNIVERSITY OF BURDWAN, BARDHAMAN, WEST BENGAL, INDIA [b]ENVIRONMENTAL CHEMISTRY LABORATORY, DEPARTMENT OF ENVIRONMENTAL SCIENCE, THE UNIVERSITY OF BURDWAN, BARDHAMAN, WEST BENGAL, INDIA

1 Introduction

Insecticides are potential water pollutants [1]. They are used to kill harmful insects in agriculture, gardens, and households. Furthermore, these chemicals can contaminate the nearest water body due to rain, wind flow, and many other sources [2]. This fact is supported by the result found by U.S. Environmental Agency (EPA) in 2004. According their study, insecticides are commonly found in ground and surface water. Contamination of ground or surface water is a major concern for human life as these waters are consumed for drinking purposes [3].

Carbaryl is a commonly used insecticide today. Many researchers have shown that carbaryl is highly toxic to humans, animals, and bees [4]. The U.S. EPA in 2004 found that carbaryl is a frequently found insecticide in groundwater. Carbaryl belongs to the carbamate family. The chemical name, 1-naphthyl methylcarbamate, is sold under the market name Sevin, and the formula is $C_{12}H_{11}NO_2$.

Molecular structure of carbaryl

Intelligent Environmental Data Monitoring for Pollution Management. https://doi.org/10.1016/B978-0-12-819671-7.00009-9

Chlorination [5], UV-photolysis [6], Fenton oxidation [7], Ozonation [8,9], Reverse osmosis [10] and Adsorption [11,12] are generally used to remove insecticide from water. Among these methods, the adsorption process is simple, easy, its initial cost is low, and it can be used to remove pesticides from dilute solutions [2,13].

For exclusion of pesticides from water, substances like activated carbon [14–16], straw [17], lignocellulosic substrate [18], fly ash [19,20], charcoal [21] and barks [22], pretreated-coffee waste [12], and Ayous wood saw dust [23] have used.

Use of biodegradable adsorbents is a new and unconventional method to eliminate insecticide from contaminated water. From an economical point of view, biomasses are also beneficial. Thus insightful research is paying attention to cost-effective, biodegradable, and easily available adsorbents such as *Rhizopus oryzae* biomass [24], fly ash [25], waste jute carbon [26], pretreated-coffee waste [12], and Ayous wood saw dust [23] to achieve effective deduction of these chemicals from the aqueous system.

But there is little work on the removal of carbaryl insecticide by suitable adsorbents. Therefore, in this chapter several easily available and low-cost biodegradable adsorbents including alluvial soil, *Pistia stratiotes* biodust (PSB), *Lemna major* biodust (LMB), neem bark dust (NBD), and eggshell powder have been utilized to remove carbaryl from water.

The studies are related with a series of batch adsorption experiments to review the effectiveness of the adsorbents for removal of the insecticide from water. The operating factors were solution pH, dose of adsorbent, stirring rate, size of the particle, contact time, temperature, and initial concentration of the solution. Adsorption isotherms and kinetic study were done using experimental data. To confirm the spontaneity of the process, thermodynamic parameters were also evaluated. The response surface model (RSM) and artificial neural network (ANN) model was applied to batch experimental data to find optimum conditions and confirm the validity of the adsorption process.

Overall, a batch study for the less costly, simple, and green process for water pollution control has been done. This study can be applied in the future for designing a simple flat-bed column operation for agriculture and household purposes in affected areas. The results acquired from these studies are presented and discussed.

2 Characterization of adsorbents

The characterization of the adsorbents was done by scanning electron microscope (SEM) and Fourier transform infrared spectroscopy (FTIR). Adsorption occurs at the surface of a substance. So, information about the surface and composition of the adsorbent is needed only, which is why TEM study was not required. Other physiochemical properties such a BET surface area, specific surface area, point of zero charge (pH_{zpc}), pH, conductivity, particle size, moisture content, specific gravity, bulk and particle density, porosity, cation exchange capacity, amount of clay, sand, Na^+ and K^+ and Ca^{2+}, PO_4^{-3} were also determined. Characterizations of the adsorbents are already published by the author [27–31].

3 Adsorption study

3.1 Adsorption of carbaryl by alluvial soil

The retention capacity of carbaryl onto alluvial soil was explored. The influence of the operating variables like initial concentration of carbaryl, solution pH, size of particle, adsorbent dose, contact time, and temperature have been determined through a batch process. Optimum pH of the process was 6. The data found in the batch study was fitted well with pseudo-first-order kinetic model, and Freundlich isotherm [32] was fitted well with carbaryl equilibrium adsorption data (obtained at different temperature). Standard enthalpy change (ΔH^0), standard change in entropy (ΔS^0), standard change in free-energy (ΔG^0), and activation energy (E_a) data confirms that the process was spontaneous in nature [32]. For soil characterization, the FTIR, SEM, and X-ray diffraction spectrum (XRD) have been used. Finally, the validity of the experimental data was tested by using ANN model, which supports the compatibility of carbaryl adsorption process onto alluvial soil. Moreover, the results suggest that the capability of carbaryl adsorption by alluvial soil is reasonable and depends on pH, particle size, and organic matter contents [32].

3.2 Removal of carbaryl by *Pistia stratiotes* biodust (PSB) by adsorption

Application of PSB as an adsorbent was investigated in this topic. The adsorbent was prepared by dead *Pistia stratiotes* aquatic plant. The outcome of the operating parameters such as pH, initial concentration, dose of adsorbent, and contact time on the adsorption of carbaryl were explored by following a batch study. Experimental data were designed through a model known as central composite design (CCD) in RSM. The ANOVA result was best fitted with quadratic model. To determine the interaction effects of main factors and optimum conditions of process response, surface plots were used. The adsorption process was also fitted with the Freundlich isotherm and pseudo-second-order kinetic model [27], respectively. Changes of free energy ($\Delta G°$) were negative at all temperatures, hence the adsorption process was spontaneous. RSM proposed that best possible conditions of the process were initial carbaryl concentration of 15.57 mgL^{-1}, pH 2.01, adsorbent dose 0.72 g, and contact time of 30 min. The highest adsorption capacity of PSB for carbaryl was 3.1 mg g^{-1}. In conclusion, this work demonstrates the feasibility of employing the dead plant biodust as an inexpensive and biodegradable adsorbent that can eliminate carbaryl from the water stream [27].

3.3 Adsorptive removal of carbaryl by *Lemna major* biodust (LMB)

In this study, biosorption of carbaryl from water sources on dead LMB was discussed. The operating parameters of the work were initial carbaryl concentration, pH of carbaryl solution, contact time, and dose of adsorbent. Two level four factor Box-Behnken design (BBD; in RSM) was used to optimize data. The ANOVA result followed the quadratic model for BBD. Perturbation plot and 3D response surface plots shows the interaction between the factors and optimum conditions of the process. The same dataset was put to the ANN model. Results were compared by both the models. Finally, both models confirm validity of the work. The kinetic data followed pseudo-second-order and adsorption isotherm data both with both Langmuir and Freundlich isotherm models with reasonable R^2 value [28]. Spontaneity of the process was confirmed by thermodynamic parameters. The adsorbent achieved maximum adsorption capacity of 6.21 mg g^{-1} [28]. Finally, the present study has predicted the feasibility of adsorption by both RSM and ANN model.

3.4 Removal of carbaryl through adsorption by neem (*Azadirachta indica*) bark dust

The application of NBD was used as an adsorbent for exclusion of carbaryl from water. Initially the most advantageous conditions of contact time, particle size, and rate of stirring was selected by batch method. Thereafter, RSM was engaged to inspect the effects of operating factors. The factors were pH, adsorbent dose, and initial concentration. A two level three factor BBD was used to locate the best condition, outcome, and interactions of the operating factors. Based on the adsorption ability with profitable use of adsorbent, the operating factors were optimized by two procedures. The desirability of first and second optimization procedures were 1.00 and 0.822, which illustrated that the obtained results were fitted well to the experimental model [31]. The second optimization procedure was more effective as the maximum adsorption capability of NBD for carbaryl was 146.3 mg g^{-1} at pH 5.86, at minimum level of adsorbent dose (0.01 g) and maximum level of initial concentration (18.29 mg L^{-1}). Finally, the result obtained from RAMP plots exposed that the NBD is an efficient adsorbent for carbaryl removal from water sources [31].

3.5 Removal of carbaryl by eggshell powder through adsorption technology

In this study, eggshell powder was used for exclusion of carbaryl from aqueous medium through adsorption process. A regression analysis was performed by using response surface methodology via BBD for percentage adsorption of carbaryl. The regression models developed represent responses as functions of initial concentration (A), pH (B), adsorbent dose (C), and contact time (D). A desirability function was used to find the best possible conditions for the exclusion of carbaryl using eggshell powder by choosing the operating factors. The removal percentage of the insecticide by eggshell powder at optimum conditions is 92.2%. The experimental data followed the pseudo-second-order kinetic model and Freundlich adsorption isotherm most effectively [29]. In addition, thermodynamic values suggest that the adsorption was endothermic and spontaneous. Overall study explored that eggshell powder is an alternative adsorbent to take carbaryl out of the water matrix [29].

3.6 Comparative study of different adsorbents

A comparative study of carbaryl adsorption ability (mg g^{-1}) by different workers are reported. NBD has highest adsorption capacity (146.3 mg/g) for carbaryl (Table 1). So,

Table 1 Comparative study on different adsorbents of carbaryl adsorption.

Adsorbent	Maximum adsorption capacity (mg g^{-1})	References
Porogen-treated banana pith carbon	45.9	[33]
Clay	10.75	[34]
Alluvial soil	0.133	[30]
Lateritic soil	0.061	[30]
Forest soil	0.941	[30]
Saline-alkali soil	0.065	[30]
Pistia stratiotes	3.10	[27]
Lemna major	6.21	[28]
Neem bark dust	146.3	[31]
Eggshell powder	105.6	[29]

it is a promising adsorbent due to its availability and maximum carbaryl uptake from aqueous phase. However, all adsorbents shown have the potential for carbaryl adsorption, but neem bark surface interacted better with carbaryl molecules. In neem bark, the major components are lignin, hemicellulose, celluloses, and carbon-related α-linkage polymer, which interacted with carbonyl and amine group of carbaryl molecules [35]. Therefore, carbaryl molecules can be easily adsorbed, and NBD has shown suitable adsorption capacity.

4 Conclusion

This chapter describes the adsorption capacity of carbaryl onto different natural adsorbents. The chapter also provides the optimum condition and interaction between the operating factors of the adsorption process by using response surface and artificial network software models. Thermodynamics and kinetics nature of the adsorption process are also provided. Moreover, the chapter provides a critical revision of carbaryl adsorption into various adsorbents. In conclusion, the chapter showed that several low-cost adsorbents have good capacity to adsorb carbaryl from an aqueous environment. In the future, these adsorbents can be further improved and used to remove carbaryl from an aqueous environment on a large scale.

References

[1] S. Cosgrove, B. Jefferson, P. Jarvis, Pesticide removal from drinking water sources by adsorption: a review. Environ. Technol. Rev. 8 (1) (2019) 1–24, https://doi.org/10.1080/21622515.2019.1593514.

[2] J.M. Salman, V.O. Njoku, B.H. Hameed, Adsorption of pesticides from aqueous solution onto banana stalk activated carbon, Chem. Eng. J. 174 (2011) 41.

[3] K.P. Singh, A. Malik, D. Mohan, S. Sinha, Persistence organachlorine pesticide residues in alluvial groundwater aquifers of gangetic plains, India. Bull. Environ. Contam. Toxicol. 74 (2005) 162–169.

[4] U.S. EPA, Interim reregistration eligibility decision for carbaryl, (2004) pp. 3–4. www.epa.gov/oppsrrd1/REDs/carbaryl_ired.pdf.

[5] R.L. Droste, Theory and Practice of Water and Wastewater Treatment, Wiley and Sons, Canada, 1997.

[6] L. Muszkat, L. Feigelson, L. Bir, Reaction patterns in photooxidative degradation of two herbicides, Chemosphere 36 (1998) 1485–1492.

[7] H. Fallmann, T. Krutzler, R. Bauer, S. Malato, J. Blanco, Applicability of the photo-Fenton method for treating water containing pesticides. Catal. Today 54 (1999) 309–319, https://doi.org/10.1016/S0920-5861(99)00192-3.

[8] M.I. Maldonado, S. Malato, L.A. Perez-Estrada, W. Gernjak, I. Oller, X. Domenech, J. Peral, Partial degradation of five pesticides and an industrial pollutant by ozonation in a pilot-plant scale reactor, J. Hazard. Mater. 38 (2006) 363–369.

[9] C. Ozdemir, S. Sahinkaya, M. Onucyildiz, Treatment of pesticide wastewater by physiochemical and fenton processes, Asian J. Chem. 20 (2008) 3795–3804.

[10] E.S.K. Chian, W.N. Bruce, H.H.P. Fang, Removal of pesticides by reverse osmosis, Environ. Sci. Technol. 9 (1) (1975) 52–59.

[11] A.H. Al-Muhtaseb, K.A. Ibrahim, A.B. Albadarin, O. Ali-khashman, G.M. Walker, M.N.M. Ahmad, Remediation of phenol-contaminated water by adsorption using poly (methylmethacrylate) (PMMA), Chem. Eng. J. 168 (2011) 691–699.

[12] N. Yeddou Mezenner, H. Lagha, H. Kais, M. Trari, Biosorption of diazinon by a pre-treated alimentary industrial waste: equilibrium and kinetic modeling, Appl Water Sci 7 (2017) 4067–4076.

[13] M.D. Martino, F. Sannino, D. Pirozzi, Removal of pesticide from wastewater: contact time optimization for a two-stage batch stirred adsorber, J. Environ. Chem. Eng. 3 (2015) 365–372.

[14] C.S. Castro, M.C. Guerreiro, M. Gonclaves, L.C. Oliveira, A.S. Anastacio, Activated carbon/iron oxide composites for the removal of atrizane from aqueous medium, J. Hazard. Mater. 164 (2009) 609–614.

[15] K. Ohno, T. Minami, Y. Matsui, Y. Magara, Effect of chlorine on organophosphorus pesticides adsorbed on activated carbon: desorption and oxon formation, Water Res. 42 (2008) 1753–1759.

[16] M.J. Salman, F.H. Hussein, Batch adsorber design for different solution volume/adsorbate mass ratios of bentazon, carbofuran and 2,4-D adsorption on to date seeds activated carbon. J. Environ. Anal. Chem. 2 (2014) 120, https://doi.org/10.4172/jreac.1000120.

[17] M. Akhtar, S.M. Hasany, M.I. Bhanger, S. Iqbal, Low cost sorbent for the removal of methyl parathion pesticide from aqueous solution, Chemosphere 66 (2007) 1829–1838.

[18] H.E. Bakouri, J. Usero, J. Morillo, R. Rojas, A. Ouassini, Drin pesticides removal from aqueous solutions using acid-treated date stones, Bioresour. Technol. 100 (2009) 2676–2684.

[19] N. Singh, Adsorption of herbicides on coal fly ash from aqueous solutions, J. Hazard. Mater. 168 (2009) 233–237.

[20] U. Traub-Eberhard, K.P. Hensche, W. Kordel, W. Klein, Influence of different field sites on pesticide movement into subsurface drain, Pestic. Sci. 43 (1995) 121–129.

[21] Y. Sudhakar, A.K. Dikshit, Adsorbent selection for endosulfan removal from waste water environment, J. Environ. Sci. Health Part B 34 (1999) 97–118.

[22] S. Boudesocque, E. Guillon, M. Aplincourt, F. Martel, S. Noael, Use of a low cost biosorbent to remove pesticides from waste water, J. Environ. Qual. 37 (2008) 631–638.

[23] F.T. Kamga, Modeling adsorption mechanism of paraquat onto Ayous (*Triplochiton scleroxylon*) wood sawdust. Appl Water Sci 9 (1) (2019), https://doi.org/10.1007/s13201-018-0879-3.

[24] S. Chatterjee, S. Das, K. Chakravarthy, R. Chakrabarti, A. Ghosh, A.K. Guha, Interactive of malathion, an organophosphorus pesticide with *Rhizopus oryzea* bio-mass, J. Hazard. Mater. 174 (2010) 47–53.

[25] V.K. Gupta, C.K. Jain, I. Ali, S. Chandra, S. Agarwal, Removal of lindane and malathion from wastewater using bagasse fly ash-a sugar industry waste, Water Res. 36 (2002) 2483–2490.

[26] S. Senthilkumaar, S.K. Krishna, P. Kalaamani, C.V. Subburamaan, N. Ganapathi Subramaniam, Adsorption of organophosphorous pesticide from aqueous solution using "waste" jute fiber carbon, Mod. Appl. Sci. 4 (2010) 67–83.

[27] S. Chattoraj, N.K. Mondal, B. Das, P. Roy, B. Sadhukhan, Biosorption of carbaryl from aqueous solution onto *Pistia stratiotes* biomass. Appl Water Sci 4 (1) (2013) 79–88, https://doi.org/10.1007/s13201-013-0132-z.

[28] S. Chattoraj, N.K. Mondal, B. Das, P. Roy, B. Sadhukhan, Carbaryl removal from aqueous solution by lemna major biomass using response surface methodology and artificial neural network. J. Environ. Chem. Eng. 2 (4) (2014) 1920–1928, https://doi.org/10.1016/j.jece.2014.08.011.

[29] S. Chattoraj, N.K. Mondal, K. Sen, Removal of carbaryl insecticide from aqueous solution using eggshell powder: a modeling study. Appl Water Sci 8 (6) (2018), https://doi.org/10.1007/s13201-018-0808-5.

[30] S. Chattoraj, B. Sadhukhan, N.K. Mondal, Predictability by Box-Behnken model for carbaryl adsorption by soils of Indian origin. J. Environ. Sci. Health B 48 (8) (2013) 626–636, https://doi.org/10.1080/03601234.2013.777283.

[31] S. Chattoraj, N.K. Mondal, B. Sadhukhan, P. Roy, T.K. Roy, Optimization of adsorption parameters for removal of carbaryl insecticide using neem bark dust by response surface methodology. Water Conserv. Sci. Eng. 1 (2) (2016) 127–141, https://doi.org/10.1007/s41101-016-0008-9.

[32] N.K. Mondal, S. Chattoraj, B. Sadhukhan, B. Das, Evaluation of carbaryl sorption in alluvial soil, Songklanakarin J. Sci. Technol. 35 (2013) 727–738.

[33] M. Sathishkumar, J.G. Choi, C.S. Ku, K. Vijayaraghavan, A.R. Binupriya, S.E. Yun, Carbaryl sorption by porogen-treated banana pith carbon, Adsorpt. Sci. Technol. 26 (2008) 679–686.

[34] M.E. Ouardi, S. Alahiane, S. Qourzal, A. Abaamrane, A. Assabbane, J. Douch, Removal of carbaryl pesticide from aqueous solution by adsorption on local clay in Agadir, Am. J. Anal. Chem. 4 (2013) 72–79.

[35] K.D. Kwon, H. Green, P. Bjoorn, J.D. Kubicki, Model bacterial extracellular polysaccharide adsorption onto silica and alumina: quartz crystal microbalance with dissipation monitoring of dextran adsorption. Environ. Sci. Technol. 40 (24) (2006) 7739–7744, https://doi.org/10.1021/es061715q.

10

Artificial neural network: An alternative approach for assessment of biochemical oxygen demand of the Damodar River, West Bengal, India

Tarakeshwar Senapati[a], Palas Samanta[b], Ritabrata Roy[c], Tarun Sasmal[d], and Apurba Ratan Ghosh[e]

[a]DEPARTMENT OF ENVIRONMENTAL SCIENCE, DIRECTORATE OF DISTANCE EDUCATION, VIDYASAGAR UNIVERSITY, MIDNAPORE, WEST BENGAL, INDIA [b]DEPARTMENT OF ENVIRONMENTAL SCIENCE, SUKANTA MAHAVIDYALAYA, DHUPGURI, WEST BENGAL, INDIA [c]HYDRO-INFORMATICS ENGINEERING DEPARTMENT, NATIONAL INSTITUTE OF TECHNOLOGY AGARTALA, AGARTALA, TRIPURA, INDIA [d]DEPARTMENT OF GEOGRAPHY, PANSKURA BANAMALI COLLEGE, PURBA MEDINIPUR, WEST BENGAL, INDIA [e]DEPARTMENT OF ENVIRONMENTAL SCIENCE, THE UNIVERSITY OF BURDWAN, BURDWAN, WEST BENGAL, INDIA

1 Background

Biochemical Oxygen Demand (BOD) is judicially applied for an interpretation of the organic load in wastewater. Additionally, BOD is also useful to check the efficiency of Water Treatment Plants. The conventional method of BOD estimation is 5-day BOD Test, which demands much time and resources. Recently, analytical estimation of water quality parameters (WQP) is gradually replaced by sensor-based devices for the sake of simplicity and rapidity. However, sensor-based device for BOD estimation is very expensive and not easily available. Moreover, a prolonged estimation time is required for conventional methods; on the other hand, it is not affordable in the situations where rapid mitigation is essential. A rapid, easy, and inexpensive method to determine the BOD level is, therefore, very useful.

Artificial Neural Network (ANN) is a type of machine learning algorithm, which requires no human supervision. This computational method, inspired by biological neural network, can be conducted with the help of different inputs and subsequently output is used for interpretation [1]. ANN provides to find out the best mathematical relationship between the inputs and related output [2] by utilizing adaptive weighing in different layers

Intelligent Environmental Data Monitoring for Pollution Management. https://doi.org/10.1016/B978-0-12-819671-7.00010-5

of neuron [3]. Apart from this, ANN also helps to find out the outputs through regression analysis in situation where input and output relationship is unknown [2]. ANN has huge applicability like forecasting of water quality [4], prediction of water parameter [5–8], and prediction of water quality [9–11].

This study attempts to develop a rapid and convenient method of BOD prediction from some other parameters by ANN through a case study from River Damodar. Conventional method of BOD assessment requires prolonged period (5 days) and sophisticated water quality laboratory setup. However, the model developed in this study can predict BOD instantly with fewer resources. Thus, it helps the decision makers to implement efficient actions to maintain desired BOD level in no time. Finally, the model can be very useful in industries, where wastewater is generated as it can predict the BOD effectively without delay and without any expensive monitoring devices. This model is developed exclusively, based on a field survey.

2 Material and methods

2.1 Study area and data collection

Sampling was performed at six selected points on River Damodar, which originates from the Chhotnagpur plateau of Palamou district in the Jharkahand state, and it flows through Jharkhand and West Bengal of India (Fig. 1). The River Damodar is an important tributary of the River Ganga. The river passes through a number of industrial cities, e.g., Bokaro, Dhanbad, Asansol, and Durgapur of both the States. Therefore, it receives a large amount of industrial wastewater from these industrial complexes, which are situated on the both of the banks of this river. Sampling was performed every month during the two consecutive years—2012–13 and 2013–14. Six sampling sites over 130-km stretches of the River Damodar from Dishergarh (Paschim Bardhaman District) to Sadarghat (near Burdwan town of Purba Bardhaman District) were selected (Table 1). Sampling sites were selected based on areas in relation to industrial, urban, and agricultural activities. The following sites were selected for sampling:

(i) Dishergarh is a coalmine area, situated downstream of the confluence point between the Damodar and Barakar river.

Table 1 Detailed description of the sampling sites of the study area.

Name of sampling site	Latitude	Longitude
Dishergarh	23°41′6.93″N	86°49′21.49″E
Dhena	23°37′36.93″N	86°58′21.09″E
Narayankuri near Raniganj Town	23°37′50.41″N	87°05′37.69″E
Majher mana	23°28′47.90″N	87°18′18.66″E
Shilampur	23°24′46.70″N	87°25′54.19″E
Sadarghat near Burdwan town	23°12′42.90″N	87°50′49.7″E

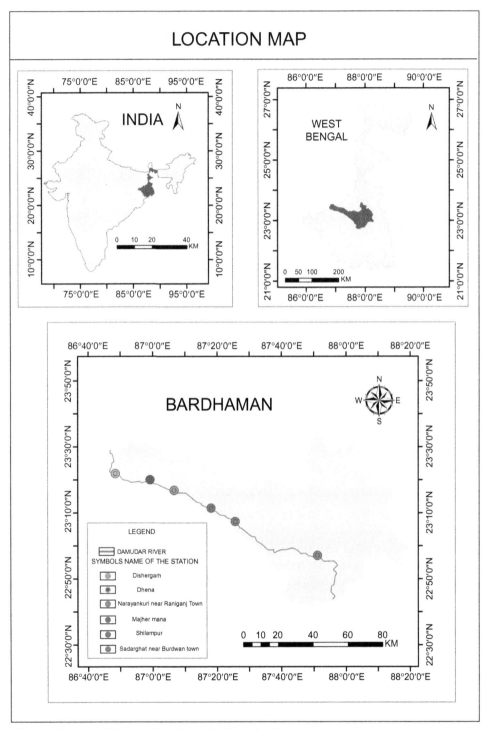

FIG. 1 The location map of the sampling sites in the Damodar River.

(ii) Dhena is situated downstream of the IISCO steel plant effluent discharge point.

(iii) Narayankuri site is situated downstream of the Nunia Nalluah discharge point.

(iv) Majhermana site is situated downstream to the discharge point of Tamla Nalluah, which carries industrial and municipal wastewater of the Durgapur city and Durgapur industrial complex area.

(v) Shilampur representing agricultural area cultivating mainly rice.

(vi) Sadarghat site representing mainly municipal area along with agricultural influence. Apart from this, this site collects wastewater of the Barddhaman city and Barddhaman Municipal Corporation (BMC).

Twenty-four different WQP were estimated from six sampling sites of the River Damodar for a period of 24 months. The pH, water temperature, electrical conductivity (EC), and total dissolved solids (TDS) were estimated with sensor-based multiparameter water quality analyzer (PCS Testr 35 Multiparameter) in situ, while Chemical oxygen demand (COD), biochemical oxygen demand (BOD), total hardness, total alkalinity, total nitrogen (TN), total phosphorus (TP), ammoniacal nitrogen, nitrate nitrogen, nitrite nitrogen, carbonate, bicarbonate, sulfate, chloride, fluoride, sodium, potassium, calcium, magnesium, total cation, and total anion were analyzed in laboratory as per the standard methods [12].

2.2 Selection of WQP

The predictor WQP was selected on the basis of two criteria, namely, high correlation (correlation coefficient: 0.9 or more) with BOD and availability of quick estimation method. These criteria ensured the accuracy of the ANN model developed and verified by actual field data.

2.3 Prediction model development

ANN was used as prediction model in this study as the relation between the parameters are complex and nonlinear. Moreover, actual field data may not follow any distinct distribution pattern, so simple regression methods are not suitable for such prediction model. Different algorithms like **GMDH Neural Network** (GNN), **Stepwise Mixed Selection** (SMS), **Stepwise Forward Selection** (SFS), and **Combinatorial** (COM) were applied for developing the prediction models. In particularly, COM and SFS are feed forward model, while SMS and GNN are back propagation model.

The COM model is performed based on formulation of different polynomial functions from a set of linear parameters [13]. Contrarily, the SFS model followed regression analysis and considered optimized when it is significant statistically [14]. The model is basically optimized by inclusion of individual variables only. On the other hand, the SMS model is optimized by both inclusion and exclusion of variables individually [13]. The GNN model basically creates different neurons layers iteratively and finally it is optimized by COM algorithm [13]. Furthermore, all these algorithms are automatized through regression analysis for rapid development of optimized ANN models inattentively. Moreover, different statistical methods are used based on actual data set to check the ANN models accuracy such as ***Mean Absolute Error*** (MAE) (Eq. 1a), ***Root Mean Square Error*** (RMSE) (Eq. 1b), ***Coefficient of Determination*** (R^2) (Eq. 1c), and ***Correlation Coefficient*** (*r*) (Eq. 1d).

■ ■ ■ ▬▬

Eq. 1 Different statistical parameters for checking ANN models accuracy

$$\text{Mean Absolute Percentage Error (MAPE)} = \frac{100}{n}\sum_{i=1}^{n}\left|\frac{A_i - P_i}{A_i}\right| \tag{1a}$$

$$\text{Root Mean Square Error (RMSE)} = \sqrt{\frac{\sum_{i=1}^{n}(P_i - A_i)^2}{n}} \tag{1b}$$

$$\text{Coefficient of Determination}\left(R^2\right) = 1 - \frac{\sum_{i=1}^{n}(A_i - P_i)^2}{\sum_{i=1}^{n}\left(A_i - \overline{A}\right)^2} \tag{1c}$$

$$\text{Correlation Coefficient}(r) = \frac{\sum_{i=1}^{n}\left(A_i - \overline{A}\right)\left(P_i - \overline{P}\right)}{\sqrt{\sum_{i=1}^{n}\left(A_i - \overline{A}\right)^2 \sum_{i=1}^{n}\left(P_i - \overline{P}\right)^2}}, \tag{1d}$$

where

$$\overline{A} = \frac{1}{n}\sum_{i=1}^{n} A_i \quad \overline{P} = \frac{1}{n}\sum_{i=1}^{n} P_i$$

n = sample number; A = actual values; P = predicted values.

▬▬▬▬▬▬▬▬▬▬▬▬▬▬▬▬▬▬▬▬▬▬▬▬▬▬▬▬▬▬▬▬▬▬ ■ ■ ■

The model with predicted values closest to the actual values was considered as the best prediction model in this study.

2.4 Validation of models

Twenty percent (20%) of the field data preserved solely for testing, not for the development of the models. After development of the models, these 20% data were used to predict the BOD value and compared with the actual BOD values. The parity among the sets of predicted and actual values indicated that the models were valid.

3 Results and discussion

3.1 Selection of predictor WQP

Among the 24 WQP estimated during the study period, only four were found highly correlated to BOD (correlation coefficient > 0.9). Those four WQP (viz. chemical oxygen demand, electrical conductivity, turbidity, and chloride) were, therefore, considered as the predictor WQP for the models (Table 2).

3.2 Prediction model development

The predictions from various models were close to the actual values (Table 3) and display similar patterns (Fig. 2). Among the prediction models, M5 had highest correlation (0.98) and least deviation (Table 3) with respect to the actual values. Additionally, M6 was found very close to M5 with same correlation and a little more deviation from the actual values (Table 3). Therefore, both the M5 and M6 models were found fairly accurate in the prediction of BOD value.

3.3 Validation of model

Twenty percent (20%) of the actual data were kept for testing of the models, developed by ANN. Those data were compared with the prediction's values from the models for

Table 2   Different combination of predictors for the prediction of the BOD value.

Model	Predictor WQP
M1	COD
M2	EC
M3	T
M4	C
M5	COD, EC, T, C
M6	EC, T, C

COD, chemical oxygen demand; EC, electrical conductivity; T, turbidity; C, Chloride.

Table 3   Closeness between actual and predicted BOD values during ANN model training.

Model	Mean absolute error (MAE)	Root mean square error (RMSE)	Coefficient of determination (R^2)	Correlation coefficient (r)
M1	0.57	1.04	0.85	0.92
M2	0.69	1.08	0.84	0.92
M3	0.76	1.31	0.77	0.87
M4	0.61	0.88	0.90	0.95
M5	0.34	0.46	0.97	0.98
M6	0.43	0.57	0.96	0.98

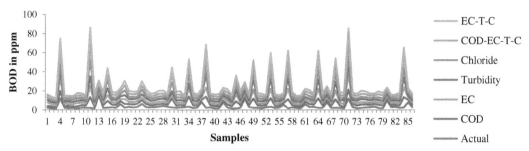

FIG. 2 Pattern of BOD prediction using different water quality parameters as predictors during ANN model training.

Table 4   Closeness between actual and predicted BOD values during testing the field data.

Model	Mean absolute error (MAE)	Root mean square error (RMSE)	Coefficient of determination (R^2)	Correlation coefficient (r)
M1	0.92	1.99	0.67	0.83
M2	0.83	1.26	0.87	0.93
M3	1.05	1.83	0.72	0.85
M4	1.19	2.46	0.50	0.71
M5	0.78	1.35	0.85	0.92
M6	0.74	1.17	0.89	0.94

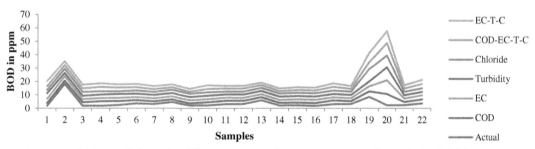

FIG. 3 Pattern of BOD prediction using different water quality parameters as predictors during field data.

validation. Based on the comparison, the predicted values were very close to actual values with high correlation and little deviations (Table 4) and they follow similar patterns (Fig. 3).

3.4 Selection of best model

During testing of the ANN with field data for the development of the models, the M5 and M6 were found to be about equally efficient in the prediction of BOD as they had the highest correlation with the actual values and least deviation from the actual values among all the models developed (Tables 3 and 4). However, the deviations of M6 predictions from the actual values showed a little higher than those of M5 (Table 3). During the testing of the models, the M6 showed highest correlation with the actual values and least deviation (Table 4). Accordingly, M6 can be considered as a better model than M5 for the prediction of BOD value.

Moreover, as M5 includes COD as a predictor, it involves more time and effort to predict BOD. The predictors of M6 (EC, T, and C), on the other hand, can be determined instantly with the aid of sensor-based devices. Prediction of BOD by M6 is, therefore, much more convenient and instantaneous. Thus, the **M6** (i.e., using EC, T, and C together as predictors) model (Eq. 2) can be considered as best model for effective BOD prediction.

■ ■ ■ ▬▬▬▬▬▬▬▬▬▬▬▬▬▬▬▬▬▬▬▬▬▬▬▬▬▬▬▬▬▬▬▬▬▬▬▬▬

Eq. 2 The M6 model to predict the BOD value

$$BOD = 0.0211036 - N185*0.374264 + N13*1.36815,$$

where

$N13 = -0.0118406 + N23*0.449672 + N51*0.553761$
$N51 = -1.05175 - N469*N91*0.0360947 + N91*1.49831$
$N91 = 1.11967 + \text{"Chloride, cubert"}*N154*0.22775$
$N154 = 6.94542 + N409*1.24069 + N409*N456*0.234695 - N456*3.48173$
$N456 = 2.53675 + Turbidity*N494*0.00165463$
$N469 = 0.806655 + \text{"EC, cubert"}*N494*0.10447$
$N494 = 26.5337 - \text{"EC, cubert"}*4.2361 + \text{"EC, cubert"}*\text{"Chloride, cubert"}*1.73976 - \text{"Chloride, cubert"}*9.88851$
$N23 = 0.0393994 + N79*1.96225 - N211*0.973678$
$N211 = 0.0589932 - N365*1.91167 + N306*2.89456$
$N306 = 1.14511 + \text{"Chloride, cubert"}*N324*0.2257$
$N365 = 1.9121 + N488*N492*0.0906233$
$N492 = 1.30874 + Chloride*\text{"Chloride, cubert"}*0.0352577$
$N488 = -8.22741 + Turbidity*\text{"EC, cubert"}*0.000885354 + \text{"EC, cubert"}*1.6556$
$N79 = 1.09573 + \text{"Chloride, cubert"}*N256*0.230684$
$N256 = 3.74622 + N409*2.77383 + N409*N419*0.133677 - N419*3.57223$
$N419 = 1.95202 + N487*N490*0.0852973$
$N490 = 0.847873 + Turbidity*0.0062723 + \text{"Turbidity, cubert"}*0.799037$
$N487 = -0.00251729 + EC*\text{"Chloride, cubert"}*0.00379555$
$N409 = 0.912347 + N421*0.525526 + N421*N495*0.0407886.$
$N185 = 0.041055 - N324*2.04798 + N247*3.03608.$
$N247 = 1.16129 + N421*0.406202 + N421*N433*0.0496239$
$N433 = 2.01096 + Chloride*N495*0.013999$
$N495 = -59.5201 - EC*0.366347 + EC*\text{"EC, cubert"}*0.0274454 + \text{"EC, cubert"}*17.4989$

■ ■ ■

As the prediction model is based upon case study, it may not be valid for all data sets. To minimize such uncertainty a large number of data were collected throughout a prolonged period of 24 months, covering all the seasons, in triplicate sets. However, if the established equilibrium of the water quality of the river in this segment gets changed due to some environmental or anthropogenic factors, the model may become less accurate. Therefore, this prediction model cannot totally replace the conventional method. It is rather suitable for occasions when BOD must be determined quickly, but accuracy may be compromised. This model can be used most effectively if it is validated in regular intervals by BOD determination by conventional method.

4 Conclusion

Present study deals with development of ANN model for quick prediction of the BOD level of water from electrical conductivity, turbidity, and chloride. The model was found fairly accurate by statistical analyses when tested with large spatial and temporal data. As this model can predict BOD quickly using less resource, it can save time, money, and effort. As BOD can be predicted almost instantaneously with this model, it is much more useful over 5-day-long conventional method for the managers to manage BOD level in water. Accordingly, the model is successful in quick and effective BOD prediction. However, the mentioned model is based on case study; therefore, there may be sampling errors and the model may not be valid for data sets. Moreover, the interactions among the natural components are extremely very complex. Furthermore, the model may be less accurate or even invalid in future because of dynamic nature of ecosystem. This model, therefore, would be an alternative substitute of existing conventional method if it is calibrated periodically with the conventional method.

Acknowledgments

The authors like to thank the Department of Environmental Science, the University of Burdwan, Burdwan, West Bengal, India, for providing the laboratory and library facilities during the course of research. We also like to thank Mr. P. K. Barai, Senior Scientist, West Bengal Pollution Control Board, West Bengal, India, for his guidance and cooperation.

References

[1] J.L. McClelland, D.E. Rumelhart, G.E. Hinton, The Appeal of Parallel Distributed Processing, MIT Press, Cambridge, MA, 1986.

[2] D.F. Specht, A general regression neural network, IEEE Trans. Neural Netw. 2 (6) (1991) 568–576.

[3] S. Agatonovic-Kustrin, R. Beresford, Basic concepts of artificial neural network (ANN) modelling and its application in pharmaceutical research, J. Pharm. Biomed. Anal. 22 (5) (2000) 717–727.

[4] R.R. Goyal, H. Patel, S.J. Mane, Artificial neural network: an effective tool for predicting water quality for Kalyan-Dombivali municipal corporation, Int. J. Sci. Res. 4 (6) (2015) 2863–2866.

[5] E. Dogan, B. Sengorur, R. Koklu, Modelling biological oxygen demand of the Melen River in Turkey using an artificial neural network technique, J. Environ. Manag. 90 (2) (2009) 1229–1235.

[6] M. Khandelwal, T.N. Singh, Prediction of mine water quality by physical parameters, J. Sci. Ind. Res. 64 (2005) 564–570.

[7] R. Roy, M. Majumder, A quick prediction of hardness from water quality parameters by artificial neural network, Int. J. Environ. Sustain. Develop. 17 (2/3) (2018) 247–257.

[8] L. Swathi, B. Lokeshappa, Artificial neural networks application in prediction of water quality, Int. J. Innov. Res. Sci. Eng. Technol. 4 (8) (2015) 6911–6916.

[9] S. Chatterjee, S. Sarkar, N. Dey, A.S. Ashour, S. Sen, A.E. Hassanien, Application of cuckoo search in water quality prediction using artificial neural network, Int. J. Comput. Intell. Stud. 6 (2–3) (2017) 229–244.

[10] S. Chatterjee, S. Sarkar, N. Dey, S. Sen, Non-dominated sorting genetic algorithm-ii-induced neural-supported prediction of water quality with stability analysis, J. Inform. Knowl. Manag. 17 (02) (2018) 1850016.

[11] R. Roy, M. Majumder, Empirical modelling of total suspended solids from turbidity by polynomial neural network in north eastern India, Desalin. Water Treat. 132 (2018) 75–78.

[12] A.D. Eaton, L.S. Clesceri, A.E. Greenberg, M.A.H. Franson, Standard Methods for the Examination of Water and Wastewater, Twenty first ed., American Public Health Association, Washington, DC, 2005.

[13] L. Anastasakis, N. Mort, The development of self-organization techniques in modelling: a review of the group method of data handling (GMDH), in: Research Report 813, Department of Automatic Control & Systems Engineering, The University of Sheffield, UK, 2001.

[14] L.J. Lancashirea, R.C. Reesb, G.R. Ball, Identification of gene transcript signatures predictive for estrogen receptor and lymph node status using a stepwise forward selection artificial neural network modelling approach, Artif. Intell. Med. 43 (2) (2008) 99–111.

Further reading

P. Sirisha, K.N. Sravanti, V. Ramakrishna, Application of artificial neural networks for water quality prediction, Int. J. Syst. Technol. 1 (2) (2008) 115–123.

11

Codesign to improve IAQ awareness in classrooms

Bradley McLaughlin[a], Stephen Snow[b], and Adriane Chapman[a]

[a]UNIVERSITY OF SOUTHAMPTON, SOUTHAMPTON, UNITED KINGDOM [b]UNIVERSITY OF QUEENSLAND, BRISBANE, QLD, AUSTRALIA

1 Introduction

Indoor air quality (IAQ) within school classrooms can be far poorer than government guidelines recommend. Multiple studies have recorded maximum classroom CO_2 concentrations more than double the UK Guidelines [1–5]. Inadequate ventilation can be a problem in naturally ventilated buildings during the cooler seasons when windows are typically closed to retain warmth [6], and in mechanically ventilated classrooms [7] if air is recirculated. Increasing requirements for energy efficiency in building standards [8] means newer buildings are increasingly airtight, which can be problematic if ventilation is not managed adequately.

Low ventilation rates have been linked to decrements in human performance and health, including elevated prevalence and intensity of health-related symptoms such as nausea, fatigue [9], asthma [10], headache, eye and throat irritation [11], poor concentration [12], allergies in certain climates, [13] as well as impaired academic performance in the classroom [14,15] and 6%–9% lower productivity in the office [16]. Given that people spend 90% of their time indoors [17], exposure to IAQ is a salient issue for research. The effects of poor IAQ are of particular concern to children, whose respiratory systems are not yet fully developed [18,19]. Furthermore, recent studies suggest that due to acclimatization to a room (sensory fatigue), decrements in human performance can begin to occur prior to—or in the absence of—occupants' awareness of the declining air quality [20,21], underscoring the value of a means for early warning.

Yet despite the well-known negative effects of poor ventilation on health and performance, and despite these effects beginning to occur prior to awareness of poor IAQ, there is currently no mandatory requirement to provide classrooms or school teachers with real-time information on IAQ [22]. Teachers are key to the ventilation of classrooms, particularly in primary school classrooms, due to their role in opening windows and operating air conditioning or heating. In this study, we focus on technology to enable teachers to affect positive change to their learning environment.

A system called Airlert was created to assist teachers and students in managing IAQ in the classroom by visually displaying IAQ and sending an alert when the air quality is poor. The system consisted of a Wi-Fi-connected sensing unit linked to a school-issued iPad running the Airlert app. By doing this, teachers and students are educated on factors affecting IAQ in the classroom and are able to detect declining air quality at an early stage. This provides a basis for better management of opening and closing windows and doors to improve IAQ as far as practicable to reduce illness, absence, and to improve student performance.

After a review of related work, the contributions of this study include:

1. An in-depth understanding of primary school teachers' experiences and understanding of IAQ in classrooms in "Methodology."
2. The user-centered design of Airlert, an IAQ sensor built specifically to engage primary school teachers and students with air quality and ventilation in classrooms, in "System Design and Implementation."
3. Deployment of the device in a set of primary school classrooms, and analysis of the effectiveness of the user-centered design process in "Understanding System Use."

We then highlight issues for consideration when designing and implementing similar devices in "Discussion" and finally conclude.

2 Related work

2.1 Ventilation, CO_2, and indoor air quality

Ventilation in indoor spaces removes indoor pollutants and maintains indoor air quality (IAQ). In naturally ventilated buildings, ventilation also substantially affects indoor temperature. Measuring actual ventilation rate in buildings is costly and time consuming [23]. Carbon dioxide (CO_2) is often used as an approximation of ventilation [24,25]; being a product of human respiration, CO_2 concentrations increase in occupied spaces when ventilation is insufficient to replace the air. The decreasing cost of sensor technology means estimating ventilation rates using CO_2 concentration is now far simpler and cheaper than measuring actual ventilation rate [2] and measurements can be taken in real time and used to inform room occupants in offices [26] and in schools [27,28]. Accordingly, CO_2 concentrations (measured in parts per million (ppm)) are increasingly becoming specified in operational building regulations and guidelines, alongside actual ventilation rate (measured in liters per second per person (l/s/p)) [29,30]. US classroom ventilation standards stipulate 7 l/s/p at the rated occupancy of the room [31], approximating to peak CO_2 concentrations above 1000 ppm [7]. The guidelines for UK classrooms stipulate average CO_2 concentrations in a naturally ventilated (with no use of mechanical ventilation systems) should remain below 1200 ppm throughout a school day and should not exceed 1500 ppm for more than 20 consecutive minutes in any day [22]. The figure is consistent with the ASHRAE (American Society of Heating, Refrigerating, and Air-Conditioning Engineers)

61.1-2016 ventilation standards for acceptable indoor air quality in buildings, stating that maximum average CO_2 concentration should not exceed 1000–1200 ppm [31].

2.2 CO_2 in classrooms

The negative health and cognitive performance effects of poor ventilation are well documented [14,32]. Within classrooms, specifically multiclassroom studies, high mean CO_2 concentrations (>1800 ppm) caused by poor ventilation are linked to poorer performance on academic test results and increased absenteeism [18,33]. Each 1000 ppm increase in average classroom CO_2 concentrations has been equated to a 10%–20% (statistically significant) relative increase in student absence [2].

Concerningly, many studies have found measured classroom CO_2 to be far worse than guideline values [1,7,11]. An extensive monitoring program of 62 classrooms in Athens found more than half (52%) to have average CO_2 concentrations above the ASHRAE guideline of 1000 ppm [1]. Fisk [7] conducted a metareview of 28 separate published classroom ventilation studies, where each of the individual studies had measured IAQ in 20 or more classrooms. Average and median CO2 concentrations in classrooms exceeded 1000 ppm in all studies, with many average values exceeding 2000 ppm and maximum peaks up to 6000 ppm [7].

2.3 CO_2 monitoring and ventilation behavior

A number of studies have deployed and tested different means of communicating real-time air quality to occupants in classrooms. In each study, the research aim was to understand whether CO_2 feedback leads to improved ventilation behavior. Table 1 summarizes studies which have deployed IAQ feedback in classrooms.

Across the studies in Table 1, there is consensus that IAQ feedback in classrooms offers promise in improving ventilation behavior. All of the studies recorded an improvement in ventilation behavior following the deployment of CO_2 feedback. This is despite a wide variance in the type of visual feedback given, e.g., a box with a single red light, which started flashing if CO_2 exceeded 1200 ppm [28]; a portrait of Albert Einstein's face turning greener (more sickly) with increasing CO_2 levels, analogous to a great mind becoming fogged by poor IAQ [35]; and a thermometer-style feedback, with pictures of a student becoming progressively less alert at higher levels of CO_2 [27].

A further defining feature of each of the studies in Table 1 is the lack of attention to the end users of the technology. Despite all studies researching the impact of CO_2 feedback on behavior change, none of the studies involve the pupils or the teachers in the design of the artifacts deployed, or in the evaluation of the system's performance. In all of the studies behavior change is measured only through quantitative air quality measurements, and none of the studies involve teachers or students in discussions or interviews evaluating the system.

The lack of end-user input into the design of CO_2 feedback in each of these studies of ventilation behavior change in classrooms was a key factor in the present study's wish to codesign Airlert with the user who would be using the technology themselves.

Table 1 Summary of classroom IAQ monitoring studies.

Study	Intervention/ Artefact deployed	Type of feedback	School type	Deployment details	Overall findings
Geelan et al. [28]	(1) Ventilation advice only (2) Ventilation advice + CO_2 warning device (3) Ventilation advice + a teaching package (4) No intervention (control) CO_2 concentration measured, follow-up questionnaires.	Situated box with sensor, digital reading and red LED light, which illuminates >1200 ppm (Dutch guideline).	Middle school	• 20 classes per intervention type • 1-week intervention, av. CO_2 levels and behavior measured 2–3 before, and 0- and 6-weeks post intervention	• 95% of teachers considered CO_2 warning device useful. • Significant reduction in CO_2 concentration of classrooms with found with ventilation advice + CO_2 warning intervention at 0 and 6 weeks after intervention, compared to control classrooms.
Heebøl et al. [34]	(1) Mechanical ventilation retrofit (2) Automated window opening (3) Automated window opening + heat recovery (4) A visual display of CO_2 (5) No intervention (control)	Scale of LEDs from 250 to 5000 ppm CO_2. Yellow >1000, red >1600, instructions to open windows once LEDs became red.	Elementary school	• 5 classrooms • 1 classroom per intervention type • Effect of CO_2 feedback monitored after 4 months	• Visual CO_2 feedback increased window opening behavior relative to control condition in the short term. • Yet a re-visit to the school 12 months later found the behavior change had not lasted.
Rigger et al. [35]	(1) Deployment of an art-based visualization of indoor CO_2 concentration (2) No intervention (control) Postdeployment questionnaires	A portrait of Albert Einstein, which turns progressively green with increasing indoor CO_2.	Secondary school	• 8-week deployment • 10 classrooms • 255 classes held during deployment • A/B testing. A = treatment, B = control	• Treatment group relative to control group recorded: higher subjective wellbeing, higher *"perceived level of engagement in room climate"* and lower average CO_2 concentration.

	Intervention scenarios	Feedback description	Setting	Study design	Results
Wargocki et al. [27]	(1) CO_2 feedback system deployed in a classroom (2) No intervention (control)	Thermometer-style CO_2 indicator depicting a student (adjacent to the readings) who becomes increasingly sleepier as CO_2 increases.	Elementary school	• Four pairs of classrooms monitored • One week per intervention	• Visual CO_2 display increased window opening frequency leading to reduced CO_2 levels.
Toftum et al. [36]	(1) CO_2 feedback + pupils made to open windows when $CO_2 > 1000$ ppm (2) CO_2 feedback + pupils recommended to open windows when $CO_2 > 1000$ ppm (3) Pupils to open windows for 5 min during lecture (4) pupils to open all windows before leaving classroom for break	Visual CO_2 indicator (few details given).	Elementary school	• Eight classes • One scenario per class (see first column for scenarios) • Two-week intervention period • One-week baseline monitoring before intervention	• Visual CO_2 display unit led to modified behavior and 40%–60% reduction in the time that CO_2 concentrations were above 1000 ppm.

2.4 Participatory design and human-computer interaction

Participatory design is an approach to design, based upon the premise that those who might use or be affected by a new technology should be included in its design [37]. Participatory design seeks to include a range of end users and stakeholders in the design process using a human-centered design process similar to complementary fields such as Human-Computer Interaction (HCI). The human-centered design process is formalized in the International Standard: ISO 9241-210:2010 [38] and involves an iterative process of problem definition, contextual enquiry (defining user needs), prototyping solutions and evaluation. Accordingly, we involved teachers both early in the design process, by gathering their opinions on existing air quality problems in the classroom, on early Airlert prototypes and again in the deployment of the system itself.

3 Methodology

To understand the effect of classroom IAQ feedback on ventilation actions, a custom-made, low-cost device called Airlert was built. Airlert followed a user-centered design process [39] involving: (1) predeployment interviews, (2) week-long deployments of Airlert in six classrooms, and (3) postdeployment interviews.

Six primary school teachers in classrooms across four schools in the South of England (Table 2) were involved in the Airlert design and evaluation process.[a] All were responsible for classes of 24 to 30 children between Reception and Year 5 (ages 4 to 10). The participants who took part in the deployment and postdeployment interviews were the same as those who had taken part in the preinstall interviews.

3.1 Predeployment

Semistructured qualitative interviews lasting approximately 20–30 minutes were undertaken with all six teachers. This structure allowed for questions to be asked, while allowing the participant freedom to openly discuss their opinions and ideas [40]. In contrast to previous studies of IAQ feedback where users themselves have not been a focus [27,28,36], this study gathered rich accounts of use and in-depth opinions from a small group of participants, allowing interviews to be tailored to each participant.

Predeployment interviews consisted of collecting: (1) personal perceptions of IAQ in the classroom and information on current ventilation practices and habits, and (2) views and opinions on various prototype user-interface (UI) designs and their features. Following Vines et al. [41] who used divergent designs and provocation in design to seed discussion and ideation, we designed paper prototypes for three potential IAQ interfaces, each of which had intentionally very different features such that the teacher could consider and discuss various different elements of the device design. This allowed for elicitation of the

---

[a]Ethical approval was obtained for the study through the University of Southampton's Ethics and Research Governance Office (ERGO: 46809). Following informed consent from each participant, interviews were audio recorded.

Table 2 Participant information.

Participant	Gender	Year group (ages)	Class size	Room ventilation	Heating control
P1	Female	Year 5 (Ages 9–10)	28	External door and windows on one side of the classroom. The opposite side had an internal door and several external windows.	Radiators not controlled by the teacher.
P2	Female	Year 2 (Ages 6–7)	30	External windows on two adjoining sides of the classroom. Internal door on one of the sides without window.	Closed-loop air conditioning/ heating system (reverse cycle), fully controllable by the teacher.
P3	Male	Year 3 (Ages 7–8)	28	External door and small windows on one side of the classroom and an internal door on the opposite side. Classroom had a low ceiling.	Radiators not controlled by the teacher.
P4	Female	Reception (Ages 4–5)	29	External door and external windows on one side of the room and an internal door on the other side. Between the two doors, there was another door leading to the adjacent Reception class which was often left open.	Radiators fully controlled by the teacher.
P5	Male	Year 1 (Ages 5–6)	24	Internal door on one side of the classroom and an external door on the opposite side, with external windows also on this wall.	Radiator not controlled by the teacher, working intermittently.
P6	Female	Year 3 (Ages 7–8)	30	Internal and external door.	No radiators but underfloor heating which was not controlled by the teacher.

needs and ideas of teachers regarding the app, maximizing the fitness for purpose of a final, high-fidelity tablet application.

Interview responses from both the predeployment and postdeployment exit interviews were analyzed firstly according to the question topic. Additionally, a thematic analysis was conducted according to Braun and Clarke [42], in order to extract themes and personal values emergent from the data additional to those asked directly in the interview questions.

3.2 Feedback from initial interviews

3.2.1 Awareness of IAQ and perceptions of the classroom environment

Feedback from the analysis of the initial interviews indicated several common themes and ideas around awareness, behavior, and control over comfort in the classrooms. Most interviewees had little to no knowledge about IAQ, with participants commenting: *"I wouldn't know anything about it"* (P1), or *"it doesn't really go through my mind"* (P5). P3 stated they would be interested to see how their perceptions of the IAQ aligned with an accurate measurement of IAQ. Thermal comfort was highlighted as a problem by P1, P2, and P6, particularly over summer. P6 equated poor thermal comfort with poor air quality, mentioning that when it gets hot in the classroom the children *"switch off and go to sleep."* The confusion between heat, humidity, *"stuffiness,"* and poor IAQ highlights the rationale for a visual indication of CO_2 specifically to enable teachers to better differentiate between thermal comfort and ventilation. It was noted that bacteria picked up from touching classroom objects is widely discussed within their school, but health effects caused by poor air quality are not.

3.2.2 Existing ventilation behavior

Various responses were received of ways of regulating the air quality within the classroom, yet if good ventilation behavior was present, this was generally triggered by subjective thermal comfort and smell, rather than for ventilation in order to preserve the health, well-being, or concentration of students. These findings compare with existing research that finds windows in naturally ventilated buildings are opened and closed primarily according to comfort considerations [6]. P2 was the only participant to have air conditioning (AC) in their classroom. This meant that in warm weather, doors and windows would be kept shut to allow the AC to cool the room, rather than opening doors or windows. P4 taught in Reception (ages 4 and 5) and tried to keep their classroom fresh and ventilated as young children often *"don't realize their temperature quite as much as we [adults] do,"* and thus do not recognize when they are getting sweaty or overheated. The participant further stated that reception classes spend a lot more time outdoors than other year groups and so the external door is regularly open to allow for this.

Ventilation behavior in some instances was a product of school protocols on other matters such as safety and energy efficiency. P5 reported their school encouraged leaving windows closed and keeping blinds down to eliminate the chances of windows being left open overnight, and for saving energy. P5 noted their classroom *"hardly has any fresh air."* Despite seemingly being aware of the need to ventilate, P5 reported only occasionally opening windows. The primary triggers for opening windows during class for P1, P5, and P6 were heat, bad smells, following sport or physical education classes, or to regulate the ambient temperature. Health or maintaining air quality did not register as a reason for ventilation for any of the teachers.

3.2.3 User interface—Expectations for IAQ feedback in the classroom

All participants were interested in the idea of receiving real-time feedback on the IAQ. Opinions differed, however, regarding how this information would best be displayed in the classroom. P1 was interested in getting the class involved with an IAQ feedback device and integrating it in learning activities. This would allow for *"a bit of an investigation"* by looking at peaks in data as well as *"predicting"* and *"speculating"* what caused these peaks, which could fit into the *"working scientifically curriculum."* P2, whose class is younger (6–7 years old), saw an IAQ feedback device as useful learning device for managing air quality, but wanted to keep the children away from any device to avoid breakage. P2 considered they might use a device for IAQ feedback to monitor the air quality during lessons with higher levels of exertion, such as acting, to determine if this affects IAQ.

P5 stated that in order for any tool to be practical for use by teachers, it needed to be *"quick, easy, and efficient otherwise no one would use it… if it's not quick and easy, no one will actually continue [using] it."* This emphasized the need to provide a concise, clear, and simple display for the device allowing the interface to be understandable by all ages across the school. All interviewees responded positively to the idea of a visual mascot for IAQ, in order to contextualize readings against possible impacts upon human performance and appeal to the children. It was agreed that interactions with the device should not be time consuming and information should be continually available, interpretable at a glance, and legible from across the room (P2). The use of colors for ease of interpretability received positive feedback from participants, in particular P1 and P2.

3.2.4 Prototype user interfaces

Prototype 1 (Fig. 1A–C) shows a multipage interface with various ways to present the data, while Prototype 2 (Fig. 2) is a simple single-page app, emphasizing the use of color coding to represent data. Prototype 3 (Fig. 3A–C) uses a mascot to visualize the data by changing "mood" dependent on the pollutant levels.

Table 3 summarizes the responses from the initial interviews.

4 System design and implementation

The Airlert device was designed according to feedback gathered in the predeployment interviews, as user engagement has a positive impact on device success [43]. The choice to build a novel device rather than use an existing off-the-shelf multiparameter air quality monitor (e.g., uHoo [44] or Netatmo [45]) was based upon four factors:

1. **Direct relationship between values and actions:** Specific focus on CO_2 and VOCs because both parameters are directly affected by ventilation. Parameters that users could not directly affect, such as outdoor pollutants, were avoided. Previous research finds users can become frustrated if they cannot themselves affect measured environmental parameters [46].
2. **Children:** Few existing user feedback systems for IAQ are designed for children specifically. Being in control of the prototype (and future iterations) allowed Airlert to incorporate features such as a mascot, and affected choice of colors and shapes.
3. **User feedback:** Airlert was codesigned with end users, which meant being able to incorporate feedback gathered from interviews when evaluating the final device.
4. **Low cost:** For realistic adoption in schools, given each school has tens of individual classrooms, the cost of any monitoring device should be a key consideration.

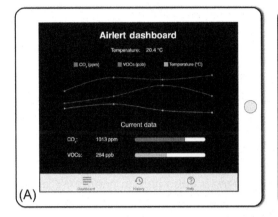
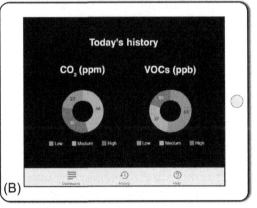

FIG. 1 Prototype user interface 1, showing three tabs: (A) Dashboard, (B) History, and (C) Help.

The Airlert design sought to find a balance between the low cost but limited functionality and usability of devices such as Wargocki et al. [27], Geelen et al. [28] and Toftum et al. [36] with the functionality, aesthetics, and utility of more expensive devices such as Uhoo (~$299 USD [47]). The total component cost of the Airlert prototype was ~£80 (~$100 USD), which would decrease substantially with mass production.

While a product can be developed with limited research, it is key to ensure the product is not only usable, but also engaging [48]. We incorporated participants' wishes for a mascot for the device who responds to the IAQ readings and color coding (e.g., green, amber, orange, red) to provide an indication of the IAQ at a point in time.

4.1 System design

The final prototype consisted of a sensor unit, with the results displayed in real time on a custom-built iPad app, for use on the school-issued iPads. The background color (green to red) is displayed behind the reading, and the posture (alertness/sleepiness) of

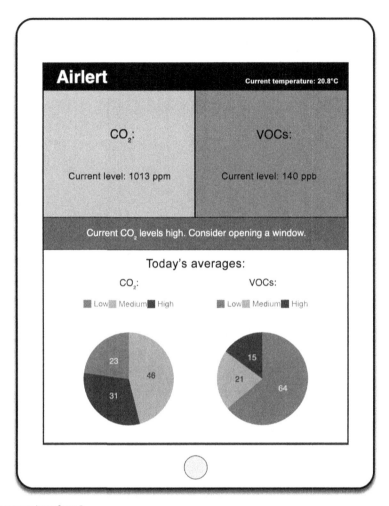

FIG. 2 Prototype user interface 2.

the mascot (called Ringo) changes at preset CO_2 concentrations informed by the UK Guidelines [22].

4.1.1 System overview

A Raspberry Pi Zero W was used to connect to the IAQ sensors due to its Wi-Fi connectivity (for database access) and small physical size. This is connected to a Google Firebase database, which then pushed data through to an iPad app (also connected to Wi-Fi).

The sensors used were an AMS CCS811 VOC and CO_2 sensor and Bosch BME280 temperature and humidity sensor. The accuracy of the CCS811 sensor is improved by providing it temperature and humidity values; hence, the BME280 chip was used. The sensors both communicate via I^2C, allowing easy integration with the Raspberry Pi. Fig. 4 shows an overview of the system.

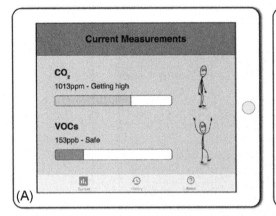

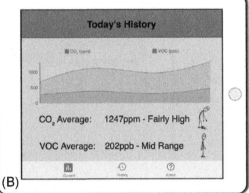

FIG. 3 Prototype user interface 3, showing three tabs: (A) Current, (B) History, and (C) About.

4.1.2 App design

The iPad app visualizes the air quality data of the classroom and provides push notifications when air quality is poor, prompting the teacher to act on this. The app was built using Xcode, as this is the official Developer's IDE (Integrated Development Environment) for Apple products, using the Swift 4.2 programming language.

The Raspberry Pi communicated to the sensors and database via the Python 3 programming language. Sensor values were taken every 15 seconds and updated on the iPad. Data were collected daily from 7 am to 7 pm to allow a sufficient window of time to collect data, while not requiring data values overnight when schools are typically closed. When a sensor reading was retrieved by the iPad, it would ascertain which "mood" the value fell within (Fig. 5). The color values were based on the UK guidelines for IAQ in schools [22]. The delimiters for the color of the device and behavior or Ringo are shown in Table 4.

A limit of a maximum of one notification every 15 minutes was set up to prevent alerts occurring too frequently, and thus disturbing the dynamics of the class. An alert was set to

Table 3   Review of key features from participant interviews.

Participant ID	Previous IAQ thoughts	Device requirements	UI requirements
P1	• Heat leads to germ accumulation • Children smelly after physical exercise • Open windows when too hot	• Able to facilitate class interaction • Pollutants visualized should be limited to those for which the source can be traced • Temperature displayed • Reflection on past data	• Data visualization • Single page • Color coding • Animal mascot to represent IAQ would encourage child interaction
P2	• Does not think about CO_2 levels • When warm, close doors and windows, turn aircon on • Take children outside when too hot	• High up away from children • Temperature displayed • One main page • Device troubleshooting information useful	• Quick-look UI, clear, big font • Color coding • Large color tiles liked • Mascot to represent IAQ and mood
P3	• Outdoor AQ in area around school is poor • Poor IAQ in class as small windows so air can feel stale • Opens window to cool room and bring fresh air	• Description of pollutant sources	• Data visualization
P4	• Children not aware of temperature • Windows open to cool children and remove smells	• Temperature not useful to them but would be to others • Device troubleshooting information helpful	• Data visualization • Not overcrowded • Likes comparison of data • Single page preferred • Mascot liked as child friendly
P5	• Lots of pollutants in air but does not cross their mind • School discourages opening windows to allow projector to be seen • Rarely opens windows	• Quick, easy, efficient	• Not overcrowded
P6	• When classroom is hot, children misbehave and do not concentrate • Open windows when hot or smelly	• Associate with guidelines • Real-time data • Temperature displayed	• Data visualization • Line graph useful • Child friendly

appear when the CO_2 level exceeded 1500 ppm (orange) or VOC levels exceeded 300 ppb (red). Also, if the CO_2 levels exceeded 2000 ppm (red), another "urgent" alert would appear, providing it had been more than 15 minutes since the last notification. The alerts produced a short audible noise and the message read *"Consider opening a window to let some fresh air in,"* or *"Open a window. The CO_2 level is far too high,"* when the "urgent" CO_2 alert was presented.

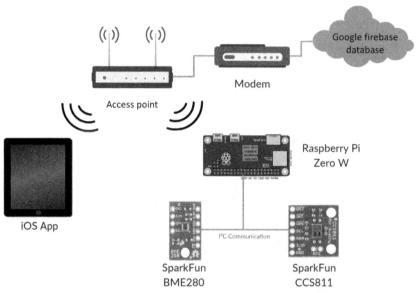

FIG. 4 System overview diagram.

The app consists of two pages, one called the "Dashboard" (Fig. 6), where the data are provided, and the app is expected to remain for the majority of the time. The second page is the "About" page (Fig. 7). This provides information on the app for the teacher, or any new teachers in the classroom, to familiarize themselves with the app and its functionalities.

In Fig. 6, Marker 1 shows the current temperature reading for the teacher's reference. Markers 2 and 3 show the CO_2 and VOC values. These are presented on two large tiles, which change color in correlation to pollutant concentration. The app mascot, Ringo, also changes mood depending on the color of the "tile" color coding, as shown by Marker 2 on page 2 of the app (Fig. 7).

In Fig. 6 Marker 4 shows a graph of the day's CO_2 and VOC values against time.

Marker 5 is used to update the graph in the morning, to clear the previous day's data. Marker 6 shows the "Tab Bar," used to navigate between the two pages of the app.

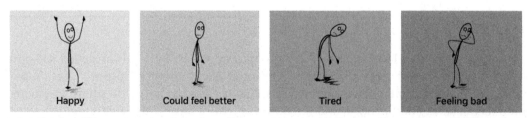

FIG. 5 The four different "moods" of the air quality mascot Ringo, with corresponding background colors.

Table 4   Colors associated to air quality data, relating to Fig. 5.

Color	CO_2 value (ppm)	VOC value (ppb)
Green	<1200	<100
Amber	1200–1499	100–199
Orange	1500–1999	200–299
Red	>2000	>300

Fig. 7 gives more information about the device. Marker 1 tells the user about Ringo; while Marker 3 provides some brief information on CO_2 and VOCs and what the side effects of high concentrations of these can be. Marker 4 provides troubleshooting information and device help and Marker 5 again outlines the "Tab Bar."

4.1.3 Casing

To house the Raspberry Pi and sensors, a 3D-printed case was designed and manufactured (Fig. 8A–D). The case was designed with multiple air gaps to maximize air flow around the sensors, and to prevent any interference from the heat generated by the Raspberry Pi on the temperature sensor. The thickness and robust nature of the casing was designed to protect the device from falls, to prevent touching of the circuitry by children, and to give the device a neat appearance.

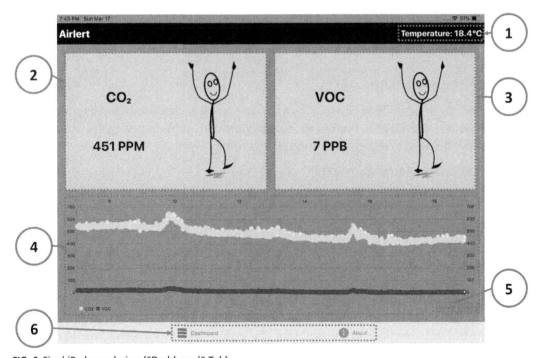

FIG. 6 Final iPad app design ("Dashboard" Tab).

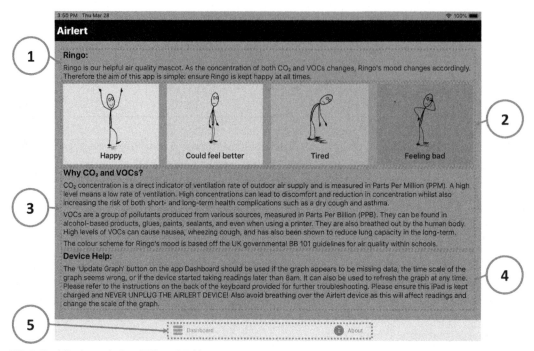

FIG. 7 Final iPad app design ("About" Tab).

4.2 System deployment

Airlert was deployed into four classrooms for 5 days in each classroom during the heating season (February–April 2019). Four of the six teachers who participated in the initial interviews trialed the device in their classrooms, including:

- P1—Year 5, 28 children (ages 9–10)
- P2—Year 2, 30 children (ages 6–7)
- P3—Year 3, 28 children (ages 7–8)
- P4—Year R (Reception), 29 children (ages 4–5)

In the first 2 days of each deployment, no visual feedback was given, allowing the sensors calibration time. The sensor in each classroom was located at seated height (as recommended by the UK guidelines for ventilation in schools [22]). It was also kept away from areas where it could have been breathed on directly, whether purposefully or not. This was to reduce numbers of sudden "spikes" in sensor readings due to CO_2 in breath exhalation, and to prevent false readings of the ambient classroom data. From days three to five, the iPad was displayed in the classroom, allowing the IAQ to be visually monitored. The iPad was in an open location chosen with the teacher to avoid risk of damage, typically out of reach of, but still visible to the children.

FIG. 8 Airlert casing shown from angles: (A) front, (B) back, (C) top, (D) bottom.

5 Understanding system use

The focus of this study is on users' perceptions, use, and understanding of the device, and we do not provide an in-depth quantitative analysis of sensor data. However, Fig. 9 shows indicative data collected from one of the days during a classroom trial, demonstrating how pollutant levels can vary throughout the day in a classroom.

The timings given below were taken from the class the timetable followed on the day of this trial, and the corresponding air quality data are reflected in Fig. 9.

(1) 9:00–11:05: Mathematics, Spelling, and Grammar
(2) 11:05–11:40: Break followed by assembly
(3) 11:40–12:20: Writing
(4) 12:20–13:05: Lunch break
(5) 13:05–14:15: Physical Education
(6) 14:15–15:15: Art/Design followed by class book
(7) 15:15 onwards: End of school day

Visible from Fig. 9 is the effect of occupancy on pollutant levels, and the speed at which IAQ deteriorates when the classroom is occupied, e.g., during lessons, and improves (i.e., pollutant levels decrease) when the class vacates the room. The deployment in winter means the classroom windows were mostly kept closed during occupation. The large

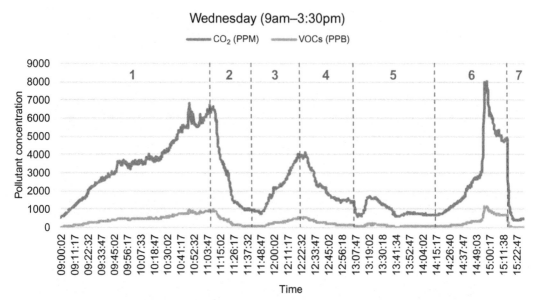

FIG. 9 Screenshot of a graph showing the air quality for a typical day in the classroom.

spike between 14:15 and 15:15 may be due to the use of glues or paints used during the Art/ Design class, affecting the true readings.

5.1 Exit interviews

Upon completion of the deployments, further interviews were conducted with classroom teachers. Exit interviews focused on use of and engagement with the device and whether it provided sufficient information, changed awareness of IAQ, changed ventilation habits during the deployment and thoughts on long term use. Interviews were semistructured, approximately 30 minutes in length, and audio recorded to allow for natural conversation and full interviewer interaction. Participants 5 and 6 did not partake in deployments. Table 5 shows a summary of the findings from the exit interviews.

5.1.1 System engagement

All participants agreed that the device was clear and easy to understand with some of the children becoming involved in monitoring and discussing readings across all year groups, from Year 5 to Year R (Reception). Examples of the comments made by the children were *"Oh, it's red, we need to open the windows,"* followed by *"Phew, we can breathe again!"* once the windows were opened and the device went back to green. The device providing a distraction to children was a concern in design; however, the teachers in three out of four trials confirmed that the device did not serve as a distraction to the children and they did not find the device alerts irritating. P3 differed by saying that on the first day the app was a little distracting as it was a *"novelty,"* although after the first day, the alerts were not distracting. Ringo, the device mascot, was said to

Table 5   Review of feedback from Airlert deployments.

Participant ID	Day-to-day usage	General comments
P1	• Discuss the data with the class • Integrate into academic curriculum	• Reflected on graphs and led regular group discussions • Class very involved • Felt in control of IAQ • Shocked how quickly IAQ improves through increasing ventilation • More aware of IAQ
P2	• Monitor changing pollutant levels across day • One child to be in charge of monitoring • Could display data on TV	• Some pupils in class very involved • Concerns over children touching device • Only used main page of app • Did not reflect at end of day • Found out how air-conditioning does not improve IAQ • More aware of IAQ—encouraged "brain breaks" outside to refresh • Concerns over poor IAQ creating negative publicity of the school
P3	• Relate IAQ to guidelines	• Investigate effects of outdoor air on IAQ • Interested to investigate use of air-conditioning on IAQ • Remove allocating an IAQ with a mood • Levels were often poor—could not improve it • Trial to improve IAQ through other methods (e.g. switching off projector)
P4	• Easy to integrate into learning • Monitor changes over the school day	• Would like to see data over longer timespan • Caused classroom to get cold by trying to ventilate more • Would like more education on the side-effects • A fixed unit would be preferred instead of iPad • Would use long-term—lots of teaching possible through using the device
P5	• Monitor changes throughout school day • Could integrate into learning	N/A: Participant did not take part in deployment
P6	• Incorporate into teaching of child independence	N/A: Participant did not take part in deployment

make IAQ more relatable and the colors allowed the device to be understood from across the classroom.

P3 also noted that Airlert negatively affected the class behavior, where the children claimed at one point that they were feeling tired when the color turned orange: *"Because the system says that's what orange means. It means you're tired!"* (P3). In this instance, the teacher felt the IAQ had only just reached the threshold for that color and felt the children were using the device as an excuse for poor behavior, rather than actually being affected by the IAQ. It was noted that a better option may be to avoid assigning a mood or feeling with a certain pollutant concentration so not to bias students as to what they *"should"* be feeling like at given concentrations of pollutants. P3 also found that sometimes the children became frustrated when the windows were open, but the readings were still red. P1 used the app to reflect on the data throughout the trial. They analyzed the graph with the children, holding various short group discussions about what could have caused peaks or troughs in the graph. P2 however chose not to reflect on the graph and was mostly interested in real-time data.

5.1.2 Increase in air quality awareness

P1, P2, and P4, through trialing this device, reported becoming more aware of indoor air quality and mentioned they would continue to consider it in the future. Participants usually responded to Airlert readings through opening the external windows, increasing ventilation of outside air into the classroom. P1 felt in control of their classroom environment with the device and found it *"really amazing"* how quickly pollutants can leave the classroom when providing more ventilation. P2 stated that they were surprised by how quickly air quality deteriorates and so they will be *"a lot more conscious"* of IAQ in the future. P3 felt that when the air quality was bad there was *"no going back,"* especially when it was a cold day, and the external door had to be closed for warmth. P3 felt that using Airlert did not make them more aware of IAQ as they knew their classroom environment was bad before the study, and the device simply confirmed this. Nevertheless, they were unsure what a "healthy" level of IAQ was and the device allowed them to understand this. P3 also mentioned that VOC sources were difficult to understand by the children.

These findings suggest that IAQ offers great potential for healthier classrooms if the occupants are able to affect IAQ themselves; however, the deployment may be counterproductive if the occupants find physical characteristics of the building limit the effect of any actions taken. Design implications for information dissemination are revisited in the discussion.

5.1.3 Changes to ventilation behavior

Following the deployment, P1 and P2 further associated stuffiness with a build-up of CO_2 and reduced oxygen. By providing more ventilation, they felt they were providing more fresh air to *"wake them [the children] up."* P2 had air conditioning in their classroom and found it interesting to see how using this led to a build-up of CO_2 in the room, as they found it recirculated the indoor air rather than providing fresh air. P2 noted that *"a barrier was taken away"* as they used the fresh air positively to encourage the children, and themselves as a teacher, to get on with work. They also took ten-minute *"brain breaks"* when the device was showing very poor air quality. This was a quick break outside which reportedly made the students refreshed and ready to do their work, highlighting the potential for IAQ feedback to instigate active breaks. These breaks *"worked really well for the students and the teachers."*

P2 realized that the air quality within the classroom had a *"bigger impact and a bigger role to play on concentration and in the ability to concentrate for a longer period of time, especially for young children when it's hard for them to concentrate at the best of times. [Air quality] could be a factor that could be helped and help [the children]."* P2 mentioned they would now use the air conditioning less and open the windows more, knowing the air conditioner does not introduce fresh air. P4 additionally claimed that their teaching assistants would ventilate the room more often than previously; that Airlert has *"massively pointed out [to the teaching assistants] the implications of not having the windows open."*

5.1.4 Misplaced analogies

Teachers described how Airlert had changed their awareness of IAQ during the exit interviews and caused them and the children to reflect on the determinants of IAQ. Yet it also became clear that certain assumptions over the factors affecting IAQ in the classroom were misplaced, for example equating changes in CO_2 to factors, which could not conceivably affect CO_2. P3 mentioned the children in her classroom equated rising CO_2 concentrations to: (1) increased talking: *"they did start to notice that the less they talked, the less carbon dioxide, so the less they are expelling out their mouths."*, and (2) the nearby airport: *"…and then the kids go: 'Oh, that's because that plane's just taken off.'"*. P1 mentioned children in her class had equated spikes in CO_2 to classmates passing gas. In reality, these factors would have little or no effect on CO_2 in the classrooms.

These findings highlight the great potential pedagogical value of IAQ sensors in schools, but equally, that teaching packages require careful consideration in order to prevent the misdiagnosis of IAQ issues and potential spread of misinformation.

5.1.5 Long-term use

The Airlert deployment was not of sufficient length to report quantitatively on engagement over time. However, reports from participants indicated a willingness to engage with such a device longer term. P1 was receptive to having Airlert continuously and found it a *"really positive"* experience, which provided *"good discussion points"* with the children. P2 and P4 would *"definitely"* use the product in the long term and could incorporate IAQ monitoring into everyday life.

5.1.6 Considerations on Airlert as a tool for teaching

P1 felt that the children can benefit from small-scale learning from this device in the long term, and P2 expressed that it *"would link to so many topics in the national curriculum."* The importance of maintaining good indoor air quality is *"not taught at home and [it] doesn't come up in curriculum so [it has been] quite an eye opener."* This could be slowly taught throughout Year R (Reception) to Year 6, with the expectation that *"by the time they leave school, they have got a really good idea of air quality and can relate that to the wider world."*

P3 taught in a school near the airport in a city known for its poor outdoor air quality and expressed interest in learning more about the effects the outdoor air quality on IAQ. They mentioned this would be a good learning opportunity for the children, investigating whether opening windows is actually beneficial to health if the outdoor air quality is poor.

P2 thought about the acceptance of Airlert from their school's senior management or higher authorities: *"it would point out the problems in the air in the schools, parents would hear about it [poor classroom air quality] from their children."*, therefore the device could *"generate more problems,"* leading to a negative impact on the school.

6 Discussion

Previous literature identifies that CO_2 feedback in classrooms can positively affect ventilation behavior toward healthier classrooms [27,28,34–36]. Findings from the Airlert deployments support this research. By employing a specifically user-centered design process and gathering teachers' reflections on the device, this research additionally highlights the potential values of networked IAQ sensing beyond basic behavioral modification toward opening windows. The Airlert device caused teachers to: (1) suggest active breaks for the purpose of refreshing the children, (2) identify factors affecting IAQ, e.g., realizing the air conditioning unit recirculates air rather than exchanges indoor air with outdoor air, and (3) acted as a tool for reflection upon IAQ and opportunistic learning opportunities.

Additionally, the specific focus on end users has raised a number of further salient considerations for the broader context of IAQ monitoring devices, which need to be considered in design. For example: (1) without guidance, IAQ devices may also foster frustration if occupants are unable to positively affect sensors readings themselves, (2) older children may "game" the system by using IAQ as an excuse for poor behavior, and (3) IAQ monitoring may face challenges to adoption by schools, if authorities believe the data may adversely affect their reputation. Below we outline some considerations for the design of IAQ feedback based on the findings from our deployments.

6.1 Design considerations

6.1.1 Human analogies for IAQ

In the preinstall interviews, all participants were keen on having a mascot embedded in the display, whether an animal or human, which responded to the IAQ readings by becoming progressively sleepier as the air quality deteriorated. Similar human analogies have been used in previous classroom displays for IAQ, which recorded positive results on ventilation [27,35].

Yet the Airlert deployments highlighted the potential for older students to take advantage of the system, highlighting the need to consider the age-appropriateness of IAQ feedback and how the value of mascots or human analogies may depend upon age. In future iterations of Airlert to be deployed in classrooms of children aged older than 9–10 years, it may be more appropriate to limit IAQ feedback to the teachers only. If visible to the whole class, IAQ feedback devices in classrooms with older children, might avoid prescribing what a pupil or teacher "*should*" be feeling to allow students to reflect upon this individually.

6.1.2 Information dissemination

The key benefit of Airlert is to inform occupants of their IAQ and encourage healthier ventilation behavior. Yet the value of situated informatics is contingent upon the ability of occupants to meaningfully change the devices readings through their actions [46]. The experience of P3 shows that providing IAQ feedback to occupants who are unable to effectively control their IAQ due to poor building design or other physical factors can frustrate users.

One suggestion for future deployments is for building managers to deploy such devices as sensors only (without the iPad display activated) for 2 weeks initially to determine the capacity of human actions to improve IAQ. Then, deploy the iPad visualizations only to those classrooms in which change can be affected. Classrooms where IAQ remains persistently poor irrespective of human actions signify the need for an engineering intervention, rather than a social intervention such as Airlert. Instead of IAQ feedback in these circumstances, building managers may instead seek to rectify the problem using engineering solutions first, before deploying IAQ feedback once users become better able to affect the IAQ themselves.

In this way, the value of IAQ feedback could be extended in future work as a means of diagnosing IAQ problems and enabling occupants and building managers alike, to characterize issues either as: (1) structural issues requiring an engineering solution, or (2) social issues, where behavior change is sufficient to produce better IAQ outcomes.

6.1.3 Fostering long-term use

Our findings support previous literature that CO_2 feedback in classrooms can inform healthier ventilation habits [27,28,34–36]. Yet the persistence of the behavior change long term is a knowledge gap [34]. The five-day deployment of Airlert is insufficient to speculate on its potential in the long term. Yet participants noted the time and attention-constrained nature of classroom teaching and the need for any device to be unobtrusive in order for continued persistence in the long term: *"…if it's not quick and easy, no one will actually continue [using] it"* (P5).

Our findings highlight that: (1) long-term use is contingent on user acceptance of the device, and not becoming distracted or annoyed by it, and (2) the potential value of minimalism and glanceability. IAQ feedback should be continually visible, yet interactions should be simple and not time consuming. Participants appreciated the color coding of the display, affording the possibility of determining the IAQ from anywhere in the classroom without the need to read the numbers on the system. We suggest that while data interrogation functionalities such as the ability to review averages and historic graphs are important for the use of IAQ feedback in learning activities, the dashboard itself should be kept simple: positioned in a place to alert users to poor IAQ, while unobtrusive enough that teachers and students do not find a reason to switch off or remove the device.

6.2 Data management considerations

This work is focused upon the user-centered design process to develop a low-cost and contextually appropriate system that fits within the routines, practices, and organized chaos of the classroom and meets the needs of users in this space. We would be remiss in this work if we did not also review the data management considerations for such a system.

6.2.1 Data quality

Understanding the quality of the gathered data helps understand and reuse the data appropriately. Data quality is considered according to the dimensions of: *completeness, timeliness, accuracy,* and *interpretability* [49]. For a situated device providing immediate IAQ feedback to the users, teachers, and students, *timeliness, accuracy,* and *interpretability* were central concerns in design.

By designing Airlert for glanceability and providing alerts only when IAQ deteriorates, we maximized the *timeliness* of the data. The choice of the color coding (green to red) and limiting the number of pollutants measured to CO_2 and VOCs were decisions taken to maximize the *interpretability* of the information. It should be noted that while Airlert may have limitations in terms of data quality, such as *precision* or *accuracy*, relative to far more expensive precision sensors, this does not invalidate the device for use within the classroom as it was designed. CO_2 is nontoxic at realistic indoor concentrations (up to 5000 ppm) [50], and the purpose of the device is simply to indicate when IAQ is deteriorating as a reminder to the occupants to ventilate the room. Future work may focus more specifically on acceptable sensor accuracies for IAQ monitoring in non-safety critical applications.

6.2.2 Information disclosure

While no personal information is captured by classroom air monitoring, the devices themselves impart some information about a group of people. Every classroom has a set of children and staff associated with it. Information gathered by an IAQ feedback/monitoring device can ultimately be linked to these individuals as a collective. Our research suggests IAQ monitoring in classrooms may face challenges of acceptance from school authorities, e.g., concerns over the IAQ data causing parent concerns or negative press. We suggest future researchers consider the appropriate extent of the dissemination of IAQ data in classrooms and these types of social factors as potential barriers to adoption of IAQ monitoring technology in classrooms.

7 Conclusion

Airlert was deployed to understand the potential roles of IAQ feedback devices in classrooms, and the potential for affordable sensor technology to bring about healthier indoor environments in classrooms. While previous work measures behavior changes from IAQ feedback quantitatively through measured changes to air quality [27,28,34–36], this research has taken a specifically human-centered focus, designing a device for and with teachers, evaluating the display with teachers and children in a number of classroom deployments.

The research concludes that awareness of indoor air quality and healthy ventilation behaviors can be increased by providing classroom teachers with a device to monitor classroom air quality. But also, that careful consideration is required into the design of such devices to ensure the timeliness, interpretability of information and to overcome limitations of accuracy in low-cost monitoring devices. For IAQ feedback devices, clear color coding for IAQ levels and permanent, yet unobtrusive displays enable information at a glance for time- and attention-constrained teachers. In future work in this space, careful consideration is required around: (1) the use of human analogies, and how appropriate solutions are likely to differ with age, (2) ensuring all information is actionable, and that occupants can meaningfully affect IAQ through ventilation actions, (3) how IAQ monitoring could be leveraged into science curriculums such as learning about outdoor air quality and the effect on indoor air quality, and (4) broader social considerations in deployment, e.g., information dissemination and the potential for IAQ information to be used for the profiling of schools, or for purposes beyond simply instigating healthier ventilation actions.

References

[1] M. Santamouris, et al., Experimental investigation of the air flow and indoor carbon dioxide concentration in classrooms with intermittent natural ventilation, Energy Build. 40 (10) (2008) 1833–1843.

[2] D.G. Shendell, R. Prill, W.J. Fisk, M.G. Apte, D. Blake, D. Faulkner, Associations between classroom CO_2 concentrations and student attendance in Washington and Idaho, Indoor Air 14 (5) (2004) 333–341.

[3] M. Simoni, et al., School air quality related to dry cough, rhinitis and nasal patency in children, Eur. Respir. J. 35 (4) (2010) 742–749.

[4] J. Toftum, B.U. Kjeldsen, P. Wargocki, H.R. Menå, E.M.N. Hansen, G. Clausen, Association between classroom ventilation mode and learning outcome in Danish schools, Build. Environ. 92 (2015) 494–503.

[5] S.R. Tortolero, et al., Environmental allergens and irritants in schools: a focus on asthma, J. Sch. Health 72 (1) (2002) 33–38.

[6] L. Bourikas, et al., Camera-based window-opening estimation in a naturally ventilated office, Build. Res. Inf. 46 (2) (2018) 148–163.

[7] W.J. Fisk, The ventilation problem in schools: literature review, Indoor Air 27 (6) (2017) 1039–1051.

[8] GOV.UK, Conservation of fuel and power: Approved Document L, [Online]. Available: https://www.gov.uk/government/publications/conservation-of-fuel-and-power-approved-document-l, 2018. Accessed 2 July 2019.

[9] T. Vehviläinen, et al., High indoor CO_2 concentrations in an office environment increases the transcutaneous CO_2 level and sleepiness during cognitive work, J. Occup. Environ. Hyg. 13 (1) (2016) 19–29.

[10] Y.H. Mi, D. Norbäck, J. Tao, Y.L. Mi, M. Ferm, Current asthma and respiratory symptoms among pupils in Shanghai, China: influence of building ventilation, nitrogen dioxide, ozone, and formaldehyde in classrooms, Indoor Air 16 (6) (2006) 454–464.

[11] J.M. Daisey, W.J. Angell, M.G. Apte, Indoor air quality, ventilation and health symptoms in schools: an analysis of existing information, Indoor Air 13 (1) (2003) 53–64.

[12] O.A. Seppänen, Association of ventilation rates and CO_2 concentrations with health and other responses in commercial and institutional buildings, Indoor Air 9 (4) (1999) 226–252.

[13] J. Sundell, et al., Ventilation rates and health: multidisciplinary review of the scientific literature, Indoor Air 21 (3) (2011) 191–204.

[14] M.J. Mendell, G.A. Heath, Do indoor pollutants and thermal conditions in schools influence student performance? A critical review of the literature, Indoor Air 15 (1) (2005) 27–52.

[15] A.N. Myhrvold, E. Olsen, O. Lauridsen, Indoor environment in schools–pupils health and performance in regard to CO_2 concentrations, in: Proceedings of the 7th International Conference on Indoor Air Quality and Climate, 1996, pp. 21–26.

[16] D.P. Wyon, The effects of indoor air quality on performance and productivity, Indoor Air 14 (Suppl 7) (2004) 92–101.

[17] A. Cincinelli, T. Martellini, Indoor air quality and health, Atmos. Environ. 14 (11) (2017) 1286.

[18] U. Haverinen-Shaughnessy, D.J. Moschandreas, R.J. Shaughnessy, Association between substandard classroom ventilation rates and students' academic achievement, Indoor Air 21 (2) (2011) 121–131.

[19] J. Madureira, et al., Indoor air quality in schools and its relationship with children's respiratory symptoms, Atmos. Environ. 118 (2015) 145–156.

[20] H. Maula, V. Hongisto, V. Naatula, A. Haapakangas, H. Koskela, The effect of low ventilation rate with elevated bioeffluent concentration on work performance, perceived indoor air quality, and health symptoms, Indoor Air 27 (6) (2017) 1141–1153.

[21] R. Maddalena, et al., Effects of ventilation rate per person and per floor area on perceived air quality, sick building syndrome symptoms, and decision-making, Indoor Air 25 (4) (2015) 362–370.

[22] GOV.UK, BB 101: Ventilation, thermal comfort and indoor air quality 2018, [Online]. Available: https://www.gov.uk/government/publications/building-bulletin-101-ventilation-for-school-buildings, 2018. Accessed 2 July 2019.

[23] D. Etheridge, A perspective on fifty years of natural ventilation research, Build. Environ. 91 (2015) 51–60.

[24] S. Cui, M. Cohen, P. Stabat, D. Marchio, CO_2 tracer gas concentration decay method for measuring air change rate, Build. Environ. 84 (2015) 162–169.

[25] Z.L.X. Zhang, P. Wargocki, Literature survey on the effects of pure carbon dioxide on health, comfort and performance, in: Indoor Air 2014 – 13th International Conference on Indoor Air Quality and Climat, 2014, pp. 1009–1011.

[26] S. Snow, M. Oakley, M.C. Schraefel, Performance by design: supporting decisions around indoor air quality in offices, in: Proceedings of the 2019 on Designing Interactive Systems Conference (DIS'19), 2019, pp. 99–111.

[27] P. Wargocki, N.A.F. Da Silva, Use of visual CO_2 feedback as a retrofit solution for improving classroom air quality, Indoor Air 25 (1) (2015) 105–114.

[28] L.M.J. Geelen, et al., Comparing the effectiveness of interventions to improve ventilation behavior in primary schools, Indoor Air (2008) 416–424.

[29] IWBI, The WELL Building Standard v1 with January 2017 Addenda, (2017).

[30] Finnish Ministry of the Environment, Indoor Climate and Ventilation of Buildings Regulations and Guidelines 2003, [Online]. Available: https://www.edilex.fi/data/rakentamismaaraykset/d2e.pdf, 2003.

[31] American Society of Heating Refrigerating and Air-Conditioning Engineers, ASHRAE Standard 62.1-2016 Ventilation for Acceptable Indoor Air Quality, (2016).

[32] D.P. Wyon, P. Wargocki, Effects of indoor environment on performance, ASHRAE J. (2013) 46–50.

[33] Z. Bakó-Biró, D.J. Clements-Croome, N. Kochhar, H.B. Awbi, M.J. Williams, Ventilation rates in schools and pupils' performance, Build. Environ. 48 (1) (2012) 215–223.

[34] A. Heebøll, P. Wargocki, J. Toftum, Window and door opening behavior, carbon dioxide concentration, temperature, and energy use during the heating season in classrooms with different ventilation retrofits—ASHRAE RP1624, Sci. Technol. Built Environ. 24 (6) (2018) 626–637.

[35] P. Rigger, F. Wortmann, A. Dahlinger, Design science in practice: Design and evaluation of an art based information system to improve indoor air quality at schools, in: Lecture Notes in Computer Science (including subseries Lecture Notes in Artificial Intelligence and Lecture Notes in Bioinformatics), 2015, pp. 71–86.

[36] J. Toftum, M.M. Wohlgemuth, U.S. Christensen, G. Bekö, G. Clausen, Managed airing behaviour and the effect on pupil perceptions and indoor climate in classrooms, in: Proceedings of Indoor Air 2016, 2016.

[37] M. Brereton, J. Buur, New challenges for design participation in the era of ubiquitous computing, CoDesign 4 (2) (2008) 101–113.

[38] International Organization for Standardization, ISO 9241-210:2010, (2010).

[39] C. Spinuzzi, The methodology of participatory design, Tech. Commun. 522 (2) (2005) 163–174.

[40] A. Galletta, Mastering the Semi-Structured Interview and Beyond: From Research Design to Analysis and Publication, New York Univerity Press, 2013.

[41] J. Vines, Playing with Provocations, Springer, Cham, 2018.

[42] V. Braun, V. Clarke, Using thematic analysis in psychology, Qual. Res. Psychol. 3 (2) (2006) 77–101.

[43] M.I. Hwang, R.G. Thorn, The effect of user engagement on system success: a meta-analytical integration of research findings, Inf. Manag. 35 (4) (1999) 229–236.

[44] uHoo Limited, uHoo, [Online]. Available: https://uhooair.com/. Accessed 3 July 2019.

[45] Netatmo, Netatmo, [Online]. Available: https://www.netatmo.com/en-eu/weather/weatherstation/. Accessed 3 July 2019.

[46] A. Mathur, M. Van den Broeck, G. Vanderhulst, A. Mashhadi, F. Kawsar, Tiny habits in the giant enterprise: understanding the dynamics of a quantified workplace, in: Proceedings of the 2015 ACM International Joint Conference on Pervasive and Ubiquitous Computing, 2015, pp. 577–588.

[47] HighYa LLC, uHoo Air Sensor Reviews, [Online]. Available: https://www.highya.com/uhoo-air-sensor-reviews/. Accessed 3 July 2019.

[48] J.R. Knight, Review of Funology: From Usability to Enjoyment, Kluwer Academic Publishers, 2003.

[49] R.Y. Wang, V.C. Storey, C.P. Firth, A framework for analysis of data quality research, IEEE Trans. Knowl. Data Eng. 7 (4) (1995) 623–640.

[50] Health and Safety Executive, General Hazards of Carbon Dioxide, [Online]. Available: http://www.hse.gov.uk/carboncapture/carbondioxide.htm. Accessed 3 July 2019.

12

Data perspective on environmental mobile crowd sensing

Mariem Brahem[a], Hafsa E.L. Hafyani[a], Souheir Mehanna[a],
Karine Zeitouni[a], Laurent Yeh[a], Yehia Taher[a], Zoubida Kedad[a],
Ahmad Ktaish[a], Mohamed Chachoua[b], and Cyril Ray[c]

[a]DAVID LAB, UNIVERSITY OF VERSAILLES SAINT-QUENTIN-EN-YVELINES, PARIS-SACLAY
UNIVERSITY, VERSAILLES, FRANCE [b]LASTIG, UNIVERSITY GUSTAVE EIFFEL, EIVP, PARIS,
FRANCE [c]THE NAVAL SCHOOL, IRENAV, BREST, FRANCE

1 Introduction

Environmental pollution has been a critical concern for many decades. Pollution engenders serious health problems and causes important economic losses [1]. In some cases, it poses enormous threat to human health and claims lives. The World Health Organization (WHO) estimates that about a quarter of the disease facing mankind today occur due to prolonged exposure to environment pollution (see Ref. [2]). Although some pollutants have fallen sharply in the last two decades, yet studies such as that by Bentayeb et al. [3] show that air pollution remain a worrying problem which reduces life expectancy by several months. Indeed, an investigation conducted by Amos [4] shows that more than 5.5 million people worldwide are dying prematurely every year as a consequence of air pollution.

The risks depend on the context and the duration of exposure. The current study mainly distinguishes between indoor and outdoor spaces. The indoor air pollution has become a matter of a grave concern since it affect human health severely, in some cases, cause life threatening diseases. According to WHO, over 4 million people die prematurely from illness attributable to the household air pollution from cooking with solid fuels [5]. As for the outdoor air pollution which heavily affects human health and leads to severe consequence, it is estimated that in 2012, around 72% of outdoor air pollution-related premature deaths were due to ischemic heart disease and strokes, while 14% of deaths were due to chronic obstructive pulmonary disease or acute lower respiratory infections, and 14% of deaths were due to lung cancer [5]. The detrimental effects of different types of pollution have given rise to a significant question *how to measure our exposure to air pollution*?

This is the first step toward the deep comprehension of the individual health effects, and further risk mitigation. This may lead to personalizing the recommendation to minimize the exposure, and change the behavior along with each individual's daily activity.

Intelligent Environmental Data Monitoring for Pollution Management. https://doi.org/10.1016/B978-0-12-819671-7.00012-9
269

The advent of the new generation of low-cost lightweight and connected sensors made a paradigm shift in environmental studies. In particular, nomadic sensors allow for a very precise personalized measurement, by continuously quantifying the individual exposure to air pollution components. Moreover, a broad dissemination among volunteers of these devices, or their deployment on vehicle fleets, is becoming a credible scenario. Either in mobile setting or in multiple fixed locations, there is a great interest in such sensor deployment to densify the air quality monitoring network, which is today restricted to sparse nodes. A plethora of research activities have been conducted over the years which aims at preventing pollution, including creating awareness among people, reducing industrial waste, etc. Nowadays, Internet of things (IoT), in particular, sensors have become critically important in pollution monitoring (see Ref. [6]). Sensors enable easier and faster collection of pollution data and cover both indoor and outdoor environments. They can be devised into *mobile sensors* and *fixed sensors*. The mobile sensors are mostly mounted on transportation mediums to collect data, because most of today's sensors are still large or heavy. However, more compact models are continuously introduced, which opens the way to carry them by humans to measure their own exposure. For such hand-held sensors, data collection process may either task the users to observe some zones or actively post them, which is called *participatory sensing*; however, in several occasions, the data collection is entirely automatic—this is known as *opportunistic sensing* [7].

In addition to sensor technology, various smart systems have been developed within the scope of different projects including personalized alert systems, analytics, and visualization tools. In concert with sensor technology, these smart systems are leading a major change in pollution monitoring and paving the foundation of high-end next-generation city-wise air quality observatory targeting both citizens and decision makers. This chapter presents a survey of mobile crowd sensing (MCS) projects related to air quality monitoring, and based on the emerging low-cost mobile sensors.

The objectives and the scope of this chapter are as follows.

- *Objectives*: The core objective of our study is manifold: carry out a deeper investigation of different solutions that have been proposed or developed in a large bodies of literature; study the solutions taking different aspects into account including architecture of the proposed solution, functional components such as data management, processing, analysis, etc.; compare them over their functional capabilities to identify the strength and weaknesses of each; and prescribe a future solution that might address the shortcomings of the state of the art.
- *Scope*: Pollution monitoring is an umbrella notion including a broad spectrum of approaches. There are different types of pollution such as air pollution, sound pollution, water pollution, and so forth. The focus of our study is *air pollution*. Additionally, we confined our investigation within mobile sensor-driven analysis, meaning that, we scrutinize the solutions that involves mobile sensors for collecting data and use systems that perform exposure analysis of both indoor and outdoor pollution. It is worth noting that the fixed station sensor-based solutions is outside of

the scope of our study unless otherwise a solution involves mobile sensors as well, or citizen sensing. We mainly focus on the architectural and data processing, and mining aspects of the proposed solutions.

After providing an overview of the projects related to environmental MCS, we provide the results of our comparative analysis among different solutions and point out the remaining challenges to be addressed. We end by presenting our proposal the current collaborative project Polluscope [8]. The remainder of this chapter is organized as follows. Section 2 introduces the core terms related to our study, followed by the identification of the main challenges from the data perspective in Section 3. The survey is presented in Section 4. Section 5 introduces potential solutions inspired from our Polluscope approach. We conclude and discuss future directions of research in this area.

2 Taxonomy

Different terms are commonly used in the literature related to environmental sensing. This section, in what follows, introduces some of the core concepts that are used throughout this survey.

2.1 Sensors

American National Standards Institute defines sensor as a physical device or equipment which acquires a physical quantity and converts it into a signal suitable for processing (e.g., optical, electrical, mechanical). In recent years, the use of sensors has increased dramatically and spanned within and across various areas especially with the advent of the notion of IoT. Environment pollution monitoring systems today leverage the power of sensors to collect data of various pollutants which can seamlessly be integrated and fed into processing and analysis engine. It is worth noting that various utility softwares are embedded within sensors today. These enabled them to perform light-weight computation over collected data. Also, they can store a small amount of data. Within the context of pollution monitoring, sensors have been classified into *fixed station sensors* and *mobile sensor*. We briefly explained this classification in the previous section. There are three different types of sensors [9].

- *Physical sensors*: Physical sensors are sensors integrated in, or attached to, mobile devices such as smart phones, tablets, GPS, microphone, camera, ambient light sensor, accelerometer, gyroscope, compass, proximity sensor, and are also used to measure the temperature and humidity. It is worth noting that many of these sensors are available on advanced smart phones.
- *Virtual sensors*: Unlike physical sensors, virtual sensors are logical components; to be more specific, they are software applications that run at user devices and collect information about users, their profile and preferences, detecting their context and situation. These sensors detect information related to user communications (voice,

SMS, etc.), user activities and interaction with devices (active applications, application in focus, the type of the interaction, etc.), user preferences and profile, user-generated content (texts, speech, videos, photos, sounds), etc. Virtually sensed information is referenced in space and time and attached to a certain location, symbolic or geographic.

- *Social sensors*: Social sensors detect user social status and activities, social network, and social media interactions (tags, likes, Tweets, photos, etc.), currently connected friends and their status, connections in vicinity, etc. Some of such information can be detected by accessing social network/media services through appropriate APIs. However, social sensors are beyond the scope of this chapter.

2.2 Mobile crowd sensing

MCS is an emerging sensing model which primarily depends on the strength of the people's sensor-enabled mobile devices to sense the data for a particular sensing task [10]. MCS paradigms have become the significant source of volunteered geographic information and crowd-sourced geo-information owing to the large number of mobile devices carried by people worldwide to support their daily activities [9]. The term "mobile crowd sensing" was coined by Ganti et al. [7] refer to a broad range of community sensing mainly participatory sensing and opportunistic sensing (which were briefly explained in the previous section).

MCS empowers a large amount of mobile phones to be utilized for trading data among their clients, as well as for activities which might have an enormous societal impact. MCS permits a large amount of mobile phone clients to share native knowledge (e.g., local information, ambient context, noise level, and traffic conditions) collected by their sensor-enhanced devices (see Ref. [11]), and more information can be collected in the cloud for large-scale sensing and community intelligence mining.

2.3 Microenvironment

According to Oxford dictionary, microenvironment refers to the immediate small-scale environment of an organism or a part of an organism, especially as a distinct part of a larger environment. The notion of microenvironment is used lately, within the context of air pollution to compute personal exposure level to different types of pollutants. The notion of microenvironment has become critical as of, because people can be exposed to pollutants. Different types of microenvironment have been reported by Colbeck and Nasir [12]. They classified microenvironments into *nonindustrial buildings* (e.g., homes, schools, offices, and transport) and *industrial buildings* (e.g., the buildings of manufacturing plants or factories). The critical aspects of these different types of environment are that they have different characteristics and therefore, the concentration of particles is highly variable and environment-specific [12].

2.4 Outdoor air pollution

Outdoor air pollution commonly referred to as *ambient air pollution* caused by combustion processes from motor vehicles, solid fuel burning, and industry in open space (unlike the closed compound such as apartment or office). The most common types of pollutants of ambient air include particulate matter (PM10 and PM2.5), ozone (O_3), nitrogen dioxide (NO_2), carbon monoxide (CO), and sulfur dioxide (SO_2).

2.5 Indoor air pollution

Very often people are prejudiced that among all pollutants indoor ones are much safer since pollution inside the building is lower than that outside. However, in reality, concentration of pollutants within a confined area in particular, *house*, may rise to unacceptable levels—in which case it is treated as *indoor air pollution* also known as *household air pollution*. Tens of sources are available from where a wide variety of pollutants emit including inorganic chemical substances (*radon*), volatile organic compound (e.g., paints, aerosol, sprays, cleaning supplies), and biological pollutants (e.g., house dust).

Although, the notion of outdoor air pollution has been segregated from the indoor, yet, interestingly, outdoor air pollution can be an important contributor to the indoor air quality, especially in highly ventilated homes, or in homes near pollution sources. Similarly, indoor air pollution sources may also be important causes of outdoor air pollution, especially in cities where many homes use biomass fuels or coal for heating and cooking.

3 Challenges of MCS

The main challenges of MCS applications will help the community to better understand and explore key research problems related to MCS in order to enable new research opportunities. The core challenges that are dealt with include the following (i) heterogeneity and variety of sensor equipments, measurements, and data analysis, (ii) the end-to-end data quality, (iii) supporting and exploiting mobility of sensors as well as context awareness, or even context inference, and (iv) involving the community in a trusted, fair, and transparent manner into the monitoring activity.

The strength of MCS relies on the usage of different types of sensors designed by different manufacturers that may vary in their sensitivity, sampling frequency, and noise immunity. Data collected from all sensing object should be merged, which could lead to measurements at irregular time intervals and missing data problems. We could observe time stamps that are closely spaced or too sparse in different cases. In fact, some sensors may be offline for hours or stay idle when the device is static (some sensors use the accelerometer to control the sampling rate), they can switch to a burst mode in some situations (increasing the sample frequency more than the normal rate) or stop transmitting the data if the variation is less than a predefined threshold, we could also get different sensor position resolutions. As the measurements of the different sensors are not synchronized by a common clock, to compare these measurements, it is necessary to perform a gap fill process that can be complex. In addition, the distribution of data may be skewed depending on the geo-location of moving sensors. For example, in some regions, participants may be not available so the collected data can contain some blank regions. Thus, data recovery techniques to infer the missing data need be used. Furthermore, advanced data management methodologies that can organize multidimensional data (e.g., streaming, geospatial) are necessary.

Besides, one of the most fundamental characteristics of mobile sensor data is the diversity of their granularity, both under temporal and spatial dimensions. The temporal domain is typically represented at different time granularities. The spatial entity can be represented using a hierarchical representation that describes the subdivision of the spatial domain into different regions or cells. Combining multiple datasets with several granularities or changing the granularity of a dataset are important analysis tasks. From a modeling point of view, a distinctive aspect of such sensor data is the spatial autocorrelation, meaning that data captured from close sensors tend to be more similar than data captured from distant sensors. The same holds for consecutive observations on the same device. As a result, collected data from moving objects cannot be modeled as independent data, and specific algorithms taking into account the correlation between observations occurring in different locations, or between different periods of time need to be considered. Huge amounts of data are being collected continuously from ubiquitous sensor-enhanced mobile devices (as many as the number of equipped holders) in different geographical areas. This requires leveraging big data processing techniques to achieve in-depth understanding, and provide useful information.

Ensuring a good data quality all along the data journey of air quality monitoring systems from acquisition to integration and analysis is an indispensable part of the process of MCS. Faulty data and data glitches will produce spurious predictions and indicators about human exposure. The data acquisition process may introduce many glitches to the data because sensors are very prone to critical failures, measurement errors, routine downtime due to maintenance or technical malfunctions (e.g., calibration and temporary power outage) which will result in missing data. Moreover, the data integration process, generally described as being composed of data parsing, cleansing, consolidation, enrichment, transformation, and storage, may also produce erroneous data. For example, useful data may be lost during parsing and cleansing. Furthermore, the storage of spatiotemporal series data, which is not a trivial task, might also deteriorate data quality. The consolidation process should also be quality-aware in the sense of exploiting every useful measured aspect to end up with a rich, comprehensive model that fosters the predictions. Other quality challenges also arise when considering the data analysis processes fed by the integrated data. Indeed, the quality of the resulting predictions and indicators is strongly dependent on the quality of the input data.

Many authors have studied different aspects of data quality for sensor data and sensors performance assessment. Fishbain et al. [13] study sensor's performance and proposed a toolkit to measure its performance. Others [14–17] focused on missing data inference and prediction of future or unobserved measurements. Approaches like those proposed by Rodríguez and Servigne [18], Hsieh et al. [16], Rahman et al. [19], and Dasu et al. [20] studied on the assessment of the quality of the data by defining the factors that affect the quality and by proposing methods for the assessment. Other approaches like Tan et al. [14], Wang et al. [21], and Cheng et al. [22] studied on data quality improvement and enhancement of data cleaning strategies. However, studies in MCS fall short of tracing the sources deteriorating the quality of the data throughout the data journey for this kind of systems. Propagating data quality across the journey of the data all the way from acquisition at sensors level to data integration till the prediction phase is crucial for detecting the points at which the data is losing part of its quality. We are interested in computing quality of indicators and predictions given, for example, a certain quality of the sensors data as we are also interested in identifying the proper constraints in order to obtain a required level of data quality. Also, and especially in the context of air quality monitoring systems, data measurements are affected by many factors that could enrich the data and be further studied in correlation with the variations of the measurements for improving inference of missing or unobserved values and prediction of future records, for example, forecast and traffic data.

The main challenge discussed by Ganti et al. [7] is local analytics. It involves data mediation such as filtering of outliers, elimination of noise, and filling in data gaps. Methods to fill missing readings in sensor data necessitate considering the temporal correlation between observations at different time stamps as well as the spatial correlation between geo time series data. Another function of data analytics is context inference to detect transportation mode (car, bus, train, on foot) or kinetic mode (walking, standing, jogging, running).

Furthermore, MCS applications potentially collect sensitive and vulnerable sensor data, this raises significant concerns about security and privacy preservation. For example, location data can be used to reveal private information about the participants, such as their home and work locations. Besides participants' locations, other private information can be obtained such as individual's health condition. Deeper knowledge about the habits of users participating in MCS applications can be inferred. The success of MCS systems requires effective methods that guarantee the privacy of their users.

The challenges discussed previously call for advanced data management and analytics models that can extract knowledge from multiple heterogeneous data generated from different types of sensors. Different goals were defined to address these challenges and to perform the pollution monitoring task efficiently.

4 Related work/projects

In the following sections, we survey related projects relying on IoT, in particular, MCS technology, and distinguish between basic experiments and more advanced "research-oriented" projects.

4.1 OpenSense I and II

The OpenSense is a research project whose major scientific goal is, as stated by Riahi et al. [23], to efficiently and effectively monitor air pollution using wireless and mobile sensors by adopting complex utility-driven approaches toward sensing and data management. The OpenSense Project system involves different sensors which are packed in a single box called *sensorbox* dedicated to collect data at both mobile and fixed stations [24].

In OpenSense, data are collected over "microwindows" of time [25] and are transmitted to a data server running global sensor network (GSN) [26]. GSN is a platform which enables building a scalable infrastructure for integrating heterogeneous sensor network technologies using a small set of powerful abstractions. As for semantic data enrichment, OpenSense developed *SeMiTri* [27] for annotating the trajectory data. The objective of OpenSense data management framework is to efficiently manage the generated environmental data. Therefore, OpenSense has introduced a framework by Cartier et al. [28] called *ConDense*. It provides a multimodel-based abstraction by condensing information generated by a GSN.

In OpenSense II, the dimension of crowdsourcing and human-centric computation were introduced to study the possibility to incentivize users to make available states based on physical measurements. They introduced a *dispersion model* to compute high-resolution air pollution maps for the cities of Zürich and Lausanne (based on land-use regression models).

4.2 CITI-SENSE and Citi-Sense-MOB

The CITI-SENSE is a research project that aimed at developing *citizens observatory* to engage mass population to contribute to and participate in environmental governance, and enable them to support and influence community and societal priorities and associated decision making [29]. In the CITI-SENSE project, sensor compliant to W3C and OGC standards are used to collect indoor and outdoor data. After processing data, they are stored in back-end server using an open-source mobile application.

In Citi-Sense-MOB, the sensors are mounted on vehicles, such as the city buses or electric bicycles. The citizens remain involved via a mobile app, and social media (see Ref. [30]). However, this project relies on a basic design of the data service architecture. Also, it does not measure the actual daily exposure of individuals since only vehicles (or bicycles) are equipped by sensors.

4.3 EveryAware

The main vision of EveryAware, as a sensitization project, is to involve the citizens not only as passive receivers of prepackaged environmental information, but also as active producers of it, by means of the networking possibilities allowed by mobile devices, pervasive Internet access, Web 2.0, and the mobile Web tools that support sharing and annotation of geo-localized contents. EveryAware technological platform combines sensing technologies, networking applications, and data-processing tool (see Ref. [31]).

An integrated platform was developed within EveryAware project to handle both subjective and objective data. The former comprises reactions of humans faced with particular environmental conditions, and the latter stems from sensors.

The platform comprises conceptual layer and implementation layer. The conceptual layer defines the basic entities and features the EveryAware system supports. The core concepts are data points with descriptions, sessions, and feeds. The implementation layer realizes the conceptual layer based on advanced storage and application structures which consists of the storage itself, the web application for receiving and retrieving data, and data processor which processes and enhances inbound data. MySQL engine has been used for implementing data storage, Apache Tomcat Servlet container has been used for implementing web application which also offers different REST end points such as WideNoise endpoint. It is worth noting that EveryAware supports measuring both sound and air pollution using WideNoise and AirProbe applications, respectively.

4.4 City Scanner

City Scanner [32] was introduced as a self-contained general purpose sensing platform that utilizes an existing fleet of vehicles, without interfering with their operations [33]. City Scanner follows a centralized IoT regime to generate a near real-time map of sensed data of thermal abnormalities and air pollutant hot spots. The individual sensing units are mounted on top of urban garbage trucks to record data and stream it to the cloud for processing and analysis. The core components of sensing units include power management, data management, and cloud streaming components. In City Scanner, data is reliably transferred from sensor nodes to the core component, as well as from the core component to the cloud. Data processing pipeline ranges from simple data storage, filtering, and visualization to more complicated services such as data analytics and machine learning.

4.5 AirCasting

AirCasting [34] is a HabitatMap Project. HabitatMap is a nonprofit environmental health justice organization whose goal is to raise awareness about the impact the environment has on human health. AirCasting is an open-source, end-to-end solution to capture real-world measurements, displaying heat maps indicating particulate matter concentrations, and sharing health and environmental data using smartphone. AirCasting proposes AirBeam and AirBeam2 sensors for PM is capable of communicating over 2G GSM, WiFi, and Bluetooth. AirBeam is based on a variant of Arduino called Teensy++. By connecting AirBeam to an Android smartphone, additional data acquisitions can be made like Zephyr HxM for Heart Rate. For CO, NO_2, temperature, and humidity, they propose AirCasting Air Monitor which is another version based on Arduino. This sensor has a Bluetooth connection to the smartphone app. The platform provides basic capabilities such as storage, pollution heat map computation, and data upload. The platform allows to query via the REST protocol. We can distinguish two kinds of query: get data set (e.g., last session), or computed average in a region for a given measurement. More advanced data processing are not covered by the platform which is aim to be a data exchange for crowd sourcing. However, these data serve to animate citizen exchanges as in schools or workshops.

4.6 hackAIR

hackAIR started in 2016 as a project to develop an open technology toolkit for citizens' observatories on air quality [35]. hackAIR is an open-source technology platform that can be used to access, collect, and improve air information on quality in Europe. It combines air quality data with a number of community data-driven sources. hackAIR proposes a home version based on Arduino which is connected to home internet, and automatically uploads a measurement every 10 minutes. A mobile version uploads its measurements through Bluetooth on the user smartphone. Citizens can check the data everyday on the map. But for advanced data analysis, as in AirCasting, user have to download data to use local computation like in Excel or use service like *data.world* or *Google Data Studio*. To import relevant data, user can express an SQL query on stored data set, or query a specific geographic area.

4.7 Other related work

While OpenSense II, CITI-SENSE, and EveryAware provide supports for monitoring and measuring both indoor and outdoor air quality, we have other projects such as INTA-SENSE [36], expAIR [37], and AirSensa [38] where measuring air quality is not provided for both environments at once. expAir, a research project, provides a cartographic model to show in great detail the black carbon concentrations in the Brussels-Capital Region as a whole. ATMO-Rennes [39], as a sensitization project by city of Rennes, aims to mobilize residents on a public environmental health problem by deploying environmental micro-sensors at low cost. Such outreach programs aim to spread a culture around air quality and mobilize citizens, via measurement tools. Many similar initiatives exist in this regard. The French environmental agency ADEME has released a recent report which surveys many of them, with the objective to assess their impact on behavioral change [40].

4.8 Summary of the projects

4.8.1 Context and main goal

	Context	Goal
OpenSense I and II	European Research Project Phase I 2009–13, phase II 2013–17	Understand the health impact of urban air pollution exposure
EveryAware	EU project from 2012 to 2014 with the consortium ISI, UCL, LUH, VITO, and PHYS-SAPIENZA	Environmental monitoring comparison with subjective opinions
AirCasting	AirCasting is a HabitatMap Project. Nonprofit environmental organization	Open-source, end-to-end solution for collecting, displaying, and sharing health and environmental data using your smartphone

Continued

	Context	Goal
CITI-SENSE	Research project from 2012 to 2016. Funded by the National Science Foundation EU project for citizens' observatories	Advance software technology to enable cyber-physical systems that allow citizens to see the environmental and health impacts of their daily activities
hackAIR	EU project from 2016 to 2018	Develop an open technology toolkit for citizens' observatories on air quality
City Scanner	MIT Senseable City Lab	Capture the spatiotemporal variation in environmental indicators in urban areas

Most of the projects leveraging low-cost environmental sensors and/or (mobile or fixed) crowd sensing have received grant from a European project. But there are also citizen initiatives such as AirCitizen and SmartCitizen, or conducted by local government (e.g., Ambassad'Air). However, the majority is restricted to fixed sensors, such as the projects luftdaten, SmartCitizen, or Ambassad'Air. We can note a notable exception with recent use in mobility, mostly based on the AirCasting platform [34].

4.8.2 Data management server

	Technology	Web portal
OpenSense I and II	GSN (global sensor networks) for stream processing and data publishing Decentralized system architecture. Each GSN instance can interact remotely with other instance using message queue UDP, HTTP REST	Hourly pollution maps
EveryAware	AirProbe, REST API web, storage on MySQL	Live track, pollution maps, and browsing other participant
AirCasting	AirCasting WebSite, MySQL Server and Redis (main memory key-value DB), REST API	Google map for individual localizing GPS and pollution trace Generate pollution maps indicating particulate matter concentrations
CITI-SENSE	Multiple data stores both for sensor data and for environmental and geospatial data Platform store data in relational and RDF	Web portal providing user track, perceptions, and comments Access the data already collected. It provides SOS, WFS, REST, SPARQL data access protocols
hackAIR	MongoDB storage SQL to export useful data, graph, compare curve Collect data from other official station, sky photo, and luftdaten data sources	REST API. PHP web server Data visualization with Node-RED Dashboard
City Scanner	Cloud servers SPI, IC, and TTL protocols for data transfer using WiFi network	Data visualization with interactive maps

All these studies share information via a client-server architecture. The server which is also the web application site loads the data from the sensor clients. The server at least display on a map the data of an individual sensor. Most produce a pollution map based on the available mobile sensor data which could be combined with fixed stations.

There is a consensus on access to the web services of a platform; they use the REST protocol which has the advantage of simplicity. For that, sensors are connected to an Android app which make the exchange with the platform via the URL.

Some projects like hackAIR offer advanced access to the server data by using high-level database languages like SQL or SparkQL in case of CITI-SENSE.

4.8.3 Advanced processing

More advanced data exploitation is expected in research projects. They follow the classic path of data analysis. This analysis can be enriched with data from questionnaires (CITI-SENSE) or social network monitoring. To enrich the analysis of the data, often these projects combine existing sources from crowdsourcing such as luftadaten.

In addition to the individual or aggregated exposure measurements provided by the platforms, the results of these analyses provide a better understanding of interactions between users and the environment (CITI-SENSE) or provide recommendations like route planning with the low pollutant.

	Data analysis	User-level feedback
OpenSense I and II	Continuous activity recognition: ECG, rate, breathing, accelerometer Air pollution dispersion modeling Privacy by cloaking of location	Health optimal route planner for pedestrians and cyclists
EveryAware	Many analysis of the data: user perception compared to real measurements for pollution, noise, etc.	Public deliverables with results of analysis
AirCasting	Not a research project. User can share its perception according to blog	Feed back to the user: Luminescent accessories (LiteBeam) or LED wearable air quality indicator. In response to the sensor measurements received by the AirCasting app
CITI-SENSE	Evaluate a set of Key Performance Indicators (KPI) from questionnaires which are published in reports	Smartphone apps share user perception and results Display locally stored data raw or in a graphical way Event notification
hackAIR	Social media monitoring, calculates a set of indicators that represent the "engagement" of social media users with the topic described by the keywords User can access to its data via data.world cloud and make data analyze, that is, building graph	The hackAIR knowledge base and reasoning framework handles both the semantic integration and reasoning of environmental and user-specific data, in order to provide recommendations to the hackAIR users LED display of air quality using Kniwwelino

Continued

	Data analysis	**User-level feedback**
City Scanner	Identify thermal abnormalities and air pollutant hot spots MSD algorithm to exclude redundant thermal images Reduce data size by using data compression	Users can browse data in both space and time dimensions A web application that offers discrete instruments to navigate and filter data in space via the interactive map and in time via a timeline scatter plot

5 Potential solutions

In this section, we sketch a general framework which addresses some limitations of state of the art in environmental MCS. It is under development within the Polluscope project [8]. Taking advantage of the technological evolution of wearable and lightweight environmental sensors, the interdisciplinary Polluscope project aims at providing methodologies, techniques, and tools expected to drastically change the way individual's exposure and exposure variability are measured, perceived, and evaluated. The first objective of Polluscope is to improve the knowledge of individual exposure anywhere at any time. It suggests a new concept, namely a *community-based participatory observatory for pollution and exposure* where citizens contribute data to the system with the purpose of sharing events of interest within the community. The measurements consider gaseous pollutants (ozone, NO_2), and particulate matter among which black carbon provides a representative overview of the air pollution. Gaining such enriched insights into individual's exposure will contribute toward reducing individual risks of some diseases by changing their behavior. This will ultimately end up in saving life and improving the individual well-being.

To achieve the aforementioned objectives, a novel infrastructure for real individual's exposure data acquisition, processing, and analysis is currently developed. It tackles several scientific and technical challenges. The data are collected at a high frequency and might be massive and noisy. Therefore, the system must be able to process them efficiently, while taking into account both their velocity and their uncertainty. More importantly, it has to offer microenvironment identification and user's activity recognition, through integration with external spatiotemporal resources. An efficient data collection and analysis will provide an insightful knowledge on individual's exposure over his/her daily life activities, and will enable conducting analytical queries, novel risk assessment modeling, mining and comparing profiles of pollution exposures, and so on. Polluscope is applied to real-world use cases. In particular, several types of population are involved in the data acquisition campaign: on the one hand, an epidemiological study is conducted by comparing diseased and healthy subjects in order to relate air pollution exposure to health; on the other hand, volunteer participants are integrated to assess the feasibility of crowd sensing scenario.

In the following we provide an overview of the proposed framework, and then focus on our proposal of a scalable data management server, including the data model and the query engine for MCS.

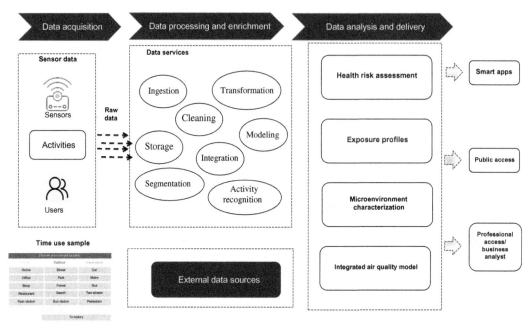

FIG. 1 Global architecture of polluscope platform.

The main structural components of the Polluscope platform include three components, centralized around data processing chain as shown in Fig. 1 [41]:

- *Data processing*: Several data processing platforms, such as Spark, R, etc., are used to perform specific tasks related to the data processing flow, such as data cleansing, enrichment, etc. For example, the collected data is notably enriched by the use of open-source tools (e.g., Open Street Map) and reference pollution data (e.g., Airparif prediction). The data services contained in data processing chain are as follows:
  - *Data ingestion*: This will enable data transformation, standardization, distribution, and loading into the data processing and analysis platform. Data ingestion is done via APIs; in a future version, the use of a distributed platform for data collection such as Apache Kafka is envisaged.
  - *Data transformation*: This will allow normalization, conversion into a target format, and transfer of output data to be used by a data analysis service or a visualization platform.
  - *Cleaning*: Sensor data are prone to errors and incompleteness, due to the imprecision or the absence of measurements notably when the device runs out of battery. This quality issue requires cleaning the data by filtering the outliers, and imputing missing data.
  - *Integration*: This service refers to the integration of heterogeneous data sources. On one hand, sensor data might be heterogeneous due to different types of sensors. On

the other hand, external data such as geographical data allow to enrich the geolocated sensor observations by their contextual information which are essential in data analysis.

- *Modeling*: Once the data are preprocessed and integrated, a high-level data model should be defined in order to capture the spatiotemporal characteristics of the data. This model should provide the necessary primitives and functions to manipulate spatiotemporal data series.

- *Storage*: One of the main services of the platform is to ensure the persistence of the data on a secondary storage. However, since the main feature of sensor data is to be ingested in a continuous manner, it is important to design a scalable architecture enabling timely storage of this data stream. Besides, the data should be organized and distributed in a way to optimize the data access and processing. To ensure the scalability, we adopted a distributed framework based on APACHE SPARK, as described in the following.

- *Segmentation*: The segmentation process is conducted within trajectory and time series data. The trajectory segmentation consists of dividing trajectories into subtrajectories called segments, in order to detect stop and move segments. A stop segment implies that the moving object is static at some location, while a move segment connects two consecutive stop segments by a mobility mean. On the other hand, time series segmentation consists of partitioning time series into homogeneous subsequences, for instance, according to changes in the microenvironment from indoor to outdoor.

- *Activity recognition*: The activity recognition process revolve around the detection of the users' microenvironment. Typical examples of the microenvironment include Home, Office, Train, Street, and Park. The users' activities are derived from trajectory data, leveraged by exploiting of the ambient air data. External data such as points of interest (POIs) and road networks are added to enhance the activity recognition process. A sample of annotated data with activities is used as a ground truth.

- *External data sources*: As indicated in the integration service, external data sources are used to semantically enrich the sensor observation by contextual information. We consider reference geographical data which will be map-matched with the geolocated sensor data. We also use the air quality data raster models from Airparif, which results from a specific simulation process. Other sources could be integrated in future, such as the weather forecast services, the road traffic data, events impacting the air quality, and so on.

- *Analysis and visualization*: This will enable the analysis and visualization of data collected and processed. An example, among others, of data analysis can relate to the semantics of the mobility of a moving object that can be represented according to various mobility dimensions and at different levels of granularity. The different semantic levels of mobility are then explicitly expressed by the annotation of the positions and subsequences of the trajectories according to the needs, the abstraction levels required, and the data layers available.

In the framework of Polluscope, we propose a scalable system [42] for managing and analyzing massive-scale air pollution data. The proposed system offers a high-level logical view of spatiotemporal data series as well as an internal physical model that combines aggressive indexing and partitioning over time and space to dissolve the heterogeneity and the variety of data. It is an extension of an existing distributed framework (Apache Spark) with the adaptation of data organization and the injection of various customized transformations rules for the optimization of spatiotemporal queries. It is a unified framework that combines batch and stream processing and tackles the unique characteristics of spatial mobile sensing data streams modeling and processing.

The data server (see Fig. 2) adopts two types of data ingestion. The first is batch data ingestion (finite datasets) that consists of loading a previously acquired data (e.g., loading the data from a previous campaign as a cold start of the data store). The second type corresponds to streaming data (infinite datasets) collected from current campaigns. We integrate the concept of indexing to achieve better query performances, and propose a model that alternates temporal and spatial indexing. We envision a two-level partitioning scheme, where the first level follows a global index, and the second depends on a second index. At the lowest level, the data will be further divided into even smaller spatial subpartitions called *buckets*. For instance, the first level can leverage a spatial index while the second follows a temporal partition, and the *buckets* could be based on the order of spatial filling curves. The way the spatial and temporal dimensions will be alternated is not yet decided and may require fine-tuning to adapt to the data and the query profile.

The queries are processed using the Spark's microbatch model, which processes data streams as a series of small batch jobs. Our query optimizer creates a series of incremental execution plans from a streaming logical plan. The idea is to inject appropriate

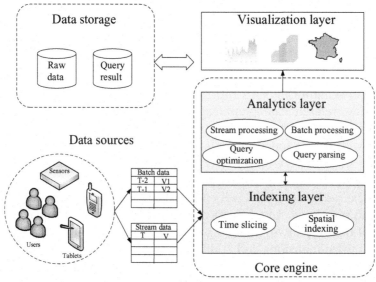

FIG. 2 Architecture of the polluscope data server.

optimization rules to obtain optimized execution plans. Rule-based optimization in our context exploits spatial indexing and time slicing to avoid scanning all the data series and eliminates records that do not contribute to the query result, avoiding Cartesian product (in case of temporal or spatial joins) which is systematically translated by the query processor in the presence of user-defined functions or performing selection on required time intervals as early as possible.

6 Conclusion and research perspectives

We have presented the state of the art related to environmental MCS, focusing on the data management and analysis aspects. We have presented our ongoing study in this domain, and sketched the architecture. An overview of the data server and the query processing has been provided.

Nowadays, we are witnessing the emergence of MCS, which will likely impact greatly the environmental monitoring. We foresee the development of the market of personal microsensors and an increase of their penetration rate in the population, as their price decreases and their quality improves. In spite of the important progress in the platforms dedicated to deal with such data, they are not sufficient to scale a large deployment. Most of the times, data cleansing, preparation, and analysis in these projects are offline and involve an important effort by experts. Therefore, much remains to be done to make the best use of the mass of MCS data, paving the way to new exciting research problems in data science.

Indeed, it is needed to automatize as much as possible the anomaly detection, the data imputation, the temporal/spatial alignment, the enrichment with the context and the knowledge extraction. Considering the opportunistic MCS, the data skew will remain a problem. Indeed, the paradox is to have frequent and dense observations per user, but the coverage also depends on their mobility. Therefore, the data should be complemented by spatiotemporal models estimated or mined from the measurement sample. The existing spatial interpolation methods should be adapted to such sparse trajectories. Privacy is another issue.

For instance, the regulation such as the General Data Protection Regulation (GDPR) in EU countries limits the data collection, and the data publishing. However, this data has a great value for several organizations as well as for the research community. How to share this data without privacy risks? Some solutions exist for privacy preserving query execution [43], but they are too specific to be deployed widely, and do not allow sharing fine-grain data. We envision a solution for privacy preserving data publishing based on extracting representative profiles from the real data, followed by simulation of synthetic data that resemble as much as possible to the original dataset [44].

Machine learning is a key technology to make sense of sensor big data. Besides mining mobility and activity profiles, which can be based on an unsupervised learning (e.g., clustering), supervised learning has a great value in predictive mining. However, it assumes the existence of training sets which are prelabeled. This is hard to achieve in the context of streams and massive data. Novel approaches using semisupervised learning or active learning are trending. There is a need to adapt them to the context of MCS, that is, to rich spatiotemporal trajectories.

The final quote is from Peter Sondergaard, Senior Vice President at Gartner: *Information is the oil of the 21st century, and analytics is the combustion engine.* In fact, without a powerful engine, the data collected from (mobile and/or crowd) sensors (the oil) is barely usable.

Acknowledgments

This study benefited from the support of the project POLLUSCOPE ANR-15-CE22-0018 of the French National Research Agency (ANR). It has been also supported by the MASTER project that has received funding from the European Union's Horizon 2020 research and innovation program under the Marie-Slodowska Curie grant agreement No. 777695.

References

[1] Air Pollution Report, What we discover when we turn urban vehicles into sensing platforms, http://www.senat.fr/notice-rapport/2014/r14-610-1-notice.html (Accessed 31 July 2019).

[2] C.G. Afrifa, F.G. Ofosu, S.A. Bamford, D.A. Wordson, S.M. Atiemo, I.J. Aboh, J.P. Adeti, Heavy metal contamination in surface soil dust at selected fuel filling stations in Accra, Ghana, Am. J. Sci. Ind. Res. 4 (4) (2013) 404–413.

[3] M. Bentayeb, V. Wagner, M. Stempfelet, M. Zins, M. Goldberg, M. Pascal, S. Larrieu, P. Beaudeau, S. Cassadou, D. Eilstein, Association between long-term exposure to air pollution and mortality in France: a 25-year follow-up study, Environ. Int. 85 (2015) 5–14.

[4] J. Amos, Polluted air causes 5.5 million deaths a year new research says, (2016). http://www.bbc.com/news/science-environment-35568249 (Accessed 19 February 2018).

[5] WHO, Household air pollution and health. (2016) http://www.who.int/mediacentre/factsheets/fs292/en/.

[6] F. Xia, L.T. Yang, V. Lizhe, A. Vinel, Internet of things, Int. J. Commun. Syst. 25 (9) (2012) 1101.

[7] R.K. Ganti, F. Ye, H. Lei, Mobile crowdsensing: current state and future challenges, IEEE Commun. Mag. 49 (11) (2011, 32–39).

[8] ATMO, Polluscope Project. (2018) http://polluscope.uvsq.fr.

[9] S. Dragan, B. Predic, N. Stojanovic, Mobile crowd sensing for smart urban mobility, in: European Handbook of Crowdsourced Geographic Information, p. 371.

[10] H. Ma, D. Zhao, P. Yuan, Opportunities in mobile crowd sensing, IEEE Commun. Mag. 52 (8) (2014) 29–35.

[11] N.D. Lane, E. Miluzzo, H. Lu, et al., A survey of mobile phone sensing, IEEE Commun. Mag. 48 (9) (2010, 140–150).

[12] I. Colbeck, Z.A. Nasir, Indoor air pollution, in: Human Exposure to Pollutants via Dermal Absorption and Inhalation, Springer, 2010, pp. 41–72.

[13] B. Fishbain, U. Lerner, N. Castell, et al., An evaluation tool kit of air quality micro-sensing units. Sci. Total Environ. 575 (2017) 639–648, https://doi.org/10.1016/j.scitotenv.2016.09.061.

[14] Y. Tan, A.L. Robinson, A.A. Presto, Quantifying uncertainties in pollutant mapping studies using the Monte Carlo method, Atmos. Environ. (2014) https://doi.org/10.1016/j.atmosenv.2014.10.003.

[15] S. Moshenberg, U. Lerner, B. Fishbain, Spectral methods for imputation of missing air quality data, Environ. Syst. Res. 4 (1) (2015) https://doi.org/10.1186/s40068-015-0052-z.

[16] H.-P. Hsieh, S.-D. Lin, Y. Zheng, Inferring Air Quality for Station Location Recommendation Based on Urban Big Data. in: Association for Computing Machinery (ACM), 2015, pp. 437–446, https://doi.org/10.1145/2783258.2783344.

[17] X. Zhao, T. Xu, Y. Fu, et al., Incorporating spatio-temporal smoothness for air quality inference, in: 2017 IEEE International Conference on Data Mining (ICDM), IEEE, 2017, pp. 1177–1182.

[18] C.C.G. Rodríguez, S. Servigne, Managing sensor data uncertainty. Int. J. Agric. Environ. Inf. Syst. (2013) https://doi.org/10.4018/jaeis.2013010103.

[19] A. Rahman, D.V. Smith, G. Timms, A novel machine learning approach toward quality assessment of sensor data. IEEE Sensors J. 14 (4) (2014) 1035–1047, https://doi.org/10.1109/JSEN.2013.2291855.

[20] T. Dasu, R. Duan, D. Srivastava, Data quality for temporal streams, (2016). Tech. Rep.

[21] L. Wang, D. Zhang, Y. Wang, C. Chen, X. Han, A. M'Hamed, Sparse mobile crowdsensing: challenges and opportunities. IEEE Commun. Mag. 54 (7) (2016) 161–167, https://doi.org/10.1109/MCOM.2016.7509395.

[22] H. Cheng, D. Feng, X. Shi, C. Chen, Data quality analysis and cleaning strategy for wireless sensor networks. EURASIP J. Wirel. Commun. Netw. 2018 (1) (2018) https://doi.org/10.1186/s13638-018-1069-6.

[23] M. Riahi, T.G. Papaioannou, I. Trummer, K. Aberer, Utility-driven data acquisition in participatory sensing, in: Proceedings of the 16th International Conference on Extending Database Technology, ACM, 2013, pp. 251–262.

[24] J.-P. Calbimonte, J. Eberle, K. Aberer, Semantic data layers in air quality monitoring for smarter cities, in: Proceedings of the 6th Workshop on Semantics for Smarter Cities S4SC 2015, at ISWC 2015.

[25] J.J. Li, B. Faltings, O. Saukh, D. Hasenfratz, J. Beutel, Sensing the air we breathe—the Opensense Zurich Dataset, in: Proceedings of the National Conference on Artificial Intelligence, vol. 1, 2012, pp. 323–325.

[26] K. Aberer, M. Hauswirth, A. Salehi, A middleware for fast and flexible sensor network deployment, in: Proceedings of the 32nd International Conference on Very Large Data Bases (VLDB)VLDB Endowment, 2006, pp. 1199–1202.

[27] Z. Yan, D. Chakraborty, C. Parent, S. Spaccapietra, K. Aberer, SeMiTri: a framework for semantic annotation of heterogeneous trajectories, in: Proceedings of the 14th International Conference on Extending Database Technology, ACM, 2011, pp. 259–270.

[28] S. Cartier, S. Sathe, D. Chakraborty, K. Aberer, ConDense: managing data in community-driven mobile geosensor networks, in: Proceedings of the 9th Annual Communications Society Conference on Sensor, Mesh and Ad Hoc Communications and Networks (SECON)IEEE, 2012, pp. 515–523.

[29] CITI-SENSE, CITISENSE—development of sensor based citizen's based observatory community for improving quality of life in cities, 2012, https://co.citi-sense.eu/.

[30] N. Castell, M. Kobernus, H.-Y. Liu, P. Schneider, W. Lahoz, A.J. Berre, J. Noll, Mobile technologies and services for environmental monitoring: the Citi-Sense-MOB approach, Urban Climate 14 (2015) 370–382.

[31] EveryAware, EveryAware: enhancing environmental awareness through social information technologies., (2007). http://www.everyaware.eu/wp-content/uploads/2011/04/EveryAware.pdf (Accessed 19 February 2018).

[32] City Scanner, What we discover when we turn urban vehicles into sensing platforms, 2017, http://senseable.mit.edu/cityscanner/.

[33] A. Anjomshoaa, F. Duarte, D. Rennings, T. Matarazzo, P. Desouza, C. Ratti, City Scanner: building and scheduling a mobile sensing platform for smart city services. IEEE Internet Things J. (2018). https://doi.org/10.1109/JIOT.2018.2839058.

[34] AirCasting, The AirCasting platform., (2018). http://aircasting.org.

[35] hackAIR, hackAIR: environmental node discovery, indexing and data acquisition., (2016). http://www.hackair.eu/wp-content/uploads/2016/12/d3.1_environmental_node_discovery_indexing_and_data_acquisition_1st_.pdf (Accessed 19 February 2018).

[36] INSTASENSE-A, INSTASENSE: integrated air quality sensor for energy efficient environment control., (2013) https://cordis.europa.eu/project/rcn/100559/factsheet/en (Retrieved 30 July 2019).

[37] B. Heene, P. Declerck, F. Beaujean, T. de Vos, G. Mendes, O. Brasseur, Evaluation de la qualité de l'air dans les parcs de la région de Bruxelles-Capitale, Bruxelles Environnement, 2016, 40 p.

[38] AirSensa, AirSensa., (2014). https://www.airsensa.com (Retrieved 30 July 2019).

[39] ATMO, ATMO Rennes. (2018) https://grafana.kabano.net/d/000000021/synthese?orgId=2.

[40] N. Saïdi, M. Planchon, Deloitte Développement Durable, L. Allard, Etude des liens entre données individuelles de la qualité de l'air, changements de comportement et mises en oeuvre de pratiques favorables à l'air. Quel apport des micro-capteurs?. (2017). http://www.ademe.fr/mediatheque (Retrieved 30 July 2019).

[41] M. Brahem, M. Chachoua, H.E. Hafyani, Z. Kedad, A. Ktaish, S. Mehanna, C. Ray, Y. Taher, R. Thibaud, L. Yeh, K. Zeitouni, Polluscope—vers un observatoire participatif de l'exposition individuelle à la pollution de l'air et de ses effets sanitaires, in: Conference on Spatial Analysis and Geomatics, SAGEO. To Appear, 2019.

[42] M. Brahem, K. Zeitouni, L. Yeh, H.E. Hafyani, Prospective data model and distributed query processing for mobile sensing data streams, in: Workshop on Multiple-Aspect Analysis of Semantic Trajectories (MASTER), in Conjunction With ECML/PKDD. To Appear, 2019.

[43] D.H.T. That, I.S. Popa, K. Zeitouni, C. Borcea, PAMPAS: privacy-aware mobile participatory sensing using secure probes, in: 28th International Conference on Scientific and Statistical Database Management, SSDBM'16ACM, 2016, pp. 4:1–4:12.

[44] N. Pelekis, S. Sideridis, P. Tampakis, Y. Theodoridis, Simulating our lifesteps by example. ACM Trans. Spatial Algorithms Syst. 2 (3) (2016) 11:1–11:39, https://doi.org/10.1145/2937753.

13

A survey of adsorption process parameter optimization related to degradation of environmental pollutants

Anindya Banerjee[a] and Avedananda Ray[b]

[a]DEPARTMENT OF ELECTRONICS AND COMMUNICATION ENGINEERING, KALYANI GOVERNMENT ENGINEERING COLLEGE, KALYANI, WEST BENGAL, INDIA [b]DEPARTMENT OF AGRICULTURAL AND ENVIRONMENTAL SCIENCES, TENNESSEE STATE UNIVERSITY, NASHVILLE, TN, UNITED STATES

1 Introduction

The booming phase of industrial expansion has led to an amplification of pollutants and contaminant dumping into the natural environment. Cosmetics, Textile, paper, plastic, leather, and industries release significant amount of contaminant impregnated wastewaters during manufacturing [1,2]. The contaminant enriched water is measured as a key problem for their toxic, nonbiodegradable character that can potentially damage human health. Some of the metals, e.g., Cu^{2+} is a necessary element for human; however higher concentration also causes severe damage. Reproductive and developmental toxicity, neurotoxicity, acute toxicity, dizziness, and diarrhea have been reported in human for higher concentration of heavy metals [3–5]. Synthetic organic dyes present in wastewater effluents can decrease sunlight penetration in water and subsequently reduce photosynthesis and raise the chemical oxygen demand (COD) [6–8]. These organic complexes are toxic in nature and it can also create unwanted effects like illnesses and diseases, such as cancer and genetic mutation in humans and other creatures [9,10]. Therefore, all these environmental contaminants must be removed from different effluent wastewaters to avoid various possible damages to living organisms [11,12]. Traditional water purification technology might not be able to entirely eliminate all the possible contaminants from water [13]. Still, it is potentially possible to remove several dissolved contaminants or pollutants from water using particular processes such as adsorption, reverse osmosis, ozonation, and nanofiltration [14,15]. Of these processes, the adsorption method of removal is beneficial because of its efficiency, low-energy-consuming behavior, and less operational cost to incur [16]. High specific surface area (high adsorption capacity) in the adsorbent is advantageous to apply in separation, adsorption of pollutants and heavy metals [17].

Intelligent Environmental Data Monitoring for Pollution Management. https://doi.org/10.1016/B978-0-12-819671-7.00013-0

As the adsorption efficiency is greatly influenced by particle size, specific surface area, and interparticle interaction affinity, the adsorption process encourages researchers to apply linear and nonlinear models, including adaptive neurofuzzy inference system (ANFIS) and artificial neural network (ANN) for the adsorption profile prediction [18–20].

Response Surface methodology (RSM) is applied to experimental design with reliable response of interest. Response surface methodology (RSM) along with artificial neural network (ANN) methods are now successfully used for both prediction and optimization purposes in adsorption processes. In the present study, various statistical repertoire has been discussed, which are used to efficiently optimize the adsorption process parameters.

2 Motivation and contributions

Extensive application of nonlinear models such as neural network and optimization design models (such as BBD, CCD, etc.) to optimize and predict adsorption of heavy metals and pollutants from aqueous solutions has increased notably. By using various neural networks, RSM model, BBD model, etc., parameters such as adsorption time, initial concentration, pH are optimized. The idea is to compare the efficiency of different type of adsorption models with different adsorption parameters.

3 Neural networks

An artificial neural network (ANN) is a combination of neurons or circuits or nodes, which is very much useful for designing algorithms and allows the computer to acquire and implement the new data by itself. The unit of neural human brain is neurons, while the basic unit of neural networks are nodes or layers. It is subpart of artificial intelligence (AI). ANNs consist of a several deep learning algorithms. Neural networks seem to have a larger applications on some complex problems like handwriting recognition, pattern recognition, automatic speech-to-text conversion, etc. ANN is extensively accepted for its adaptive nature, which means it can adapt and modify with the changes in the network accordingly. Once we design and develop the network, it generates the desired result and also acclimates with the changes in the input.

Artificial neural network (ANN) is an extensively used statistical technique, which can be improved by using heuristic optimization algorithm such as genetic algorithm (GA), imperialist competitive algorithm (ICA), etc. ANNs are also characterized based on the layers they have in between the input and output layers. These layers are then formed with various nodes, which are also interconnected. Every hidden layer is connected with output layer. This link between one node to another is represented by a number known as weight. A schematic diagram is represented in Figs. 1–3. Weight can also be negative, positive. Nodes having higher weights have more impact on the other node.

For a three-layer feed forward ANN, the nth response can be written as,

$$Y_n = \varphi_0 \left(b' + \sum_{j=1}^{k} w'_j \varphi_h \left(\sum_{i=1}^{m} w_{ij} X_i^n + b_j \right) \right) \tag{1}$$

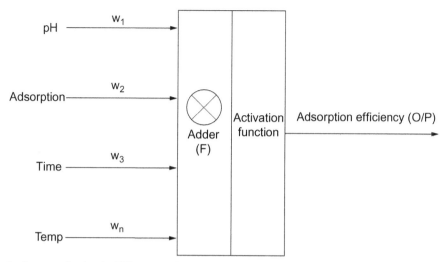

FIG. 1 Basic diagram of a simple ANN structural process.

W is defined as weighted matrix, b_j is bias matrix of the hidden layer (before applying activation function). b' denotes bias of output layer. $w_1', w_2', \ldots\ldots w_n'$ is the weighted matrix of output layer. $y^1, y^2, \ldots\ldots, y^n$ can be referred as response vector. $x^1, x^2, \ldots\ldots, x^n$ can be referred as input vector. The transfer functions of hidden layer and output layers are denoted as φ_h and φ_0, respectively [21].

Transfer functions that are commonly used are as follows,

Linear transfer function:

$$\varphi_x = x \quad -\infty < \varphi_x < +\infty \tag{2}$$

Sigmoid transfer function:

$$\varphi_x = \frac{1}{1+e^{-x}}, \quad 0 \le \varphi_x \le 1 \tag{3}$$

Each input node is multiplied with their corresponding weights and summation of these nodes is done by adder. Activation functions or threshold function like tanh function, tan 1.5 h function, sigmoid function are nonlinear in nature. The output of adder acts as input to the activation function (Fig. 4), it decides whether the output should be activated using further function (sigmoid function, tan function, etc.).

To make the neural network understand the logic cum rules it uses different set of principles like genetic algorithms, fuzzy-based logic, etc. For example, designing a facial image recognition application we can provide multiple images of a person with different postures, different poses. By using these images ANN can easily identify the person and also when there would possibly occur some changes in the person like scars on face, moustache, or jawline grows, etc. ANN will identify with the help of its previously provided inputs(this is the adaptive nature of ANNs).

ANN does not need any ideal design to build the required model. It also allows to add extra experimental data into the design set whenever required [21]. The performance of

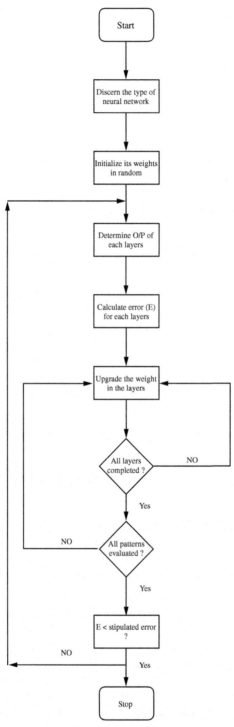

FIG. 2 Basic flowchart diagram for backpropagation algorithm of neural network.

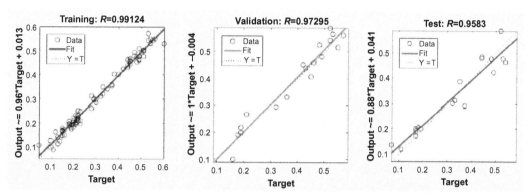

FIG. 3 Output (adsorption efficiency) prediction from input variables (pH, temperature, adsorption dosage) of ANN.

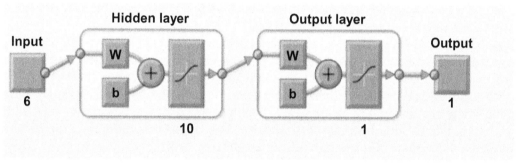

FIG. 4 Basic diagram of different ANN layers.

ANN is much better than RSM because RSM has a large number of nonlinear behavior [22]. ANN requires a wide range of experimental set of data to compute its design model. Although the training duration is much more than SVM [23], it also provides a little information about the factors and also their contribution to the response parameters [24].

ANNs learn with feedback process, known as back propagation. For example, if after designing our network we get the output, which is hugely different than actual output, in that scenario the ANN will use backpropagation algorithm. ANN is very simple and super fast technique to implement. The sensitivity analysis provides relative importance of input variables which could guide to predict better input treatments for heavy metals in the real wastewater.

A recurrent neural network (RNN) is a class of ANN where intermodal connections are directed toward graph along with temporal sequence, which shows temporal dynamic behavior. RNN basically follows feed-forward neural networks, which may use internal memory for processing variable length sequences of inputs. RNN is divided into multilayers, where every node in particular layer is directly connected with other successive layer's node. Every connection has a changeable real-value weight. The independent RNN focuses on the gradient vanishing and boosting approaches. Each node in one particular layer only receives its own past state as context information (instead of free, open connection to all other nodes in this layer) and, thus, dependency of the nodes is decreased.

4 Path analysis

Path analysis is a type of multiple regression statistical analysis, which is very much beneficial to determine relationship between an independent (adsorbent mineral content, pH of the solution, Adsorption dosage, Organic matter content) and dependent variable (Adsorption efficiency). One of many advantages of this method is we can easily determine relationship with variables and also the significance of the same.

Path diagram: The double-headed arrow in the Fig. 5 used to indicate two variables, which are connected (in the diagram) are linked or attached to each other. It is also named as unanalyzed association as this path association does not imply why the variables are connected with each other.

In path analysis we can easily determine the effect of a variable hypothesized used for an example. It also allows the decompositions of correlation of the variables, thus beneficial for determining the relation and effect patterns on the variable with another one [22]. It can also be noted determination of Path coefficients is also possible with repeated regression analysis to the variable subsets [25].

Path analysis distinguishes by correlation coefficient between independent variables (adsorbent mineral content, pH of the solution, adsorption dosage, organic matter content) and dependent variables (adsorption efficiency) into direct and indirect effects.

Disadvantages: When the collinearity increases, the capability to detect and analyze the significant effect turns out to be mitigated so path coefficient becomes inaccurate [22]. If variance inflation factor becomes less than 10, consequently R-squire needs to be less compared to R-square of the design model [26,27].

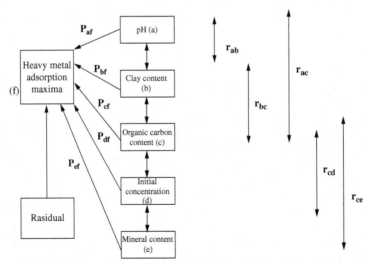

FIG. 5 Basic diagram of path analysis work flow for the prediction from input variables (pH, temperature, adsorption dosage).

However there are some drawbacks also for using path analysis:

(1) we need to keep in mind that while applying path analysis, we are assuming some conditions. Those are:
  (i) variables present in the path analysis measured on an time interval
  (ii) variables present on path model should be additive in nature, depends upon present input(causal) and linear.
  (iii) Measured variables are completely errorless.
  (iv) It should follow one way flow between variables.
(2) Error in calculation occurs when the independent variables are extremely correlated. It is known as collinearity. Collinearity is inversely proportional to rate of accuracy of path coefficient, i.e., more accurate the path coefficient, less is the collinearity [22,28–31].

5 Neurofuzzy network analysis

Neurofuzzy systems are one kind of automatic self-learning system, which is blend of techniques of fuzzy systems and ANNs. Neurofuzzy systems are very useful for solving complex problems like estimating density, pattern recognition, etc. Although there exists other methods to solve above problems, Neurofuzzy systems plays a great role when mathematical derivations are missing. It consists of a group of fuzzy rules, although it is also possible that a fuzzy system has not been learned without any initial knowledge.

Neurofuzzy network can be five-layer/three-layer feed forward neural network. Nevertheless it is also possible to convert them into a three-layer network architecture from a five-layer network architecture. Following is the example of a five-layer network architecture.

In the figure (Fig. 6), the first layer depicts input elements, i.e., Adsorption process parameters.

The second and third layers depict fuzzy rules.

The last layer depicts output elements, i.e., adsorption efficiency.

There exists different types of Neurofuzzy systems solely characterized upon multiple combinations of artificial neural networks and fuzzy logic procedures. They are briefly described as:

5.1 Cooperative neurofuzzy systems

When the neural networks are used in initial stages, it is known as Cooperative Neurofuzzy Systems. Hence the task of neural network is to configure the fuzzy system by using training data. After that only the fuzzy system is rendered keeping the neural network idle. Here both fuzzy system and artificial neural network do not depend upon each other (Fig. 7). Shows a simple architecture of cooperative neurofuzzy systems.

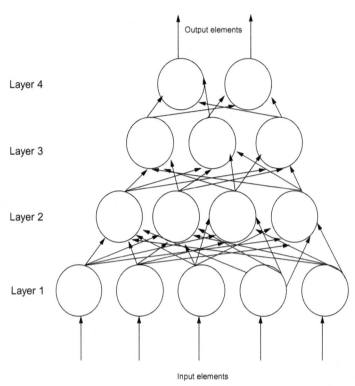

FIG. 6 Sample architecture of a neurofuzzy system.

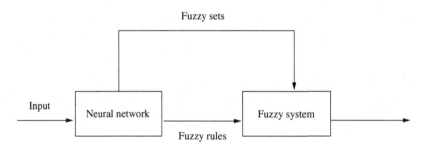

FIG. 7 Simple diagram of a cooperative neurofuzzy systems.

5.2 Concurrent neurofuzzy systems

Neural network and fuzzy systems act together in this system. If the Input enters in fuzzy systems then output of the same becomes input to neural network. The basic sample diagram is represented in Fig. 8. Reverse way is also possible. But from this system, we can't really set forth the result. This is main disadvantage of this kind of system.

FIG. 8 Simple diagram of a concurrent neurofuzzy systems.

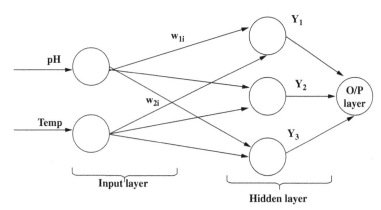

FIG. 9 Diagram of a hybrid neurofuzzy systems.

5.3 Hybrid neurofuzzy systems

It uses an algorithm solely based upon neural network algorithms to determine fuzzy sets by using its input and output There exists numerous ways to design hybrid neurofuzzy systems (see Fig. 9). Nowadays use of hybrid neurofuzzy systems is drastically increasing in applications like trade systems, medical application, engineering design, etc. [32–34].

Neurofuzzy networks had been used to design nonlinear processes with various properties in several regions. It's easier to explain rather than classical neural-network models. It allows utilization of data with process knowledge. The benefit of using this method over time-consuming methods like uncertainty in convergence to the global optimum and classical numerical optimization methods are avoided [13].

6 Response surface methodology

Response surface methodology (RSM) is the combination of various mathematical, statistical tools and techniques, which is efficient for simulating problems as well as evaluating them. RSM deals with the relationship between various explanatory and response variables. The primary objective of this method is to get an optimized result by using a sequence of designed experiments [35]. RSM design extensively used to optimize the multivariance system where a set of independent variables impacts the group of dependent variables; the relationship between both can be evaluated by a quadric equation [21].

It also allows analysis of different quadric response models due to influence of different parameters [36]. It is also a fast approach and cost effective as it needs small number of experimental trials. RSM is widely used to develop the mathematical design in the form of multiple regression equations [25].

RSM design is useful only for the condition where number of inputs are limited [37]. However it is also not applicable for all systems with second-order polynomial model.

Response surface methodology can be classified into two types, namely, Central Composite Design (CCD) and Box-Behnken Design (BBD).

6.1 Central composite design (CCD)

CCD is extensively used to build second-order design model. It uses a huge number of sample points (e.g., center point, axial point, corner point). In CCD, the variables at centric level (0) implies center points, commix of variables at higher level (+1) or lower levels (−1) and also with another variables of centric level forms star points. It is a useful substitute to classical design models as we can collect huge number of data by using less number of experiments. To determine the effects of variables/parameters second-order polynomial CCD model can be formed according to the equation [38].

$$y = \beta_0 + \sum_{i=1}^{k} \beta_i X_i + \sum_{i=1}^{k} \beta_{ii} X_{ii}^2 + \sum_{i<j=1}^{k} \beta_{ij} X_i X_j + \varepsilon \tag{4}$$

Here y is the target response of the experiment X_i is the value of ith process parameter β is coefficients of squared effects, linear effect, ε is error in the experiment [39].

CCD contains twice the star points of the total factors present in the design. There are three types of CCD designs based upon the presence of star points. Those are (1) CCC design, (2) CCI design, (3) CCF design.

6.1.1 CCC design

The most primary form of CCD design is CCC design. CCC designs enable precise predictions along with the entire design space. Based upon number of factors present and design characteristics, star points exists at some distance from the center. CCC must need factors outside the range of the factorial part factors. The sample diagram is presented in Fig. 10.

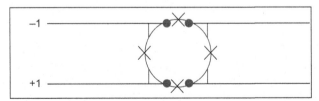

FIG. 10 Pictorial representation of starting points in CCC design.

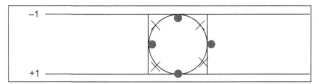

FIG. 11 Pictorial representation of starting points in CCI design model.

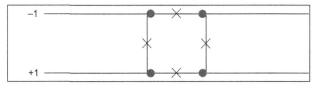

FIG. 12 Pictorial representation of starting points in CCF design model.

6.1.2 CCI design

In Central Composite Inscribed (CCI) design, the start points are inscribed within the circle. It computes the factor setting of the design within the boundaries. It is also known as scaled-down CCC design [35] (Fig. 11).

6.1.3 CCF design

CCF design is similar to CCD design. This design model is useful where second-order response requires to fit within the model. Factorial points, center points, and axial points (Fig. 12) are used in this design model as well [40].

6.2 Box-Behnken design

Box and Behnken introduced a class of three-level design for response analysis. BBD is developed by combination of 2^k factorials based upon incomplete block designs. When the points lie on the center of the cube, it become very much difficult to examine as the process is much expensive; but by using BBD model we can easily overcome this problem [41]. The design is analyzed in Figs. 13 and 14 with different parameters such as time, pH, temperature, adsorption doses, etc. (Table 1).

R-squared value lies between 0 and 1. Lower R^2 value suggests that the variability of the response output (adsorption efficiency) does not lies around the dataset mean, whereas higher value suggests the maximized variability lies around the mean in the particular statistical model. An R^2 value of 1 means that all movements of dependent variable (adsorption efficiency) are completely explained by movements in the independent variable (e.g., pH, temperature, time) (Table 2).

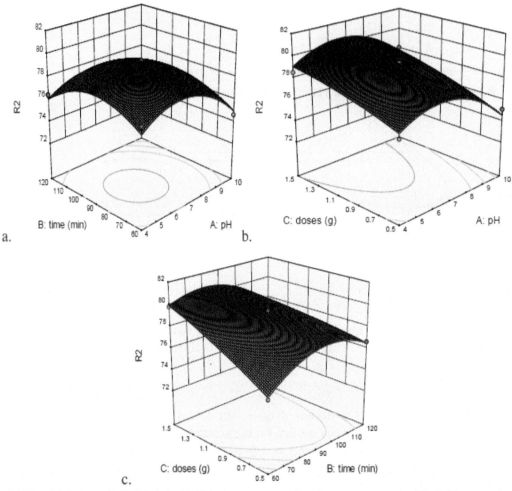

FIG. 13 Model diagram of a BBD design with different parameters (three input parameters are (A) pH, (B) contact time (min), and (C) adsorbent doses (g); R^2 is the adsorption efficiency).

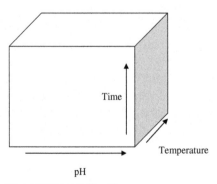

FIG. 14 Sample diagram of BBD with a center point ($k = 3$).

Table 1 Hypothetical data for different levels of inputs (temperature, pH, adsorption dosage) required for a 3D BBD model.

Experimental nos	pH	Temperature	Time
1	1	−1	0
2	−1	1	0
3	−1	−1	0
4	−1	0	−1
5	0	−1	1
6	0	1	−1
7	1	1	0
8	0	0	−1
9	0	−1	−1
10	−1	0	1
11	1	0	1
12	0	0	0
13	0	0	0
14	0	0	0

Table 2 Small-scale literature survey of Statistical methods with their R^2 value.

Adsorbent	Heavy metals/ pollutant/ contaminant	Input parameters	Computational repertoire	R^2 value	Reference
Magnetic multi-wall carbon nanotube	Microcystin	Microcystin-LR concentration ($\mu g\ L^{-1}$), contact time (min), adsorbent dosage (g), and pH	ANN-Levenberg-Marquardt backpropagation	0.9505	[42]
			ANN-Scaled conjugate gradient backpropagation	0.9413	
			ANN-BFGS quasi-Newton backpropagation	0.9146	
			ANN-One step secant backpropagation	0.8535	
			ANN-Batch gradient descent	0.6512	
			ANN-Variable learning rate back propagation	0.9121	
			ANN-Batch gradient descent with momentum	0.4488	
			ANN-Fletcher-Reeves conjugate gradient backpropagation	0.7899	
			ANN-Polak-Ribi'ere conjugate gradient backpropagation	0.8571	
			ANN-Powell-Beale conjugate gradient backpropagation	0.8961	

Continued

Table 2   Small-scale literature survey of Statistical methods with their R^2 value—cont'd

Adsorbent	Heavy metals/ pollutant/ contaminant	Input parameters	Computational repertoire	R^2 value	Reference
Activated spent tea	Methylene blue	Dye concentration (mg L^{-1}), adsorbent dose (g/L), pH, temperature (K), time (min)	ANN	0.99	[43]
Apatitic tricalcium phosphate	Fluoride	pH, adsorbent dose, temperature	ANN	0.979	[21]
			RSM-BBD	0.927	
Copper nanowares loaded on activated carbon	Malachite green	Dye concentration (mg L^{-1}), adsorbent dose (g), pH, time (min)	ANN-Levenberg-Marquardt backpropagation	0.9658	[44]
Light expended clay aggregate (LECA)	Cu^{2+}	Initial concentration (mg L^{-1}), adsorbent dose (mg), temperature (°C)	ANN-Levenberg-Marquardt back propagation algorithm (LMA)	0.962	[45]
			RSM-CCD	0.941	
PbO nanoparticle	Methyl orange	Concentration (mg L^{-1}), adsorbent dosage (g), and contact time (min).	Feed-forward architecture of ANN-PSO model with backpropagation (BP) algorithm	0.9685	[46]
Polyaniline nano-adsorbent	Methyl orange	Concentration (mg L^{-1}), adsorbent dosage (g), temperature (°C), pH, adsorption Time	ANN	0.99	[47]
ZnS nanoparticles loaded on activated carbon	METHYLENE blue dye	pH, concentration (mg L^{-1}), adsorbent dosage (g), sonication time (Min)	ANN-Levenberg-Marquardt (LMA) algorithm	0.99	[48]
			CCD	0.99	
Co2O3-NP-AC	Eosin Y dye	Adsorbent dosage (g), contact time (min) and initial dye concentration (mg L^{-1})	ANN-GA	0.99	[49]
Graphene oxide nanoplatelets	Synthetic azo dye	pH, temperature, and adsorbent dose	ANN-Levenberg-Marquardt backpropagation	0.96	[50]
Zn (OH)2 nanoparticles activated carbon	Sunset yellow	Dye concentration (mg L^{-1}), contact time (min), and adsorbent dosage (g)	ANN	0.9782	[44]
Chemically modified orange peel	Copper	Initial solution pH, Initial copper concentration (mg L^{-1}), time (min)	Feed-forward backpropagation neural network	0.967	[51]
			CCD	0.99	
Superheated steam activated granular carbon	Chromium (VI)	pH, GAC dose, contact time, temperature, and initial concentration	CCD	0.99	[42]
			ANN	0.99	
Gold nanoparticle-activated carbon	Reactive orange 12	Concentration, contact time (min), and adsorbent dosage (g)	Artificial neural network-imperialist competitive algorithm	0.9443	[52]

7 Random forest model

Random forest is an ensemble learning method for classified regression performs by building multidecision trees for training and outputting the modal class (classified) or prediction of mean (regression) of the individual decision trees. Random forest has following tuning parameters: n_{tree}, m_{try}. The relative importance of each variable can be corresponding by a mean decrement in Gini index and accuracy. For the mean decrease in accuracy, a variable is validated by the out-of-bag error computation, an error computation-based absent samples during training. Mean decrease in Gini index is employed for homogeneity of the nodes and leaves. Huge mean decreases in accuracy of some variables are most important for the training of data, whereas large mean decreases in Gini index show greater purity (Fig. 15).

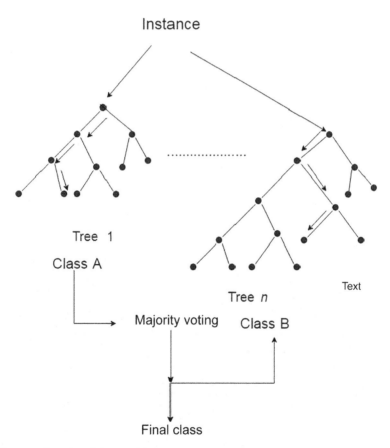

FIG. 15 Basic diagram of random forest model.

8 Stochastic gradient boosting

Stochastic gradient boosting, a hybrid bagging-boosting procedure, works at each itera-tion by least squares develops that regression models by additive sequentially method and deploying a simple parameterized function to current "pseudo"-residuals. The pseudo-residuals are being minimized with respect to each current step's training data. Execution speed and accuracy can be potentially improved by including randomization. At each iteration, random subsample from training dataset is used in place of overall sam-ple to compute the recent iteration. This randomization increases robustness of the model (Fig. 16).

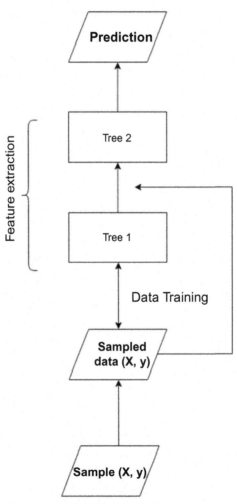

FIG. 16 Basic flowchart of stochastic gradient boosting.

9 Conclusion

Adsorption data mentioned in various literature indicated the strong efficacy of statistical repertoire to predict the adsorption efficiency. The influences of process parameters such as adsorbent characteristics, physiochemical nature of heavy metal, initial contaminant concentration, contact time, initial pH of the aqueous solution, amount of adsorbent, and temperature on the contaminant removal percentage were easily investigated, but still other parameters and other kinds of prediction, e.g., time-series analysis should be focused on future research work. The optimal adsorption boundary conditions to achieve highest removal percentage could be found by trained neural network as fitness function. Results of the optimum stage corroborate the applicability of this novel computational repertoire for process parameters optimization in the adsorption process. The accurate prediction for adsorption efficiency of heavy metals, organic dyes, complex molecules based on the biochars and nanoadsorbents characteristics are valuable to avoid redundancy in repetitive experiments.

References

[1] A.A. Babaei, A. Khataee, E. Ahmadpour, M. Sheydaei, B. Kakavandi, Z. Alaee, Optimization of cationic dye adsorption on activated spent tea: equilibrium, kinetics, thermodynamic and artificial neural network modeling, Korean J. Chem. Eng. 33 (4) (2016) 1352–1361.

[2] M. Rafatullah, O. Sulaiman, R. Hashim, A. Ahmad, Adsorption of methylene blue on low-cost adsorbents: a review, J. Hazard. Mater. 177 (1–3) (2010) 70–80.

[3] T.G. Chuah, A. Jumasiah, I. Azni, S. Katayon, S.T. Choong, Rice husk as a potentially low-cost biosorbent for heavy metal and dye removal: an overview, Desalination 175 (3) (2005) 305–316.

[4] A. Papandreou, C.J. Stournaras, D. Panias, Copper and cadmium adsorption on pellets made from fired coal fly ash, J. Hazard. Mater. 148 (3) (2007) 538–547.

[5] B. Yu, Y. Zhang, A. Shukla, S.S. Shukla, K.L. Dorris, The removal of heavy metal from aqueous solutions by sawdust adsorption—removal of copper, J. Hazard. Mater. 80 (1–3) (2000) 33–42.

[6] A.S. Franca, L.S. Oliveira, M.E. Ferreira, Kinetics and equilibrium studies of methylene blue adsorption by spent coffee grounds, Desalination 249 (1) (2009) 267–272.

[7] L.S. Oliveira, A.S. Franca, T.M. Alves, S.D. Rocha, Evaluation of untreated coffee husks as potential biosorbents for treatment of dye contaminated waters, J. Hazard. Mater. 155 (3) (2008) 507–512.

[8] S. Chowdhury, P.D. Saha, Artificial neural network (ANN) modeling of adsorption of methylene blue by NaOH-modified rice husk in a fixed-bed column system, Environ. Sci. Pollut. Res. 20 (2) (2013) 1050–1058.

[9] I. Ali, AL-Othman, Z. A., & Alwarthan, A., Molecular uptake of congo red dye from water on iron composite nano particles, J. Mol. Liq. 224 (2016) 171–176.

[10] N. Daneshvar, D. Salari, A.R. Khataee, Photocatalytic degradation of azo dye acid red 14 in water on ZnO as an alternative catalyst to TiO_2, J. Photochem. Photobiol. A Chem. 162 (2–3) (2004) 317–322.

[11] C.P. Sekhar, S. Kalidhasan, V. Rajesh, N. Rajesh, Bio-polymer adsorbent for the removal of malachite green from aqueous solution, Chemosphere 77 (6) (2009) 842–847.

[12] T.A. Khan, R. Rahman, I. Ali, E.A. Khan, A.A. Mukhlif, Removal of malachite green from aqueous solution using waste pea shells as low-cost adsorbent–adsorption isotherms and dynamics, Toxicol. Environ. Chem. 96 (4) (2014) 569–578.

[13] M.P. Gatabi, H.M. Moghaddam, M. Ghorbani, Efficient removal of cadmium using magnetic multi-walled carbon nanotube nanoadsorbents: equilibrium, kinetic, and thermodynamic study, J. Nanopart. Res. 18 (7) (2016) 189.

[14] N. Tokodi, D. Drobac, Z. Svirčev, D. Lazić, Cyanotoxins in Serbia and water treatment procedures for their elimination, Geograph. Pannon. 16 (4) (2012) 155–163.

[15] D.P. Smith, V. Falls, A.D. Levine, B. MacLeod, M. Simpson, T.L. Champlin, Nanofiltration to augment conventional treatment for removal of algal toxins, taste and odor compounds, and natural organic matter, in: Proceedings of the AWWA 2002 Water Quality Technology Conference, Seattle, Washington, November, 2002, pp. 10–14.

[16] A. Asfaram, M. Ghaedi, S. Agarwal, I. Tyagi, V.K. Gupta, Adsorption of basic dye Auramine-O by ZnS: Cu nanoparticles loaded on activated carbon using response surface methodology with central composite design, RSC Adv. 5 (2015) 18438–18450.

[17] A. Mittal, A. Malviya, D. Kaur, J. Mittal, L. Kurup, Studies on the adsorption kinetics and isotherms for the removal and recovery of methyl orange from wastewaters using waste materials, J. Hazard. Mater. 148 (1–2) (2007) 229–240.

[18] M. Ghaedi, A.M. Ghaedi, F. Abdi, M. Roosta, A. Vafaei, A. Asghari, Principal component analysis-adaptive neuro-fuzzy inference system modeling and genetic algorithm optimization of adsorption of methylene blue by activated carbon derived from *Pistacia khinjuk*, Ecotoxicol. Environ. Saf. 96 (2013) 110–117.

[19] H. Karimi, F. Yousefi, Application of artificial neural network–genetic algorithm (ANN–GA) to correlation of density in nanofluids, Fluid Phase Equilib. 336 (2012) 79–83.

[20] J. Kennedy, The particle swarm: social adaptation of knowledge, in: Proceedings of 1997 IEEE International Conference on Evolutionary Computation (ICEC'97), April, 1997, pp. 303–308.

[21] M. Mourabet, E.R. Abdelhadi, M. Bennani-Ziatni, T. Abderrahim, Comparative study of artificial neural network and response surface methodology for modelling and optimization the adsorption capacity of fluoride onto apatitic tricalcium phosphate, Univ. J. Appl. Math. 2 (2014) 84–91.

[22] J. Jeon, The Strengths and Limitations of the Statistical Modeling of Complex Social Phenomenon: Focusing on SEM, Path Analysis, or Multiple Regression Models (Version 10001434). (2015), https://doi.org/10.5281/zenodo.1105869.

[23] R. Moraes, J.F. Valiati, W.P. Gavião Neto, Document-level sentiment classification: an empirical comparison between SVM and ANN. Expert Syst. Appl. 40 (2) (2013) 621–633, https://doi.org/10.1016/j.eswa.2012.07.059.

[24] A.K. Lakshminarayanan, V. Balasubramanian, Comparison of RSM with ANN in predicting tensile strength of friction stir welded AA7039 aluminium alloy joints. Trans. Nonferrous Metals Soc. China 19 (1) (2009) 9–18, https://doi.org/10.1016/s1003-6326(08)60221-6.

[25] R.J. Mitchell, Testing evolutionary and ecological hypotheses using path analysis and structural equation modelling, Funct. Ecol. 6 (2) (1992) 123–129.

[26] P.S. Petraitis, A.E. Dunham, P.H. Niewiarowski, Inferring multiple causalities: the limitations of path analysis, Funct. Ecol. 10 (4) (1996) 421–431.

[27] J. Zhang, A.J. Morris, Recurrent neuro-fuzzy networks for nonlinear process modeling. IEEE Trans. Neural Netw. 10 (2) (1999) 313–326, https://doi.org/10.1109/72.750562.

[28] J. Cohen, P. Cohen, S.G. West, L.S. Aiken, Applied Multiple Regression/Correlation Analysis for the Behavioural Sciences, third edition, Erlbaum, Mahwah, NJ, 2003.

[29] T.Z. Keith, Multiple Regression and beyond: A Conceptual Introduction to Multiple Regression, Confirmatory Factor Analysis, and Structural Equation Modeling, Allyn & Bacon, Boston, 2006.

[30] R.B. Kline, Principles and Practice of Structural Equation Modeling, Guilford Press, New York, 1998.

[31] T.H. Wonnacott, R.J. Wonnacott, Regression: A Second Course in Statistics, Wiley, New York, 1981.

[32] R. Kruse, Fuzzy neural network, Scholarpedia 3 (11) (2008) 6043.

[33] R. Singh, T. Prasad, Exploration of hybrid neuro fuzzy systems. in: National Conference on Advances in Knowledge Management, NCAKM 2010, At Lingaya's University, Faridabad, Haryana, India, 2010, https://doi.org/10.13140/RG.2.1.3570.0327.

[34] A. Abraham, N. Baikunth, Hybrid Intelligent Systems: A Review of a decade of Research, School of Computing and Information Technology, Faculty of Information Technology, Monash University, Australia, 2000, pp. 1–55 Technical Report Series, 5/2000.

[35] B. Olawoye, A comprehensive handout on central composite design (Ccd), in: NIST/SEMATECH e-Handbook of Statistical Methods, 2016, http://www.itl.nist.gov/div898/handbook/.

[36] B. Koç, F. Kaymak-Ertekin, Response surface methodology and food processing applications, Gıda 7 (2009) 1–8.

[37] S. Muthusamy, L.P. Manickam, V. Murugan, M. Chendrasekar, A. Pugazhendhi, Pectin extraction from *Helianthus annuus* (sunflower) heads using RSM and ANN modelling by a genetic algorithm approach. Int. J. Biol. Macromol. 124 (2018) 750–758, https://doi.org/10.1016/j.ijbiomac.2018.11.036.

[38] T. Rajmohan, K. Palanikumar, Application of the central composite design in optimization of machining parameters in drilling hybrid metal matrix composites. Measurement 46 (4) (2013) 1470–1481, https://doi.org/10.1016/j.measurement.2012.11.034.

[39] M. Muthukumar, D. Sargunamani, N. Selvakumar, J. Venkata Rao, Optimisation of ozone treatment for colour and COD removal of acid dye effluent using central composite design experiment. Dyes Pigments 63 (2) (2004) 127–134, https://doi.org/10.1016/j.dyepig.2004.02.003.

[40] W. Sibanda, P. Pretorius, Comparative study of the application of central composite face-centred (CCF) and Box–Behnken designs (BBD) to study the effect of demographic characteristics on HIV risk in South Africa. Netw. Model. Anal. Health Inform. Bioinform. 2 (3) (2013) 137–146, https://doi.org/10.1007/s13721-013-0032-z.

[41] S.L.C. Ferreira, R.E. Bruns, H.S. Ferreira, G.D. Matos, J.M. David, G.C. Brandão, et al., Box-Behnken design: an alternative for the optimization of analytical methods. Anal. Chim. Acta 597 (2) (2007) 179–186, https://doi.org/10.1016/j.aca.2007.07.011.

[42] G. Halder, S. Dhawane, P.K. Barai, A. Das, Optimizing chromium (VI) adsorption onto superheated steam activated granular carbon through response surface methodology and artificial neural network, Environ. Prog. Sustain. Energy 34 (3) (2015) 638–647.

[43] A.A. Babaei, S.N. Alavi, M. Akbarifar, K. Ahmadi, A. Ramazanpour Esfahani, B. Kakavandi, Experimental and modeling study on adsorption of cationic methylene blue dye onto mesoporous biochars prepared from agrowaste, Desalin. Water Treat. 57 (56) (2016) 27199–27212.

[44] M. Ghaedi, A. Ansari, M.H. Habibi, A.R. Asghari, Removal of malachite green from aqueous solution by zinc oxide nanoparticle loaded on activated carbon: kinetics and isotherm study, J. Ind. Eng. Chem. 20 (1) (2014) 17–28.

[45] T. Shojaeimehr, F. Rahimpour, M.A. Khadivi, M. Sadeghi, A modeling study by response surface methodology (RSM) and artificial neural network (ANN) on Cu^{2+} adsorption optimization using light expended clay aggregate (LECA), J. Ind. Eng. Chem. 20 (3) (2014) 870–880.

[46] S. Agarwal, I. Tyagi, V.K. Gupta, M. Ghaedi, M. Masoomzade, A.M. Ghaedi, B. Mirtamizdoust, RETRACTED: kinetics and thermodynamics of methyl orange adsorption from aqueous solutions—artificial neural network-particle swarm optimization modeling, J. Mol. Liq. 218 (2016) 354–362.

[47] M. Tanzifi, S.H. Hosseini, A.D. Kiadehi, M. Olazar, K. Karimipour, R. Rezaiemehr, I. Ali, Artificial neural network optimization for methyl orange adsorption onto polyaniline nano-adsorbent: kinetic, isotherm and thermodynamic studies, J. Mol. Liq. 244 (2017) 189–200.

[48] A. Asfaram, M. Ghaedi, S. Hajati, A. Goudarzi, Synthesis of magnetic γ-Fe$_2$O$_3$-based nanomaterial for ultrasonic assisted dyes adsorption: modeling and optimization, Ultrason. Sonochem. 32 (2016) 418–431.

[49] N.P. Assefi, M. Ghaedi, Removal of Eosin B from aqueous solution by cobalt oxide nanoparticle loaded on activated carbon, in: Iranian Chemistry Congress, vol. 16, 2013.

[50] P. Banerjee, P. Das, A. Zaman, P. Das, Application of graphene oxide nanoplatelets for adsorption of ibuprofen from aqueous solutions: evaluation of process kinetics and thermodynamics, Process Saf. Environ. Prot. 101 (2016) 45–53.

[51] A. Ghosh, K. Sinha, P.D. Saha, Central composite design optimization and artificial neural network modeling of copper removal by chemically modified orange peel, Desalin. Water Treat. 51 (40–42) (2013) 7791–7799.

[52] R.H. Nia, M. Ghaedi, A.M. Ghaedi, Modeling of reactive orange 12 (RO 12) adsorption onto gold nanoparticle-activated carbon using artificial neural network optimization based on an imperialist competitive algorithm, J. Mol. Liq. 195 (2014) 219–229.

Further reading

M.J. Amiri, M. Bahrami, F. Dehkhodaie, Optimization of Hg(II) adsorption on bio-apatite based materials using CCD-RSM design: characterization and mechanism studies. J. Water Health 17 (4) (2019) 556–567, https://doi.org/10.2166/wh.2019.039.

A. Fakhri, Investigation of mercury (II) adsorption from aqueous solution onto copper oxide nanoparticles: Optimization using response surface methodology, Process Saf. Environ. Protect. 93 (2015) 1–8.

V.K. Gupta, S. Agarwal, I. Tyagi, M. Sohrabi, A. Fakhri, S. Rashidi, N. Sadeghi, Microwave-assisted hydrothermal synthesis and adsorption properties of carbon nanofibers for methamphetamine removal from aqueous solution using a response surface methodology. J. Ind. Eng. Chem. 41 (2016) 158–164, https://doi.org/10.1016/j.jiec.2016.07.018.

N.A. Jarrah, Adsorption of Cu(II) and Pb(II) from aqueous solution using Jordanian natural zeolite based on factorial design methodology. Desalin. Water Treat. 16 (1–3) (2010) 320–328, https://doi.org/10.5004/dwt.2010.1089.

R. Myers, Classical and Modern Regression with Applications, 2nd ed., Duxbury Press, Boston, 1990.

P. Leechart, W. Nakbanpote, P. Thiravetyan, Application of 'waste' wood-shaving bottom ash for adsorption of azo reactive dye, J. Environ. Manag. 90 (2) (2009) 912–920.

A. Bhatnagar, A.K. Jain, A comparative adsorption study with different industrial wastes as adsorbents for the removal of cationic dyes from water, J. Colloid Interface Sci. 281 (1) (2005) 49–55.

K.T. Chung, G.E. Fulk, A.W. Andrews, Mutagenicity testing of some commonly used dyes, Appl. Environ. Microbiol. 42 (4) (1981) 641–648.

R. Jain, S. Sikarwar, Removal of hazardous dye congored from waste material, J. Hazard. Mater. 152 (3) (2008) 942–948.

V.K. Gupta, R. Jain, A. Mittal, M. Mathur, S. Sikarwar, Photochemical degradation of the hazardous dye Safranin-T using TiO$_2$ catalyst, J. Colloid Interface Sci. 309 (2) (2007) 464–469.

V.K. Gupta, R. Jain, S. Varshney, Electrochemical removal of the hazardous dye Reactofix Red 3 BFN from industrial effluents, J. Colloid Interface Sci. 312 (2) (2007) 292–296.

D. Kavitha, C. Namasivayam, Experimental and kinetic studies on methylene blue adsorption by coir pith carbon, Bioresour. Technol. 98 (1) (2007) 14–21.

S. Chen, J. Zhang, C. Zhang, Q. Yue, Y. Li, C. Li, Equilibrium and kinetic studies of methyl orange and methyl violet adsorption on activated carbon derived from *Phragmites australis*, Desalination 252 (1–3) (2010) 149–156.

K. Ahmadi, M. Ghaedi, A. Ansari, Comparison of nickel doped zinc sulfide and/or palladium nanoparticle loaded on activated carbon as efficient adsorbents for kinetic and equilibrium study of removal of Congo Red dye, Spectrochim. Acta A Mol. Biomol. Spectrosc. 136 (2015) 1441–1449.

Index

Note: Page numbers followed by *f* indicate figures, *t* indicate tables, and *b* indicate boxes.

Printed in the United States
By Bookmasters